Essentials of RF Front-end Design and Testing

IEEE Press
445 Hoes Lane
Piscataway, NJ 08854

Essentials of RF Front-end Design and Testing

A Practical Guide for Wireless Systems

Ibrahim A. Haroun

Published by John Wiley & Sons, Inc., Hoboken, New Jersey.
Published simultaneously in Canada.

Library of Congress Cataloging-in-Publication Data Applied for:

Hardback ISBN: 9781394210619

Cover Design: Wiley
Cover Image: © ronstik/Adobe Stock Photos

Set in 9.5/12.5pt STIXTwoText by Straive, Chennai, India

This book is dedicated to my late wife Magdalene who used to edit my publications and I have learned a lot from her. She will always be remembered.

Contents

About the Author *xiii*
Preface *xv*
Acknowledgments *xvii*

1 Introduction to Wireless Systems *1*
1.1 Chapter Objectives *1*
1.2 Overview of Wireless Communications *2*
1.3 Radio Systems Classification *3*
1.3.1 Radio Transceivers *3*
1.4 Cellular Phone Systems *6*
1.5 Terahertz (THz) for 6G Wireless Technology *9*
1.6 Multiple-Input Multiple-Output (MIMO) *10*
1.7 Basic Concept of Modulation *11*
1.7.1 Baseband Signals *11*
1.8 Modulation *11*
1.8.1 Time and Frequency Domains of Signals *12*
1.9 Radiofrequency Spectrum Allocation *13*
Review Questions *14*
References *15*
Suggested Readings *16*

2 Analog Communication Systems *19*
2.1 Chapter Objectives *19*
2.2 Overview of Analog Communications *20*
2.3 Amplitude Modulation (AM) *20*
2.4 Single-Sideband Modulation *23*
2.4.1 Filtering Method *24*
2.4.2 Phase Method *24*
2.5 AM Demodulation *26*
2.5.1 AM Envelope Detector *26*
2.6 Frequency Modulation (FM) *26*
2.6.1 Frequency Spectrum of FM Signal *27*
2.7 Noise Suppression in FM Systems *32*
2.8 FM Demodulation *32*
2.8.1 PLL FM Demodulator *33*

2.9 Phase Modulation (PM) 34
Review Questions 36
References 37
Suggested Readings 37

3 Digital Communication Systems 39
3.1 Chapter Objectives 39
3.2 Overview of Digital Communication 40
3.3 Types of Digital Signals 41
3.3.1 Non-Return to Zero (NRZ) Signal 41
3.3.2 Return to Zero (RTZ) Signal 41
3.3.3 Non-Return to Zero (Bipolar) 41
3.3.4 Return to Zero (Bipolar) 41
3.4 Data Conversion 42
3.4.1 Analog-to-Digital Conversion 42
3.4.2 Digital-to-Analog Conversion 44
3.5 Digital Modulation 45
3.5.1 Frequency-Shift Keying (FSK) 45
3.5.2 Phase-Shift Keying (PSK) 47
3.5.3 Quadrature Amplitude Modulation (QAM) 49
3.6 Spectral Efficiency and Noise 51
3.7 Wideband Modulation 52
3.7.1 Frequency-Hopping Spread Spectrum 52
3.7.2 Direct-Sequence Spread Spectrum 54
3.7.3 Orthogonal Frequency Division Multiplexing (OFDM) 56
Review Questions 58
References 59
Suggested Readings 59

4 High-Frequency Transmission Lines 61
4.1 Chapter Objectives 61
4.2 RF Transmission Lines Overview 62
4.3 Transmission Line Analysis 63
4.4 Reflection Due to Impedance Mismatch 67
4.5 Voltage Standing Wave Ratio VSWR 68
4.6 Input Impedance of a Transmission Line 70
4.7 Quarter-Wave Transformer 73
4.8 Planar Transmission Lines in Radio Systems 74
4.8.1 Planar Transmission Line Types 75
4.9 Smith Chart 76
4.10 Impedance Matching Using Smith Chart 88
4.10.1 Single-Stub Matching 89
4.10.2 Double-Stub Matching 91
4.11 ABCD Parameters 98
4.11.1 ABCD Parameters of a Lossless Transmission Line 102
4.12 *S*-parameters 102
4.13 Transmission Line Connectors 106

Review Questions 106
References 109
Suggested Readings 109

5 RF Subsystem Blocks 111
5.1 Chapter Objectives 111
5.2 Introduction to RF Building Blocks 111
5.3 Low-Noise Amplifiers 112
5.3.1 Noise Figure (NF) 112
5.3.2 Noise Figure Measurement Methods 114
5.3.2.1 *Y*-Factor Method for Measuring NF 114
5.3.2.2 Cold Noise Method for Measuring NF 115
5.3.3 Intermodulation Distortion 115
5.3.4 LNA Performance Parameters 117
5.4 RF Mixers 117
5.4.1 Frequency Conversion 118
5.4.2 Image Frequency 118
5.4.3 Image Rejection Mixers 119
5.4.4 Double-Balanced Mixers 119
5.4.5 Mixer's Conversion LOS 120
5.4.6 Mixer's Noise Figure 120
5.4.7 Mixer's Performance Parameters 120
5.5 Filters 121
5.5.1 Low-Pass Filters 121
5.5.2 High-Pass Filters 122
5.5.3 Bandpass Filters 123
5.5.4 Bandstop Filters 123
5.5.5 Cavity Resonator Filters 123
5.5.6 Duplexers 124
5.5.7 RF Filter's Specifications 125
5.6 Frequency Synthesizers 125
5.6.1 Phase-Locked Loop (PLL) Frequency Synthesizer 125
5.7 RF Oscillators 127
5.7.1 Oscillator's Frequency Stability 128
5.7.2 Voltage-Controlled Oscillator 128
5.7.3 VCO Performance Parameters 129
5.8 RF Power Amplifiers 129
5.9 Power Amplifier Linearization Techniques 130
5.9.1 Feedforward Linearization Method 131
5.9.2 Power Amplifier Performance Parameters 132
5.10 Circulators/Isolators 132
5.11 Directional Couplers 133
5.12 Power Splitter/Combiner 134
5.13 Attenuators 134
5.14 RF Phase Shifters 135
5.14.1 Digital Phase Shifters 137
5.15 RF Switches 137

5.15.1 PIN Diode RF Switch *137*
5.15.2 RF Switches Performance Parameters *138*
Review Questions *138*
References *140*
Suggested Readings *140*

6 Basics of RF Transceivers *143*
6.1 Chapter Objectives *143*
6.2 RF Transceivers *144*
6.3 Superheterodyne Receiver Architecture *144*
6.4 Receiver System Parameters *146*
6.4.1 Receiver Sensitivity *146*
6.4.2 Noise Figure *147*
6.4.3 Intermodulation and Spurious-Free Dynamic Range *150*
6.4.4 Receiver Spurious Response *154*
6.4.5 Receiver Selectivity *155*
6.4.6 Intermediate Frequency and Images *155*
6.5 Dual-Conversion Superheterodyne Receivers *156*
6.6 Direct Conversion (Zero-IF) Receiver *157*
6.7 Software-Defined Radios *158*
6.8 RF Block-Level Budget Analysis *159*
6.9 Direct-Conversion Transmitters *162*
6.10 RF Transmitters System Parameters *162*
6.10.1 Transmitter's Output Power *162*
6.10.2 Transmitter's Output Power Dynamic Range *162*
6.10.3 Transmitter's Frequency Error *162*
6.10.4 Transmitter's Error Vector Magnitude (EVM) *163*
6.10.5 Occupied Bandwidth (OBW) *164*
6.10.6 Adjacent Channel Leakage Power Ratio (ACLR) *164*
6.10.7 Operating Band Unwanted Emissions (OBUE) *164*
6.10.8 Transmitter's Spurious Emission *164*
6.10.9 Transmitter's Intermodulation *165*
Review Questions *165*
References *167*
Suggested Readings *167*

7 Antenna Basics and Radio Wave Propagation *169*
7.1 Chapter Objectives *169*
7.2 Introduction *170*
7.3 Antenna Fields *172*
7.4 Antenna Radiation Pattern and Parameters *173*
7.4.1 Antenna Gain *175*
7.4.2 Antenna Beamwidth *176*
7.4.3 Antenna Efficiency *176*
7.4.4 Antenna Aperture Efficiency *176*
7.4.5 Antenna Input Impedance *178*
7.4.6 Antenna Bandwidth *178*

7.4.7 Antenna Polarization *178*
7.5 Isotropic Antenna *179*
7.6 Fields Due to Short Antenna *180*
7.7 Received Power and Electric Field Strength *181*
7.8 Effective Radiated Power *183*
7.9 Antenna Types *184*
7.9.1 Dipole Antenna *184*
7.9.2 Microstrip Antenna *186*
7.10 Antenna Impedance Mismatch *187*
7.11 Antenna Polarization Mismatch *189*
7.12 Antenna Noise Temperature *190*
7.12.1 Antenna Gain-to-Noise-Temperature (G/T) *190*
7.13 Multielements Antenna (Array) *191*
7.14 Multipath Propagation *193*
7.14.1 Radio Wave Propagation Path Loss *195*
7.14.2 Radio Fresnel Zones *197*
7.15 Antenna Characterization *198*
7.15.1 RF Antenna Anechoic Chamber *198*
7.15.2 Compact Antenna Test Range (CATR) *198*
7.16 Antenna Measurements *200*
7.16.1 Antenna Gain Measurements *200*
7.16.1.1 Absolute Gain Method *200*
7.16.1.2 Gain-Transfer Method *200*
7.16.2 Antenna Radiation Pattern and Directivity Measurement *201*
7.16.3 Antenna Input Impedance Measurement *201*
Review Questions *202*
References *203*
Suggested Readings *204*

8 Introduction to MIMO and Beamforming Technology *207*
8.1 Chapter Objectives *207*
8.2 Overview of 5G NR Technology and Beyond *207*
8.3 5G NR Frequency Ranges *209*
8.4 5G NR Radio Frame Structure *210*
8.5 5G NR Numerology *211*
8.6 5G NR Resource Grid *212*
8.7 Massive MIMO for 5G Systems *213*
8.7.1 Simplified Mathematical Model of a MIMO Channel *214*
8.8 Beamforming Technology *216*
8.8.1 Analog Beamforming *219*
8.8.2 Digital Beamforming *219*
8.8.3 Hybrid Beamforming *219*
Review Questions *220*
References *221*
Suggested Readings *221*

9 RF Performance Verification of 5G NR Transceivers *223*
9.1 Chapter Objectives *223*
9.2 Test Instruments for Radio Performance Verification *224*

9.2.1 RF Signal Generators 224
9.2.2 Vector Spectrum Analyzer 224
9.3 RF Performance Verification of 5G NR Transmitters 225
9.3.1 Transmitter Output Power 226
9.3.2 Transmitter Total Power Dynamic Range 227
9.3.3 Transmit ON/OFF Power 227
9.3.4 Transmitter Frequency Error 228
9.3.5 Transmitter's Error Vector Magnitude (EVM) 229
9.3.6 Time Alignment Error (TAE) 231
9.3.7 Occupied Bandwidth (OBW) Measurement 232
9.3.8 Adjacent Channel Leakage Power Ratio (ACLR) 233
9.3.9 Operating Band Unwanted Emissions (OBUE) Measurement 234
9.3.10 Conducted Spurious Emission 235
9.3.11 Transmitter Intermodulation 237
9.4 RF Performance Verification of 5G NR Receivers 238
9.4.1 Receiver Reference Sensitivity 239
9.4.2 Receiver Dynamic Range 239
9.4.3 Adjacent Channel Selectivity (ACS) Measurement 242
9.4.4 In-band Blocking 242
9.4.5 Out-of-Band Blocking 245
9.4.6 Receiver Spurious Emissions 245
9.4.7 Receiver Intermodulation 246
9.4.8 In-channel Selectivity 247
9.5 Over-The-Air (OTA) Testing of Radio Systems 251
Review Questions 253
References 254
Suggested Readings 254

Index 257

About the Author

Dr. Ibrahim A. Haroun is a retired senior RF system designer, Canada. In his career, Dr. Haroun has worked for a number of companies in different capacities, from a senior radio frequency (RF)/microwave designer to an RF hardware system design manager, and worked as a Research Scientist/Engineer at the Communications Research Centre, Canada. He was a Lecturer with the Marine Institute of Memorial University, Canada, and he was a Research Adjunct Professor with the Department of Electronics, Carleton University, and the University of Western Ontario, Canada. Dr. Haroun has also published and presented several papers at international conferences and is a Senior Member of the IEEE.

Preface

The purpose of this book, *Essentials of RF Front-end Design and Testing: A Practical Guide for Wireless Systems*, is to provide the required knowledge for developing RF transceiver front-ends using commercial off-the-shelf building blocks, as well as verifying their RF performance, and developing RF prototypes for wireless applications. It covers relevant topics for underlying RF systems from baseband to transmission. In addition, the book should serve as a reference for RF systems engineers to develop and characterize RF transceiver front-ends and proof-of-concept wireless systems, and provide complementary materials for academic students (undergraduate/graduate) to enhance their learning experience on RF system development and testing. The book includes nine chapters. Each chapter has learning objectives, review questions, a list of references related to the covered topics, and suggested readings that present the up-to-date developments related to the topics covered in the chapters. The chapters are organized as follows:

Chapter 1 provides an overview of wireless communications, radio system classifications (simplex, half-duplex, and full-duplex), radio transceivers, cellular phone systems, terahertz (THz) communication, MIMO (multiple-input, multiple-output), the basic concept of modulation, and radio frequency spectrum allocation.

Chapter 2 introduces and explains the principles of analog communications, amplitude modulation AM, single-sideband modulation, AM demodulation, frequency modulation FM, noise suppression in FM, FM phase-locked loop detector, and phase modulation PM.

Chapter 3 discusses and presents an overview of digital communications, types of digital signals, analog-to-digital and digital-to-analog conversion, digital modulation including frequency-shift keying FSK, phase shift keying PSK, and quadrature amplitude modulation QAM. Spectral efficiency and noise, wideband modulation including spread spectrum frequency hopping, direct sequence spread spectrum, and orthogonal frequency division multiplexing OFDM.

Chapter 4 covers high-frequency transmission lines, transmission line analysis, reflection due to impedance mismatch, voltage standing wave ratio (VSWR), the input impedance of a transmission line, quarter-wave impedance matching, planar transmission lines, the Smith chart, single-stub and double-stub matching, S-parameters, and ABCD parameters.

Chapter 5 provides the essential background for testing and characterizing the building blocks of RF transceiver front-ends. Such blocks include low-noise amplifiers, mixers, RF filters (low-pass, high-pass, band-pass, band-stop), cavity filters, duplexers, frequency synthesis, voltage-controlled oscillators, power amplifiers, circulators and isolators, directional couplers, power combiners, RF switches, and RF phase shifters.

Chapter 6 presents different receiver and transmitter architectures, focusing on system performance and characterization rather than circuit design. It covers different architectures including superheterodyne, direct conversion (i.e., Zero-IF), low-IF, software-defined radio (SDR). The receiver parameters that are discussed include receiver sensitivity and selectivity, receiver intermodulation, and dynamic range. The transmitter system parameters that are covered include output transmitter power, spurious emission, frequency error, error vector magnitude (EVM), occupied bandwidth (OBW), adjacent channel leakage power ratio (ACLR), operating band unwanted emission (OBUE), spurious emission, and transmitter intermodulation. The impact of the transmitter and receiver parameters on the system performance is also addressed. The RF block-level budget analysis and examples of the receiver RF block-level budget analysis are presented to help the readers apply the topic to develop an RF block-level budget.

Chapter 7 covers antenna fundamentals including antenna gain, radiation pattern, antenna input impedance, antenna polarization, antenna noise temperature, antenna types, microstrip antennas, and the theory of array antennas. It also discusses multipath propagation and propagation path loss, Fresnel zones, antenna anechoic chamber, compact antenna test range, and antenna measurements including gain, radiation pattern, and input impedance.

Chapter 8 covers the basics of MIMOs and beamforming technology for wireless systems such as 5G, with a focus on the RF architecture of the beamforming subsystems. The chapter provides a brief overview of 5G NR technology including the technology evolution, 5G NR frequency bands, 5G NR frame structure, 5G numerology and subcarrier spacing, and 5G resource grid. Such an overview helps the readers understand the specifications of the RF conformance testing of 5G NR transceivers. The concept of massive MIMO, the MIMO channel, and the beamforming types including analog, digital, and hybrid beamforming are discussed.

Chapter 9 introduces and explains the test setups for characterizing the key RF test parameters of 5G NR base station transmitters and receivers. The transmitter test parameters that are discussed include the output power dynamic range, ON/OFF transmit power, error-vector-magnitude (EVM), occupied bandwidth, adjacent channel leakage power, operating band unwanted emissions (OBUE), spurious emissions, and intermodulation. Images of the measured transmitter test results are presented. The receiver test parameters that are covered include receiver sensitivity, dynamic range, adjacent channel selectivity, in-band blocking, receiver spuriousness, and receiver intermodulation. The chapter also explains the over-the-air (OTA) testing method for characterizing the performance of wireless systems.

Canada, 13 October 2023 *Ibrahim A. Haroun*

Acknowledgments

I would like to thank my former students at the Marine Institute of Memorial University, who suggested having a book that covers various radio communication disciplines in a single book. Also, I would like to thank Wiley's team for their support in completing this book project.

1
Introduction to Wireless Systems

This chapter presents a brief overview of wireless communications and cellular phone systems to lay the foundation for discussing the book's remaining chapters. It covers the basic functional blocks of a wireless communication system, radio systems classifications (simplex, half-duplex, and full-duplex), radio transceivers of wireless systems, cellular phone systems including first-generation (1G), second-generation (2G), third-generation (3G), fourth-generation (4G), and fifth-generation (5G) New Radio wireless technologies. Further topics cover the air interface technologies, including frequency division multiple access (FDMA), time division multiple access (TDMA), code division multiple access (CDMA), and orthogonal frequency division multiple access (OFDMA), terahertz (THz) communications for next-generation 6G wireless technology, multiple-input multiple-output (MIMO) technology, the basic concept of modulation, time and frequency-domain signals, and frequency spectrum allocation. These topics are discussed in further detail in the references list at the end of the chapter. The chapter also provides review questions to help the reader understand the covered topics, and suggested readings that present the up-to-date research activities that are related to the topics of this chapter.

1.1 Chapter Objectives

On reading this chapter, the reader will be able to:

- Draw a block diagram of a wireless communication model and explain the function of each block.
- Compare and contrast full-duplex, simplex, and half-duplex radio systems.
- Explain the difference between 1G, 2G, 3G, 4G, and 5G wireless technologies.
- Explain the applications of frequency range 1 (FR1) and frequency range 2 (FR2) of 5G NR (new radio) wireless technology.
- Explain the advantages and challenges of using THz frequency in the future 6G wireless technology.
- Explain the advantages of MIMO wireless systems.
- Briefly explain the difference between amplitude modulation (AM) and frequency modulation (FM) and why modulation is needed.
- Illustrate the time-domain and frequency domain of a sinusoidal signal.
- Explain the differences between FDMA, TDMA, and CDMA wireless access technologies.

Essentials of RF Front-end Design and Testing: A Practical Guide for Wireless Systems, First Edition. Ibrahim A. Haroun.

– Explain why radio transceivers' front end uses circulators and radiofrequency (RF) switches.
– Illustrate and explain the frequency spectra of an OFDM signal.

1.2 Overview of Wireless Communications

Wireless communication [1–8] is a global industry people depend on in many ways for effective business, efficiency in service (e.g., education, health, security, etc.), as well as for critical situations (e.g., lifeline emergencies), which makes it indispensable to public safety. Wireless technology was first introduced in 1901 when Guglielmo Marconi successfully achieved the first transatlantic communication. Radio broadcasting began in the 1920s and used vacuum-tube technology in transmitters and receivers. The invention of integrated circuits in 1958 enabled the design and development of small-size wireless communication systems.

The primary function of a wireless communication system is to send information, such as voice, video images, data, or other physical variables (e.g., temperature, speed, pressure, etc.), from the information source to the information destination through a radio channel (i.e., free-space transmission medium). Figure 1.1 shows a simplified block diagram of a wireless communication model.

In Figure 1.1, the source block represents a transducer that converts physical variables into electrical signals. These electrical signals are also called baseband signals that can be processed (e.g., encoding, interleaving, etc.) in the transmitter chain. Microphones and cameras are transducers that convert sound and visual images to baseband signals. Baseband signals cannot be transmitted directly over the air because such a transmission requires large antennas. To transmit baseband signals over the air, they must be combined with a higher frequency called carrier frequency, and this process is called modulation. The modulated signal gets amplified and applied to the antenna, which acts as an interface between the transmitter's output and the free space. The antenna converts the RF signals to radio waves (i.e., electromagnetic waves) that propagate over the air at the speed of light (3×10^8 m/s). Antennas and radio wave propagation are discussed in detail in Chapter 7.

At the receiver end of a wireless link, the antenna converts the radio waves into electrical signals that get down-converted to an intermediate frequency (IF) and baseband. Finally, the transducers convert the baseband signals back to the original physical variables that were sent by the transmitter.

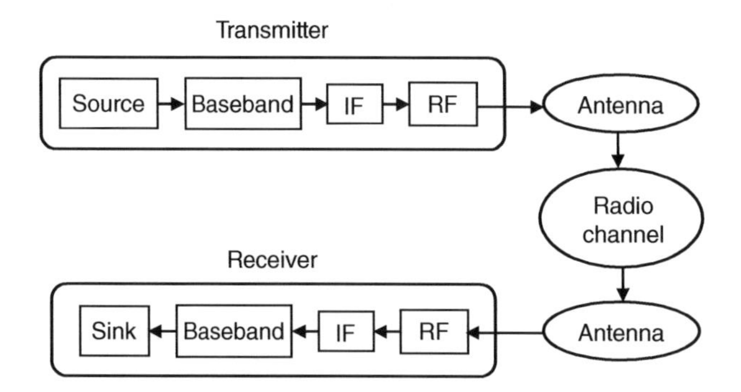

Figure 1.1 Simplified block diagram of a wireless communication model.

1.3 Radio Systems Classification

Radio communications systems are classified into simplex, half-duplex, and full-duplex. In a simplex system, communication is one-way from the transmitter to the receiver, as in radio and TV broadcasting. In a full-duplex system, transmission and reception coincide (thus creating a two-way communication) as in mobile phones. In a half-duplex radio system, the receiver and transmitter alternate (i.e., take turns). Figure 1.2 illustrates half-duplex and full-duplex transmission between two wireless systems *A* and *B*.

1.3.1 Radio Transceivers

Radio transceivers transmit and receive radio signals in wireless communication systems such as cellular phones, radars, two-way radios, and electronic navigation systems. In radio transceivers, the transmitter and the receiver share the same antenna and are packaged in the same enclosure.

When the transmit frequency differs from the receive frequency, a duplexer is used to enable the use of a single antenna for both transmit and receive operations. Figure 1.3 shows a simplified block diagram of an RF transceiver that uses a duplexer.

Half-duplex (*A*-to-*B* or *B*-to-*A* take turns)

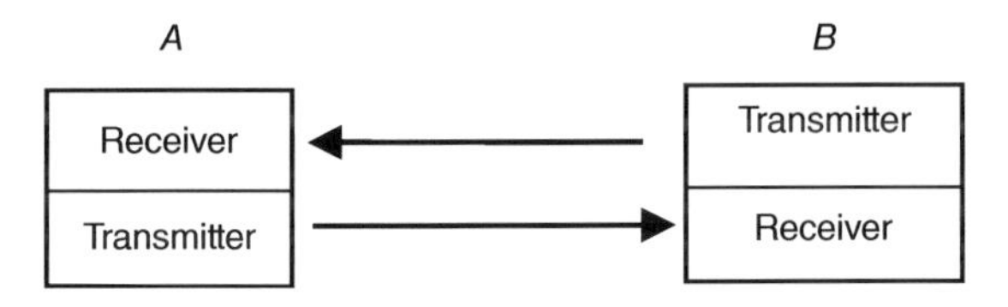

Full-duplex (*A*-to-*B* and *B*-to-*A* simultaneously)

Figure 1.2 Types of radio communication systems.

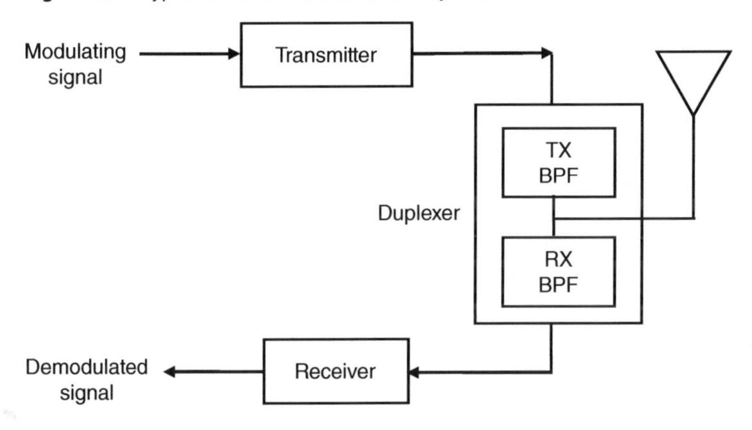

Figure 1.3 Block diagram of a radio transceiver front-end using a duplexer.

A duplexer is a device that enables bi-directional communication in wireless communications systems, and it isolates the receiver from the transmitter while permitting them to share a common antenna. In Figure 1.3, the duplexer consists of two bandpass filters: one is the transmit bandpass filter (TX BPF) that passes the transmit signal to the antenna and blocks the received signal, while the other is the receive bandpass filter (RX BPF) that passes the received signal from the antenna to the receiver and blocks the transmit signal. In this way, a single antenna can be shared by both the transmitter and the receiver.

If the transmitter and receiver operate on the same frequency, an RF switch connects the transmitter to the antenna during the transmission and isolates the receiver. The switch also connects the antenna to the receiver and isolates the transmitter during the reception to enable sharing of the antenna by the transmitter and the receiver. Figure 1.4 shows a block diagram of an RF transceiver that uses an RF switch.

To prevent transmit power leakage to the receiver, the RF switch should have sufficient isolation. The transmit (TX) leakage to the receiver's input reduces the signal-to-noise ratio, degrading the receiver sensitivity (i.e., the ability to detect weak signals). A circulator can be used to enable the transceiver to transmit and receive without constraint on the transmitter and the receiver timing and frequency. Figure 1.5 shows a radio transceiver that uses a circulator.

A circulator is a passive, nonreciprocal three-port device that enables a radiofrequency signal to pass from one port to another while isolating the signal from the other port. The performance specifications and testing of duplexers, RF switches, and circulators are discussed in further detail in Chapter 5. Figure 1.6 shows a block diagram of an RF transceiver that uses a transmit/receive (T/R) switch.

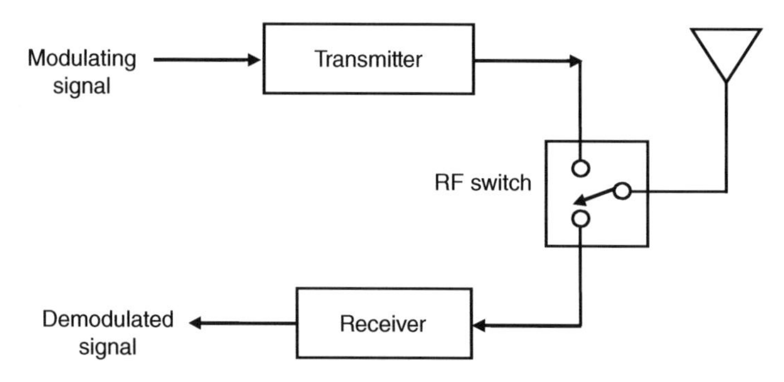

Figure 1.4 Block diagram of a radio transceiver with an RF switch.

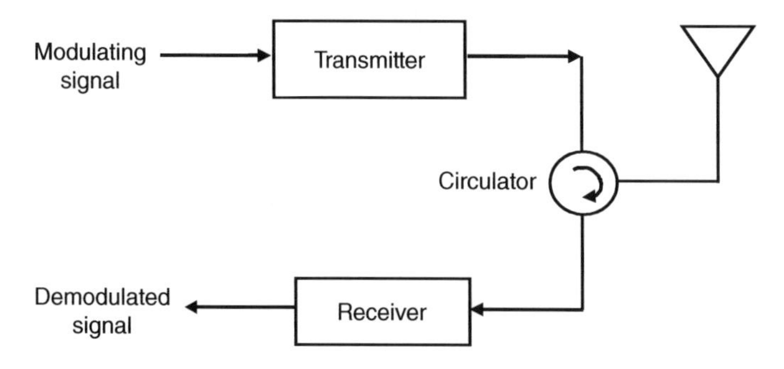

Figure 1.5 Block diagram of a radio transceiver using a circulator.

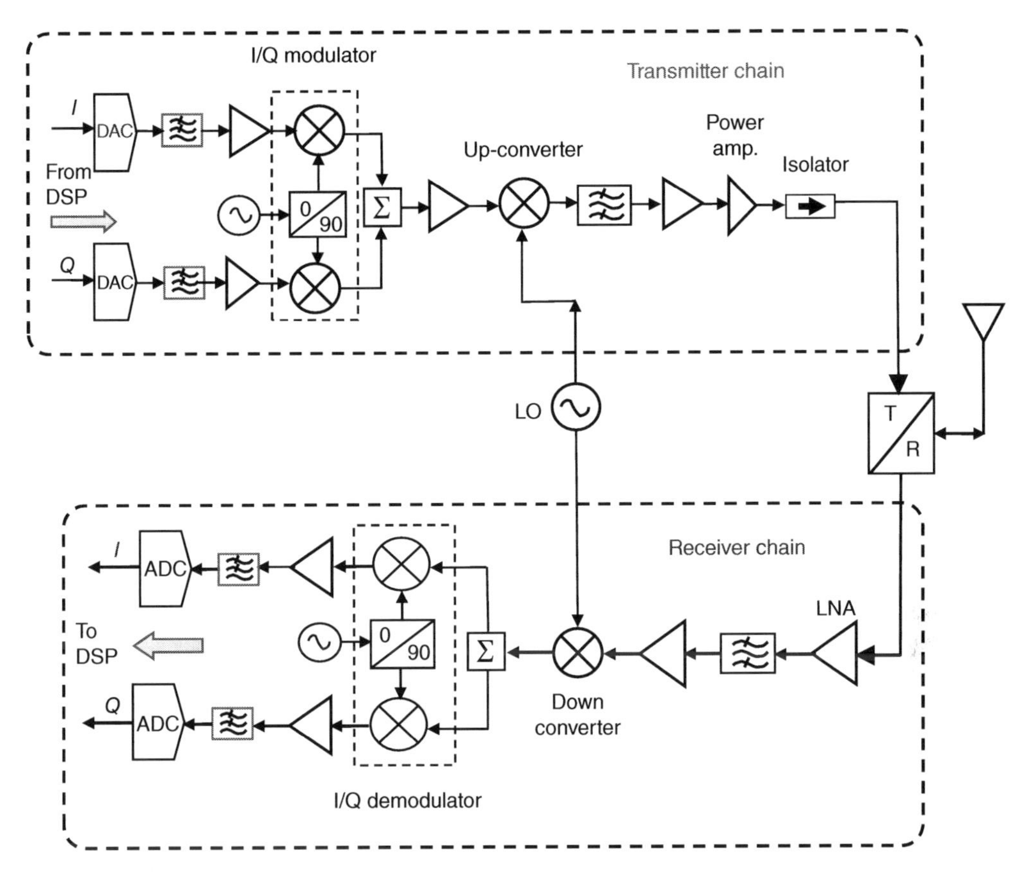

Figure 1.6 Block datagram of an RF transceiver.

Figure 1.6 shows the basic functional blocks of an RF transceiver, including a (T/R) switch, low-noise amplifier (LNA), filters, power amplifier, mixers, local oscillators (LO), analog-to-digital converter (ADC), digital-to-analog converter (DAC), modulator, and demodulator. Some of these building blocks can be used standalone or integrated into an integrated chip. Although similar RF building blocks in different types of transceivers perform the same basic function, the complexity of each implementation can considerably vary depending on the overall system requirement and operating environment for each specific application. The performance of these blocks is affected by many factors, such as the communication link range, information bandwidth, and power budget. The basic theory of operation and testing of the RF building blocks that are used in wireless communication systems are discussed in Chapter 5.

In the transmit chain shown in Figure 1.6, the in-phase/quadrature (I/Q) signals from the digital-signal-processing (DSP) block get converted by the DAC blocks and then filtered, amplified, and applied to the I/Q modulator. The signals from the I/Q modulator are combined by a power combiner and then applied to a gain block to drive the mixer that converts the signal to the required transmit frequency. The LO port of the mixer is connected to the LO block that generates the carrier signal. The output of the mixer is filtered by a bandpass filter that suppresses any undesired frequency components that are generated in the mixing process. The output of the filter is applied to a driver amplifier to compensate for the filter's insertion loss and to amplify the signal to the level that drives the power amplifier. The power amplifier amplifies the signal to the required transmit power level. The isolator between the power amplifier and the T/R switch prevents any reflection from the switch to the power amplifier. The T/R switch connects the transmit signal to the antenna and isolates the receiver from the antenna during the transmission time. The antenna converts the transmit signal to electromagnetic waves propagating in the free space at the speed of light.

In the receiver chain shown in Figure 1.6, the antenna converts the received electromagnetic waves to an electrical signal. At the same time, the T/R switch connects the antenna to the LNA and isolates the transmit chain to protect the receiver. The LNA amplifies the received RF signal for subsequent processing; the output of the LNA is then applied to a BPF filter to suppress out-of-band signals. The down converter mixer converts the RF to a lower frequency signal to be demodulated by the I/Q demodulator and converted to I and Q signals by the ADC blocks. The I and Q signals are further processed by the DSP block to optimize the receiver performance. Each block of the transceiver's blocks and their performance parameters is discussed in Chapter 6.

1.4 Cellular Phone Systems

Cellular phones are the most widely used wireless systems because they enable users to connect to the standard telephone systems, support sending text messages, checking email, accessing the Internet, finding a location, and taking high-resolution photos. The original cellular technology advanced mobile phone system (AMPS) was commercially introduced in the 1980s as a 1G (first-generation) product and used the frequency division multiple access (FDMA) technique. Figure 1.7 shows a representation of an FDMA transmission.

The FDMA technique allows multiple users to communicate with the base station (i.e., a fixed radio system that handles the cellular traffic) without interfering with each other. In this technique, the multiplexing process is achieved by assigning each user a narrow frequency band (i.e., channel), and all the channels are combined and transmitted over a broadband channel. This multiplexing type is called frequency division multiplexing (FDM). In an FDM, narrowband channels are separated by guard bands to prevent interference between the channels. Figure 1.8 shows a simplified block diagram of an FDM transmitter. A representation of FDM transmission is shown in Figure 1.9.

In Figure 1.8, the multiple baseband signals $m_1(t)$, $m_2(t)$, and so on modulate different subcarriers $f_1, f_2, \ldots, f_n$. The outputs of the subcarrier modulators are fed to a combiner that merges all the modulated signals to produce a composite baseband signal. This composite baseband signal is then fed to an RF up-conversion block (i.e., the frequency conversion to higher frequency) in the transmitter chain, where the output of the upconverter is filtered, amplified, and connected to an antenna for transmission over the air.

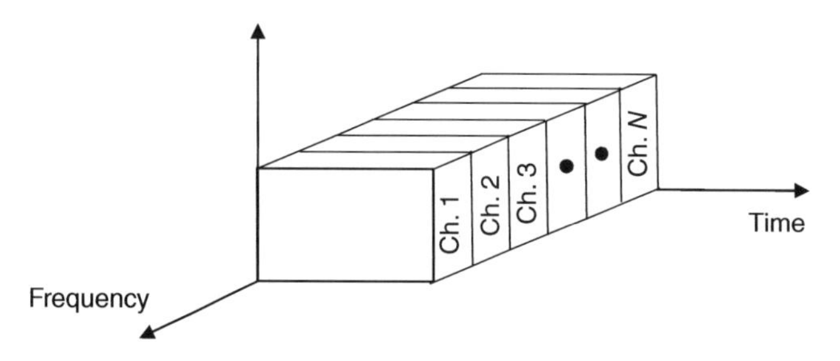

Figure 1.7 Illustration of an FDMA channels.

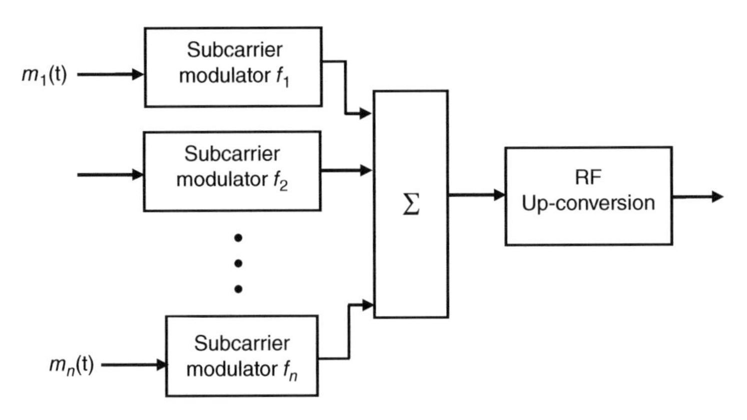

Figure 1.8 Simplified block diagram of an FDM system transmitter.

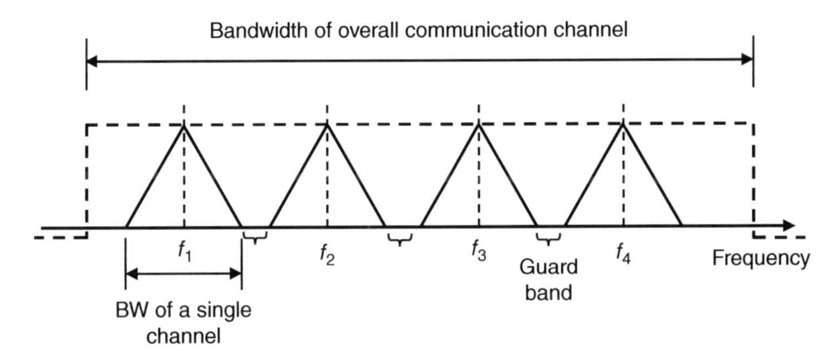

Figure 1.9 FDM transmission.

Since the AMPS technology could not meet the demand for more services, a 2G phone technology was developed. The 2G marked the transition of mobile networks from analog to digital.

The 2G technology supported basic data services such as short message services. It used time division multiple access (TDMA) multiplexing. In TDMA, multiplexing is achieved by assigning each user a time slot to transmit and receive. Figure 1.10 shows a representation of TDMA transmission.

Later, the continuous demand for more services led to the development of the 3G cellular systems. The 3G technology introduced improved mobile broadband services and enabled new applications such as multimedia message services, video calls, and mobile TV.

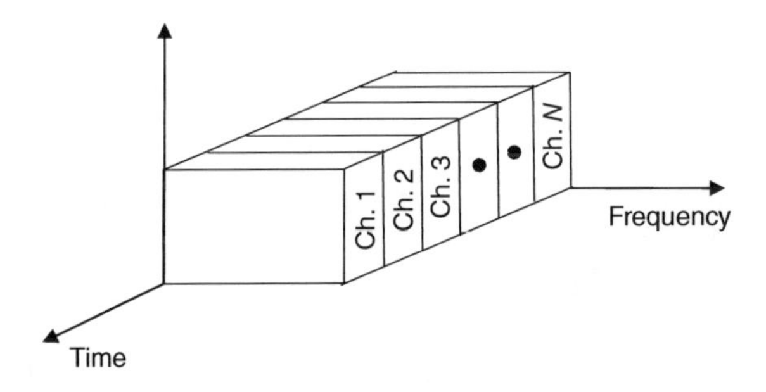

Figure 1.10 Illustration of TDMA transmission.

The 3G cellular systems used the code division multiple access (CDMA) technology. In CDMA, the narrow band message signal is multiplied by a large bandwidth signal called the *spreading signal*. Spreading signals are a pseudorandom sequence whose chip rate is larger than the data rate of the message. CDMA systems use the same carrier frequency to modulate all the users' signals and to allow the users to transmit simultaneously. In a CDMA system, each user has a unique pseudo-random code word, which is approximately orthogonal to all other code words to ensure minimum mutual interference among users. Figure 1.11 shows a representation of a CDMA transmission.

The CDMA receiver performs a time correlation function so that signals other than the desired signal appear uncorrelated. To decode the desired signal, the receiver uses the same code word that is used in the transmission.

Further, improved mobile broadband services, Voice Over Internet Protocol (VoIP), ultra-high-definition video streaming, and online gaming were introduced in the 4G systems. The 4G technology is also known as long-term evolution (LTE). It used orthogonal frequency division multiple access (OFDMA) and MIMO techniques. Figure 1.12 shows a representation of the frequency spectra of an OFDM signal.

In Figure 1.12, the subcarriers overlap and whenever there is a peak for one subcarrier, there is none for the other subcarriers.

A decade after introducing 4G systems, the fifth-generation new radio (5G NR) wireless systems were introduced. The 5G NR technology is a new standard provided by the 3rd Generation

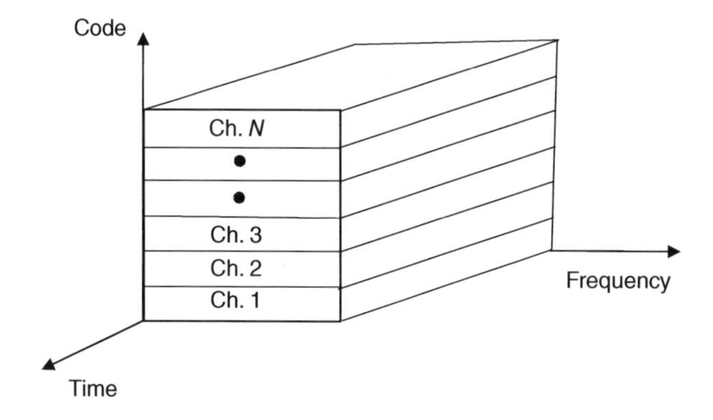

Figure 1.11 CDMA transmission.

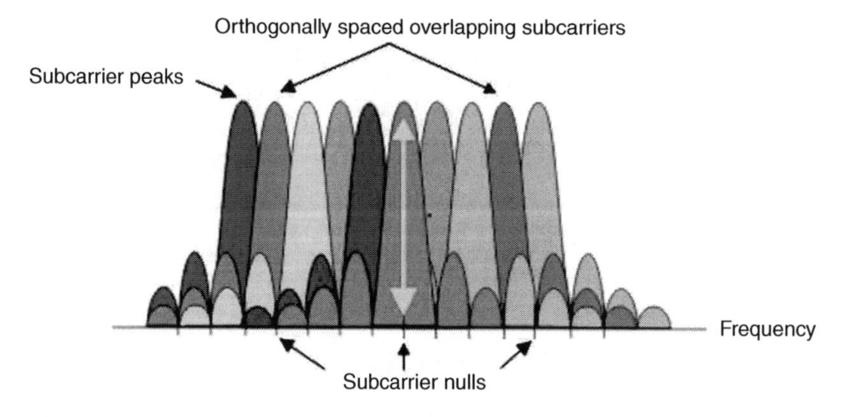

Figure 1.12 Representation of an OFDM transmission.

Partnership Project (3GPP) organization. The main goals of this technology include enhanced mobile broadband (eMBB), the downlink data rate of 10 Gbps, massive machine-type communication (mMTC) associated with Internet-of-Things (IoT), and ultra-reliable low-latency communication (uRLLC).[1] In addition, mmWave 5G systems support large bandwidth up to 400 MHz.

There are many 5G NR bands, and they vary from 600 MHz to 52.5 GHz. In the United States, the operating frequencies are designated FR1 and FR2. FR1 band ranges from 410 MHz to 7.125 GHz, which is used to carry most of the traditional cellular mobile communications traffic, whereas FR2 (24.25 to 52.6 GHz) band is focused on short-range, high data rate capabilities.

Additional space has been allocated in the mmWave spectrum from 24 up to 95 GHz. In the United States, three spectrum segments which are 28, 37–39, and 40 GHz have been dedicated to 5G NR. These segments offer sufficient bandwidth to support the Gbps data rate offered by the standard—specifically, 27.5–38.35 GHz, 37–38.6 GHz, and 38.6–40 GHz.

1.5 Terahertz (THz) for 6G Wireless Technology

Although the 5G mobile communication [9–11] networks are now widely deployed worldwide, the capabilities of the 5G technology are challenged by many new applications that are required. In this context, researchers have worked towards the next-generation 6G mobile communication network [12–14]. The THz frequency [15–17], also known as submillimeter-wave, ranges from 100 GHz to 10 THz (1 THz = 10^{12} Hz). It is becoming one of the cornerstones of the future 6G wireless systems, as this frequency has very large available bandwidths. Figure 1.13 shows the location of the THz region in the electromagnetic spectrum.

The THz technology enables high-definition video resolution radars that can provide TV-like pictures to complement radars at low frequencies that give a more extended range The technology

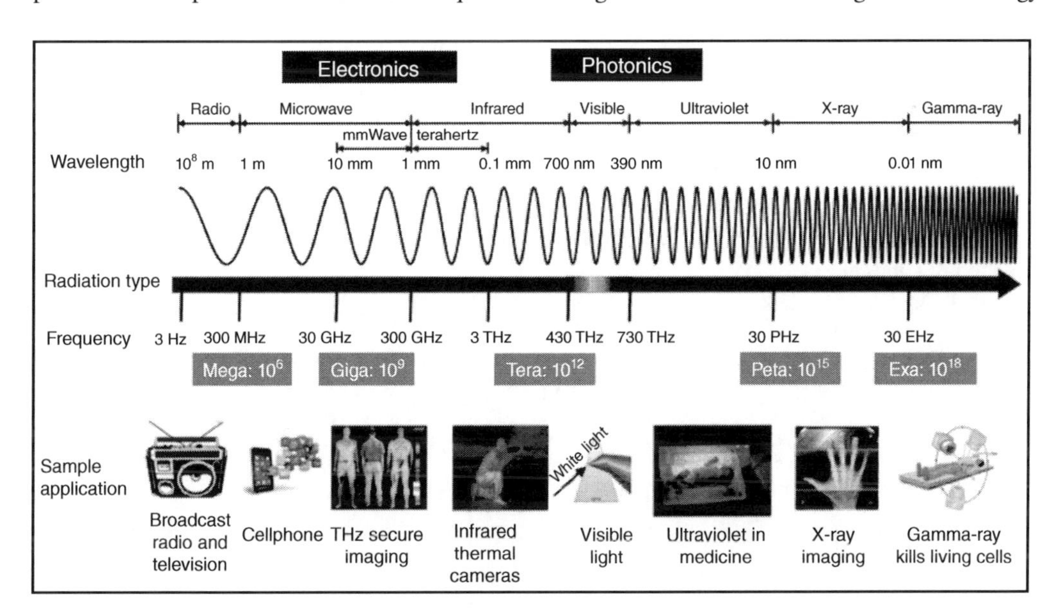

Figure 1.13 Illustration of THz region in the electromagnetic spectrum [17].

1 Latency is the time delay between initiating an action and the time the action occurs.

also makes it possible to achieve high-resolution imaging through fog, dust, and smoke. In addition, the THz technology has promising applications in security, public safety, health environment, and education.

The disadvantages of THz communications are related to the strong absorption caused by water vapor in the atmosphere and the low output power from the currently available THz sources. These disadvantages limit the achievable link distance. In addition, the high propagation losses at THz frequencies will require the simultaneous use of high-gain directional antennas at the transmitter and the receiver of a THz communication link. However, the propagation path loss of a THz link can be compensated for by using ultra-massive MIMO (UM-MIMO) systems. The THz frequencies enable the implementation of UM-MIMO systems because of the small wavelengths at these frequencies. Furthermore, having a small wavelength empowers the development of small-size antenna elements, which is essential for implementing UM-MIMO systems.

In summary, future applications of THz communications will largely depend on the availability of more efficient sources, coherent detectors, and modulators at this frequency range, as well as on the availability of relevant test equipment to validate the performance of THz systems.

1.6 Multiple-Input Multiple-Output (MIMO)

MIMO [18] is a wireless technology that uses multiple antennas at the transmitter and receiver to use multipath signals to improve the link reliability and increase the channel capacity. Common arrangements of MIMO include 2×2 MIMO (i.e., two transmit antennas and two receive antennas), 4×4 MIMO, and 8×8 MIMO. At mmWave frequencies, massive MIMO systems use 100s antenna elements.

In MIMO systems, serial data are divided into separate data streams that are transmitted simultaneously over the same channel. Figure 1.14 illustrates a basic MIMO system with multiple transmit antennas at the input of the propagation channel and multiple receive antennas at the channel's output (i.e., MIMO).

Since the antenna size is inversely proportional to the frequency, mmWave frequencies enable having small antennas to implement large antenna arrays. A large antenna array provides a high gain and narrow beam, thereby enabling interference mitigation and extending the communication range of the link. Small-size antenna elements contribute to having an integrated antenna system and removing the interconnect cables. Further details on the concept of MIMO technology are provided in Chapter 8.

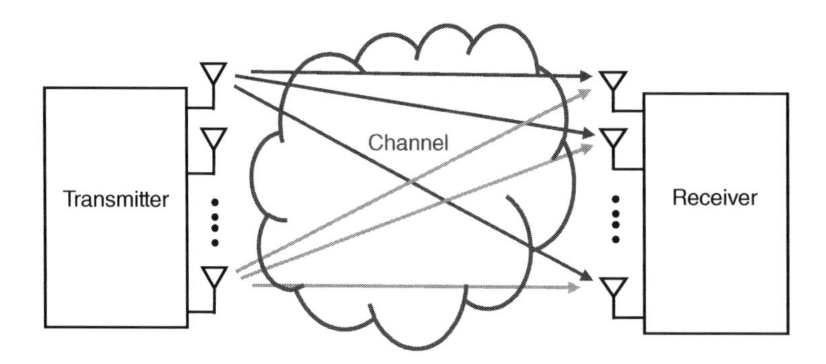

Figure 1.14 Illustration of a basic MIMO system.

1.7 Basic Concept of Modulation

1.7.1 Baseband Signals

Although this book focuses on the RF subsystem blocks of transmitters and receivers in wireless communication systems rather than on baseband subsystem blocks, it is worthwhile to describe baseband signals briefly for a more comprehensive view of a wireless communication system.

In wireless communication systems, to transmit a baseband signal (i.e., low-frequency signal) over the air via an antenna, it should be combined with a high-frequency signal called the carrier signal. The antenna structure has very small dimensions compared to the wavelength, λ, of the carrier signal. The wavelength of a carrier signal is given by

$$\lambda = \frac{c}{f} \text{m/s} \tag{1.1}$$

where,

$c = 3 \times 10^8$ m/s
f = frequency of the signal in Hz

The direct transmission of baseband signals would require very large antennas. For that reason, baseband signals need to modulate a high-frequency signal (i.e., to be mixed with RF signal) to be transmitted.

1.8 Modulation

In radio transmitters, having the baseband signal modify the carrier signal is called modulation, and the device that performs the modulation is called a modulator. Figure 1.15 shows a modulator block in a wireless transmitter system.

The carrier signal $s(t)$ of a wireless transmitter is expressed as

$$s(t) = A \cos(2\pi f_c t + \theta) \tag{1.2}$$

where

A = amplitude of the carrier
f_c = frequency of the carrier
θ = phase of the carrier

If the amplitude of the carrier changes proportionally to the amplitude of the baseband signal, this modulation is called ***AM***. Conversely, if the frequency of the carrier changes proportionally

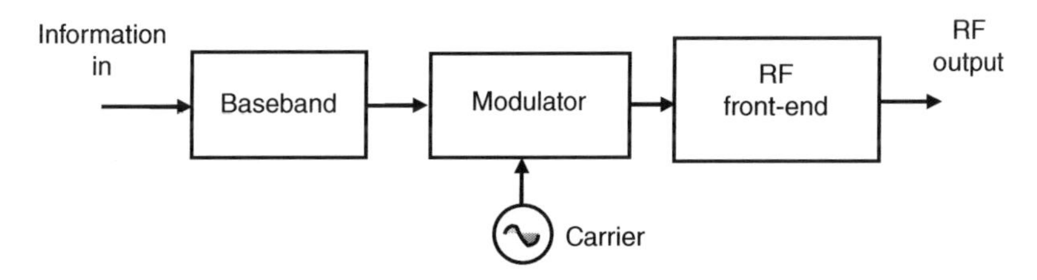

Figure 1.15 Modulator block in a wireless transmitter.

to the baseband, this modulation is called **FM**. Finally, if the phase of the carrier changes proportionally to the baseband's amplitude, such modulation is called **PM.** AM, FM, and PM modulation schemes are discussed in further detail in Chapter 2, and the digital modulation schemes are discussed in Chapter 3.

In all modulation schemes, the modulation process produces frequency components different from the carrier frequency. The receiver of a communication link demodulates the received RF signal to extract the baseband signal that was used in the transmitter.

1.8.1 Time and Frequency Domains of Signals

Electrical signals can be represented in both time and frequency domains. In a time domain, the amplitude of the signal is expressed as a function of time, whereas in a frequency domain, the amplitude of the signal is expressed as a function of frequency. Time-domain signals are displayed using oscilloscopes, and frequency-domain signals are displayed using spectrum analyzers. Figure 1.16 shows two different sinusoidal signals' time and frequency domains (A and B).

In Figure 1.16b, the frequency of signal A is smaller than that of signal B, and this is because the period (T) of signal A is longer than that of signal B (frequency = $1/T$), as shown in Figure 1.16a. Figure 1.17 shows the frequency and time domains of a signal $v(t)$, which results from mixing two signals. These two signals are $v_1(t)$ with amplitude "A" and $v_2(t)$ with amplitude "B."

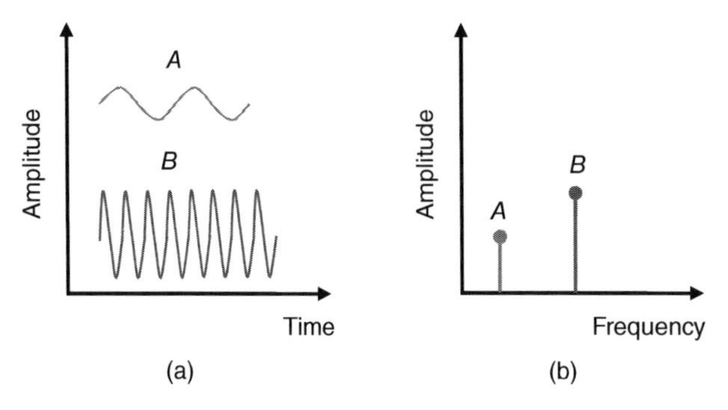

Figure 1.16 Time and frequency domains of two sinusoid signals A and B. (a) Time-domain representation. (b) Frequency domain representation.

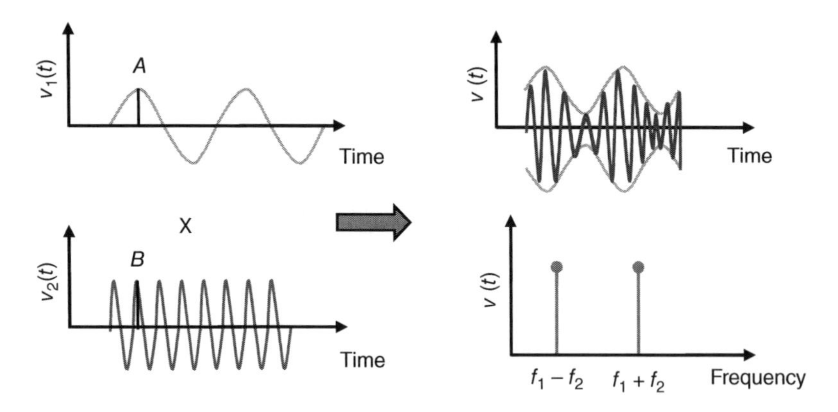

Figure 1.17 Time and frequency domains of two multiplied signals $v_1(t)$ and $v_2(t)$.

The signal $v(t)$ is expressed as

$$v(t) = v_1(t) \times v_2(t) \tag{1.3}$$

where

$v_1(t) = A \cos(2\pi f_1\, t)$
$v_2(t) = B \cos(2\pi f_2\, t)$

$$\therefore \quad v(t) = \frac{AB}{2}\cos[2\pi(f_1 + f_2)t] + \frac{AB}{2}\cos[2\pi(f_1 - f_2)t] \tag{1.4}$$

where A, B, f_1, and f_2 are the amplitudes and frequencies of $v_1(t)$ and $v_2(t)$ signals, respectively. From Eq. (1.4), the obtained signal from the multiplication process has two frequency components $(f_1 + f_2)$ and $(f_1 - f_2)$, referred to as upper and lower sidebands.

1.9 Radiofrequency Spectrum Allocation

Radio waves are controlled and regulated by various governmental and international organizations such as the International Telecommunication Union (ITU) in Geneva, Switzerland; Federal Communication Commission (FCC) in the United States; Innovation Science and Economic Development Canada (ISED) in Canada, Australia Communication and Media Authority (ACMA), and other organizations in other countries. Regulation of the spectrum is implemented to ensure that the communication systems do not interfere with each other. Table 1.1 lists standard frequency ranges in the radiofrequency spectrum. Table 1.2 specifies some spectrum allocations. The 3GPP frequency bands [19] for mmWave 5G-NR cellular phone systems that are time-division duplexing (TDD) systems are listed in Table 1.3.

Table 1.1 Standard frequency ranges.

Frequency	Designation	Abbreviation
30–300 Hz	Extremely low frequency	ELF
300–3000 Hz	Voice frequency	VF
3–30 KHz	Very low frequency	VLF
30–300 KHz	Low frequency	LF
300 KHz–3 MHz	Medium frequency	MF
3–30 MHz	High frequency	HF
30–300 MHz	Very high frequency	VHF
300 MHz–3 GHz	Ultra-high frequency	UHF
3–30 GHz	Super high frequency	SHF
30–300 GHz	Extra high frequency	EHF

Table 1.2 Some spectrum allocations.

Frequency (MHz)	Allocation
26.9–27.4	Citizens band
29–54	Land mobile
50–54	Amateur
54–88	TV low VHF
88–108	FM BCB
108–136	Aircraft
132–174	Land mobile
174–216	TV high VHF
406–512	Land mobile
470–806	TV UHF
806–947	Land mobile
806–947	Cellular AMPS
1200–1600	Amateur Land mobile GPS
1700–2000	Cellular PCS
2400–2500	ISM Bluetooth Wi-Fi

Table 1.3 Frequency range 2 (FR2) for 5G-NR mmWave systems.

Band	Frequency (GHz)	Uplink/Downlink (GHz)
n257	28	26.50–29.50
n258	26	24.25–27.50
n259	41	39.50–43.50
n260	39	37.00–40.00
n261	28	27.50–28.35
n262	47	47.20–48.20

Review Questions

1.1 Draw a block diagram of a wireless communication model and explain the function of each block.

1.2 Explain the difference between full-duplex and half-duplex radio systems.

1.3 Why baseband signals cannot be transmitted directly without modulation.

1.4 Briefly explain the difference between AM and FM and why modulation is needed.

1.5 Illustrate the time domain and frequency domain of a sinusoidal signal.

1.6 Illustrate the process of multiplexing in wireless communications.

1.7 Draw a block diagram of an FDM (frequency division multiplexing) system and illustrate the spectrum of an FDM signal.

1.8 Briefly discuss the advantages of 4G wireless technology compared to 3G wireless technology.

1.9 List the advantages of 5G wireless technology.

1.10 Explain the application of FR1 and FR2 bands of 5G NR wireless technology.

1.11 Explain the advantages and challenges of using THz frequency in 6G wireless technology.

1.12 Illustrate and explain the basic concept of MIMO, and what are the advantages of MIMO systems?

1.13 Explain why RF circulators and RF switches are used in radio front-ends of wireless systems.

References

1 Leon W. Couch, *Digital and Analog Communication Systems*, 8th ed., Pearson, 2013.
2 Masoud Salehi, John G. Proakis, *Fundamentals of Communication Systems*, Pearson Education, 2013.
3 Rodger E. Ziemer, William H. Tranter, *Principles of Communications: Systems, Modulation, and Noise*, 7th ed., Wiley, 2015.
4 Andreas F. Molisch, *Wireless Communications*, Wiley, 2010.
5 Michael Moher, Simon Haykin, *Communication Systems*, 5th ed., Wiley, 2009.
6 Theodore S. Rappaport, Robert W. Heath Jr., Robert C. Daniels, James N. Murdock, *Millimeter Wave Wireless Communications*, 1st ed., Pearson, 2014.
7 Masoud Salehi, John G. Proakis, *Communication Systems Using MATLAB*, Couch, 1998.
8 David Michael Pozar, *Microwave and RF Design of Wireless Systems*, Wiley, New York, 2001.
9 Muhanned Rabbani, James Churm et al., "26 GHz band beam-steered antenna for Mm-wave 5G systems," *51st European Microwave Conference (EuMC)*, 2021.
10 Anton Tishchenko, Ali Ali et al., "Reflective metasurface for 5G mmwave coverage enhancement," *International Symposium on Antennas and Propagation (ISAP)*, 2022.
11 Muhammed Rabbani, James Churm et al., "Enhanced data throughput using 26 GHz band beam-steered antenna for 5G systems," *16th European Conference on Antennas and Propagation (EuCAP)*, 2022.

12 Carlos de Lima, Didier Belot et al., "Convergent communication, sensing and localization in 6G systems: an overview of technologies, opportunities and challenges," *IEEE Access*, Volume: 9, pp. 26902–26925, 2021.

13 Jie Hu, Qing Wang, Kun Yang, "Energy self-sustainability in fullspectrum 6G," *IEEE Wireless Communications*, Volume: 28, Issue: 1, pp. 104–111, 2021.

14 Chamitha de Alwis, Anshuman Kalla, "Survey of 6G frontiers: trends, application, requirements, technologies and future research," *IEEE Open Journal of the Communications Society*, Volume: 2, pp. 836–886, 2021.

15 Mohamed Shehata, Ke Wang, Withawat Withayachumnankul, "Mitigating the nonlinearity of radio-over-fiber terahertz systems," *47th International Conference on Infrared, Millimeter and Terahertz Waves (IRMMW-THz)*, 2022.

16 Mohamed Shehata, Ke Wang, Withawat Withayachumnankul, "Carrierless I–Q mixing for terahertz communications," *47th International Conference on Infrared, Millimeter and Terahertz Waves (IRMMW-THz)*, 2022.

17 Theodore S. Rappaport, Yunchou Xing, et al., "Communications and applications above 100 GHz: opportunities and challenges for 6G and beyond," *IEEE Access*, Volume: 7, 2019.

18 Alice Faisal, Hadi Sarieddeen et al., "Ultra-massive MIMO systems at terahertz bands: prospects and challenges," *IEEE Vehicular Technology Magazine*, Volume: 15, Issue: 4, 2020.

19 https://www.3gpp.org/DynaReport/38104.htm.

Suggested Readings

3GPP Release 18 Overview, ATIS Webinar, 2023.

Cheng-Xiang Wang, Jie Huang et al., "6G wireless channel measurements and models: trends and challenges," *IEEE Vehicular Technology Magazine*, Volume: 15, Issue: 4, pp. 22–32, 2020.

Girija Shankar Sahoo, Anumoy Ghosh, "Antenna array design for higher spectral efficiency for hybrid beamforming 5G MIMO wireless communication system," *IEEE Wireless Antenna and Microwave*, 2022.

Grigoriy Fokin, "Channel model for location-aware beamforming in 5G ultra-dense mmWave radio access network," *International Conference on Electrical Engineering and Photonics*, IEEE, 2022.

Ian F. Akyildiz, Ahan Kak, Shuai Nie, "6G and beyond: the future of wireless communications systems," *IEEE Access*, Volume: 8, pp. 133995–134030, 2020.

Kenichi Okada, Jian Pang et al., "Millimeter-wave CMOS phased-array transceivers for 5G and beyond," *IEEE 33rd Annual International Symposium on Personal, Indoor and Mobile Radio Communications (PIMRC)*, 2022.

Latif U. Khan, Ibrar Yaqoob et al., "6G wireless systems: a vision, architectural elements, and future directions," *IEEE Access*, Volume: 8, pp. 147029–147044, 2020.

Marco Giordani, Michele Polese, et al., "Toward 6G networks: use cases and technologies," *IEEE Communications Magazine*, Volume: 58, Issue: 3, pp. 55–61, 2020.

Marcos Katz, Iqrar Ahmed, "Opportunities and challenges for visible light communications in 6G," *Proceedings of IEEE 2nd 6G Wireless Summit (6G SUMMIT)*, pp. 1–5, 2020.

Mostafa Z. Chowdhury, Md. Shahjalal, Shakil Ahmed, Yeong Min Jang, "6G wireless communication systems: applications, requirements, technologies, challenges, and research directions," *IEEE Open Journal of the Communications Society*, Volume: 1, pp. 957–975, 2020.

Patrick Marsch, Ömer Bulakci, Olav Queseth, Mauro Boldi, *5G System Design: Architectural and Functional Considerations and Long Term Research*, Wiley, 2018.

Rawan Alghamdi, Reem Alhadrami et al., "Intelligent surfaces for 6G wireless networks: a survey of optimization and performance analysis techniques," *IEEE Access*, Volume: 8, pp. 202795–202818, 2020.

Satya Prakash Rout, "6G wireless communication: its vision, viability, application, requirement, technologies, encounters and research," *11th International Conference on Computing, Communication and Networking Technologies (ICCCNT)*, pp. 1–8, 2020.

Shanzhi Chen, Ying-Chang Liang, "Vision, requirements, and technology trend of 6G: how to tackle the challenges of system coverage, capacity, user data-rate and movement speed," *IEEE Wireless Communications*, Volume: 27, Issue: 2, pp. 218–228, 2020.

Vesa Lampu, Lauri Anttila et al., "Air-induced PIM cancellation in FDD MIMO transceivers," *IEEE Microwave and Wireless Components Letters*, Volume: 32, Issue: 6, 2022.

Volker Ziegler, Harish Viswanathan et al., "6G architecture to connect the worlds," *IEEE Access*, Volume: 8, pp. 173508–173520, 2020.

Youngmin Kim, Hongjong Park et al., "High efficiency 29-/38-GHz hybrid transceiver front-ends utilizing Si CMOS and GaAs HEMT for 5G NR millimeter-wave mobile applications," *IEEE Symposium on VLSI Technology and Circuits (VLSI Technology and Circuits)*, 2022.

Zhiwen Qin, Zhiming Yi et al., "Analog beamforming for millimeter-wave communication," *IEEE International Students' Conference on Electrical, Electronics and Computer Science (SCEECS)*, 2022.

2

Analog Communication Systems

This chapter provides the basic concept of analog modulation and demodulation. It covers amplitude modulation (AM) and detection, single-sideband (SSB) AM, double-sideband suppressed carrier (DSB SC) AM, AM envelope detector, frequency modulation (FM), Noise suppression in FM, phase-locked loop (PLL) FM demodulation, preemphasis and deemphasis in FM systems, and the principles of phase modulation (PM). The covered topics are explained with a focus on the system operation rather than the circuit design. Review questions are provided to help the readers understand the covered topic and to help solve problems related to analog modulation. A list of references and suggested readings that present the recent research activities that are related to the covered topics are provided at the end of the chapter.

2.1 Chapter Objectives

On reading this chapter, the reader will be able to:

- Draw a block diagram of an analog communication system and explain the function of each block.
- Explain the disadvantages of an analog communication system.
- Write an expression for amplitude modulation (AM) and illustrate an AM signal's time and frequency domains.
- Calculate the total power P_t of an AM signal and explain the impact of the modulation index on the signal's power.
- Explain the difference between single-sideband (SSB) and double-sideband (DSB) transmissions.
- Draw a block diagram of an SSB transmitter using the filter method and explain the function of each block.
- Draw a block diagram of a phase-shift SSB transmitter and explain the function of each block.
- Illustrate the envelope detection process in an AM receiver.
- Calculate the modulation index of a frequency-modulated (FM) signal and explain the advantages of the FM technique.
- Calculate the bandwidth of a frequency signal using Carson's rule.
- Draw a block diagram of a phase-locked loop (PLL) demodulator and explain its operation.
- Compare and contrast AM, FM, and PM.

Essentials of RF Front-end Design and Testing: A Practical Guide for Wireless Systems, First Edition. Ibrahim A. Haroun.

2.2 Overview of Analog Communications

In analog communication systems [1–6], the information is transferred from the transmitter to the receiver using analog signals. An analog signal is a continuous signal which varies in amplitude, frequency, phase, or some other property with time. Figure 2.1 shows a block diagram of a simple analog communication system model consisting of a modulator, channel, and demodulator.

The combination of the modulator and demodulator is referred to as a modem. The modulator varies specific parameters of the carrier signal according to the change in the baseband signal. A baseband signal represents the information to be transmitted; such signals are the output of the transducers that convert some physical variable into an electrical signal. In a radio receiver, the demodulator (called detector) recovers the original baseband signal that was transmitted. The main disadvantage of an analog communication system is its susceptibility to noise and interference. Interference signals of large amplitudes could corrupt the transmitted information so that it cannot be recovered. In analyzing communication systems, the primary objective is to have a certain SNR at the output of the detector in the presence of interference at the receiver's input. The required SNR to achieve specific system performance depends on the modulation scheme. If the system has a high tolerance for noise, the transmitter's output power specifications can be relaxed. In communication systems, modulation schemes impact the transmission bandwidth and the type of power amplifier that is needed in the transmitter. Analog communication systems use amplitude, phase, and frequency modulation techniques.

2.3 Amplitude Modulation (AM)

In AM systems [4], the amplitude of the carrier signal varies according to the change in the amplitude of the modulating signal (i.e., Baseband). The instantaneous modulating signal $v_{\mathrm{m}}(t)$ and carrier signal $v_{\mathrm{c}}(t)$ are expressed as

$$v_{\mathrm{m}}(t) = V_{\mathrm{m}} \cos(\omega_{\mathrm{m}} t) \tag{2.1}$$

$$v_{\mathrm{c}}(t) = V_{\mathrm{c}} \cos(\omega_{\mathrm{c}} t) \tag{2.2}$$

where V_{m} and V_{c} are the peak amplitudes of the modulating and carrier signals. The angular frequencies of the carrier and modulating signals are given by

$$\omega_{\mathrm{c}} = 2\pi f_{\mathrm{c}}$$

$$\omega_{\mathrm{m}} = 2\pi f_{\mathrm{m}}$$

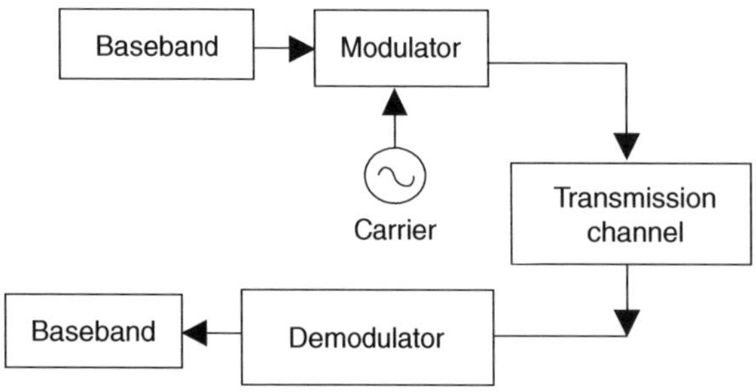

Baseband → Modulator → Transmission channel → Demodulator → Baseband; Carrier → Modulator

Figure 2.1 Simplified block diagram of an analog communication system.

where f_m and f_c are the frequencies of the modulating and carrier signals. The amplitude of an AM signal is expressed as

$$\begin{aligned} V_{AM} &= V_c + V_m(t) \\ &= V_c + V_m \cos(\omega_m t) \end{aligned} \tag{2.3}$$

$$V_{AM} = V_c\,[1 + m\cos(\omega_m t)] \tag{2.4}$$

where m is called the *modulation index* and is given by

$$m = (V_m/V_c)$$

Thus, the instantaneous AM signal can be written as

$$v_{AM}(t) = V_c[1 + m\cos(\omega_m t)] \times \cos\omega_c t \tag{2.5}$$

$$\therefore\ v_{AM}(t) = V_c \cos(\omega_c t) + \frac{m\,V_c}{2}\cos(\omega_c - \omega_m)t + \frac{mV_c}{2}\cos(\omega_c + \omega_m)\,t \tag{2.6}$$

Equation (2.6) indicates that the modulated signal has upper and lower sidebands $(\omega_c - \omega_m)$ and $(\omega_c + \omega_m)$ with peak amplitude $mV_c/2$. Figure 2.2 shows a time domain representation of an AM signal, and Figure 2.3 shows a frequency domain representation of an AM signal.

The modulation index can also be written as

$$m = \frac{V_{max} - V_{min}}{V_{max} + V_{min}}$$

where V_{max} and V_{min} are the maximum and minimum values of the AM signal. In an AM signal, if $m > 1$, the modulation is referred to as over-modulation, and it is not allowed because it produces distortion called sideband splatter. Sideband splatter transmits frequencies outside the allocated frequency range and could impact other co-located systems.

In an AM transmission, the amplitude of the sidebands changes according to the change in the modulating signal. The total power P_t of an AM signal (i.e., sidebands and carrier) is given by

$$P_t = P_c\left(1 + \frac{m^2}{2}\right) \tag{2.7}$$

Equation (2.7) indicates that the power in the sidebands is much less than the power of the carrier. Since the carrier signal does not contain information, it can be suppressed. Figure 2.4 shows AM signals with different modulation indices.

Example 2.1

Calculate the modulation index $m\%$ for an AM signal that has maximum peak-to-peak of 125 volts and minimum peak-to-peak of 35 volts.

Solution

The modulation index $m\%$ is calculated as follows:

$$m\% = \frac{125 - 35}{125 + 35} \times 100\% = 56.25\%$$

Example 2.2

Determine the total transmit power of a 500-watt carrier to be modulated to a 90% level.

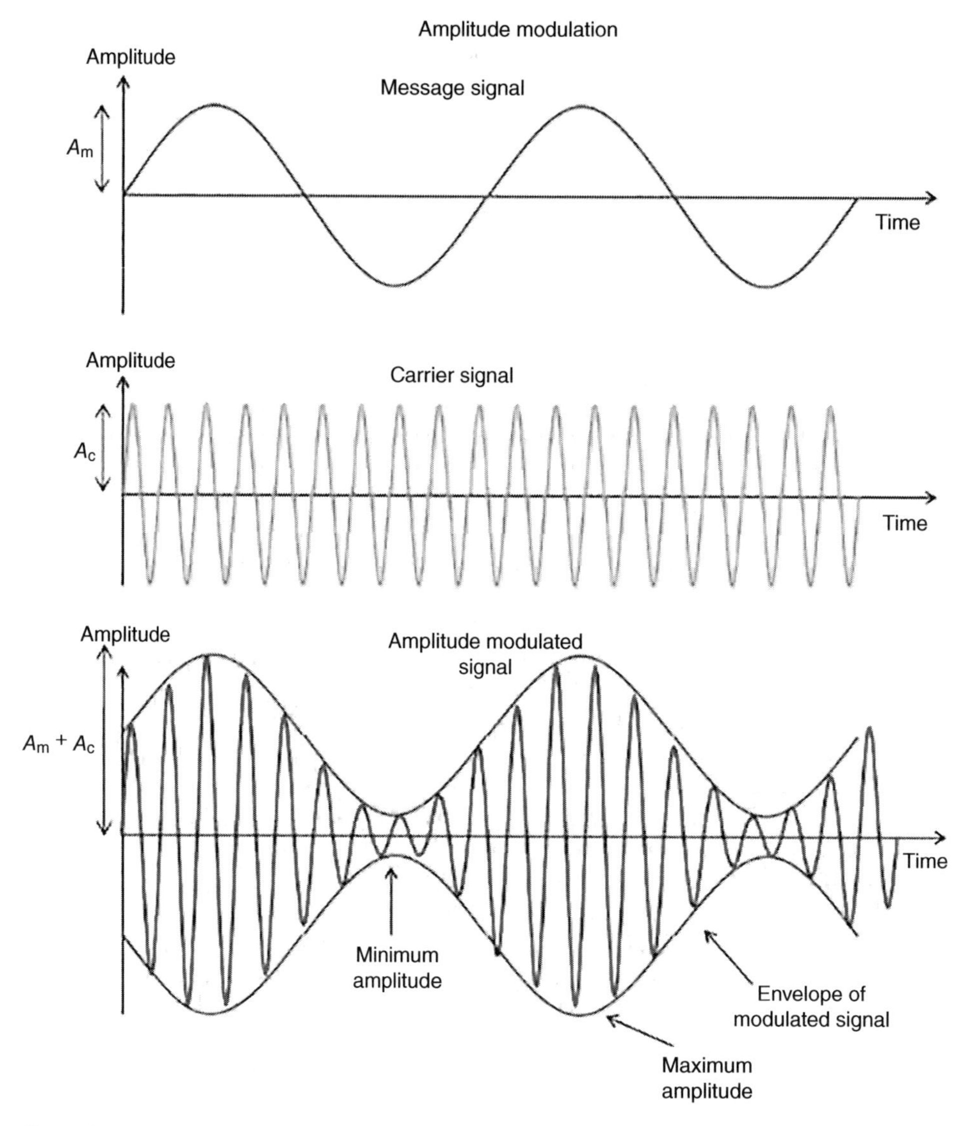

Figure 2.2 Time domain representation of an AM signal.

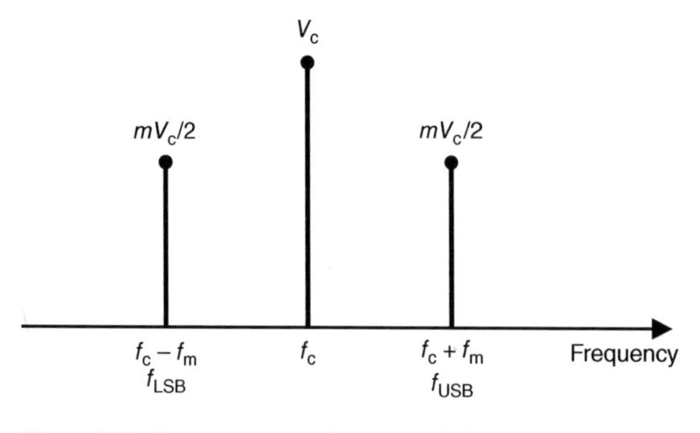

Figure 2.3 Frequency domain representation of an AM signal.

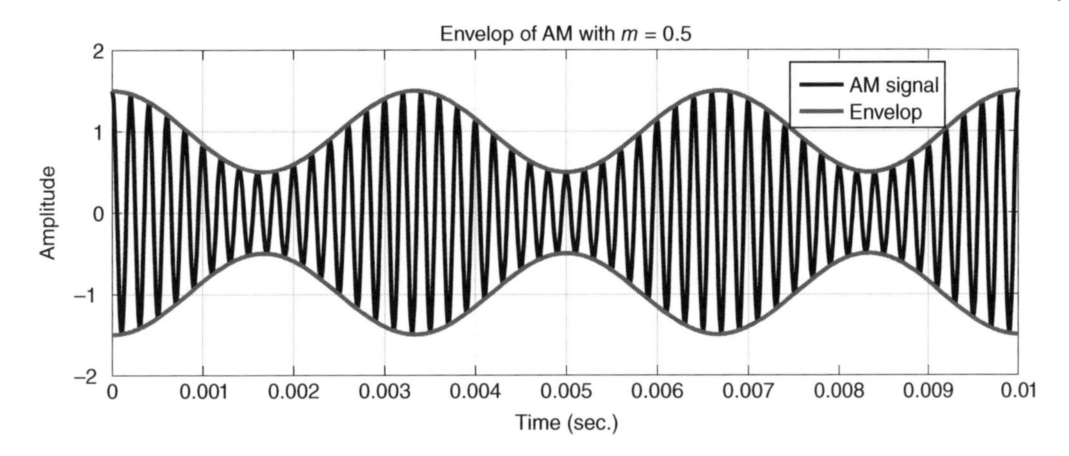

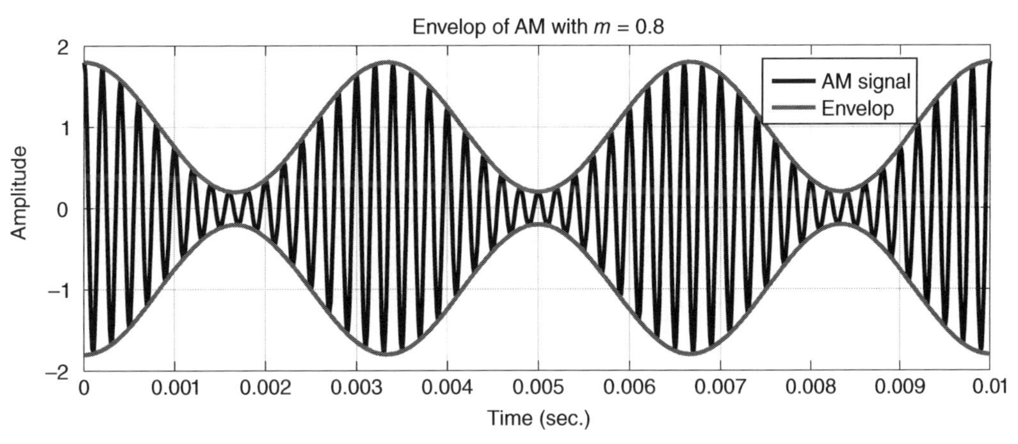

Figure 2.4 Illustration of AM signals with different modulation indices.

Solution

The total transmit power, P_t, is given by

$$P_t = P_c\left(1 + \frac{m^2}{2}\right)$$

$$P_t = 500\left(1 + \frac{(0.9)^2}{2}\right) = 702.5 \text{ watt}$$

2.4 Single-Sideband Modulation

SSB carrier is a type of AM modulation that transmits only a SSB instead of the two sidebands and the carrier. In an SSB transmission, the carrier signal does not contain any information, and the other sideband duplicates the information of the transmitted sideband. Thus, neither the carrier nor the other sideband is needed to send information. Figure 2.5 shows a frequency domain representation of an SSB AM signal.

Suppressing one of the sidebands in an SSB transmission reduces the bandwidth to half the bandwidth needed for a double-sideband AM transmission. As a result, the noise in the received signal gets reduced (i.e., better SNR at the receiver input). In double-sideband AM, the frequency in one sideband could experience a different phase shift than the opposite sideband, causing partial or complete cancellation of the two sidebands. In an SSB transmission, the carrier is first

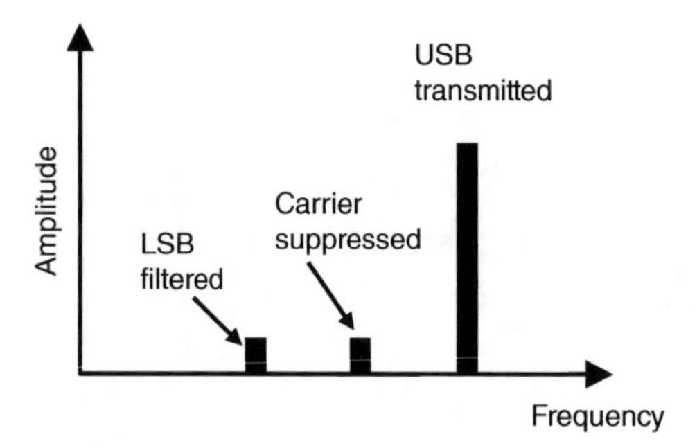

Figure 2.5 Frequency domain representation of a single-sideband AM signal.

suppressed using a balanced modulator. Once the carrier is suppressed, one of the two sidebands gets eliminated by filtering or phasing techniques.

2.4.1 Filtering Method

Figure 2.6 shows a block diagram of an SSB-SC (i.e., suppressed carrier) transmitter using the filter method.

In Figure 2.6, the output signal of the balanced modulator is a double-sideband suppressed carrier, and the filter after the balanced modulator passes only the upper sideband, resulting in an SSB-SC signal.

2.4.2 Phase Method

Figure 2.7 shows a block diagram of a phase-shift SSB modulator.

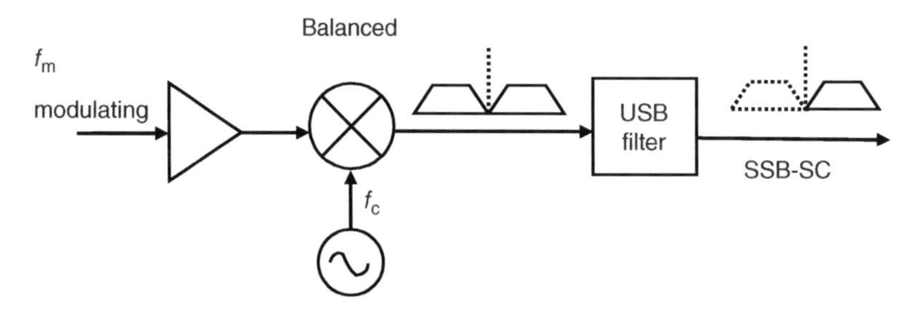

Figure 2.6 Block diagram of an SSB transmitter using filter method.

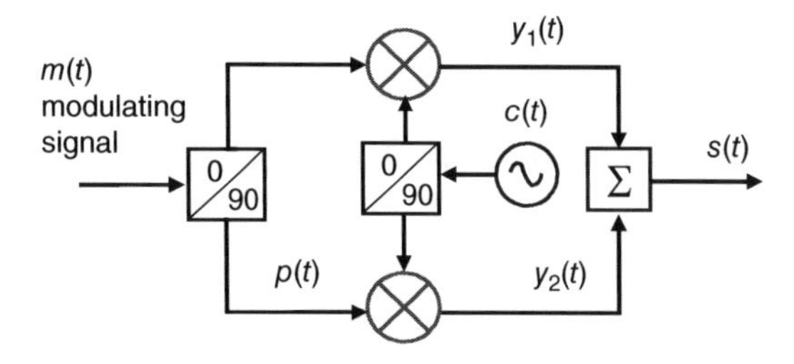

Figure 2.7 Block diagram of a phase-shift SSB modulator.

Figure 2.7 uses two balanced modulators and a pair of 90° phase shifters. This technique eliminates the need for a sideband filter. Signals $m(t)$, $p(t)$, and $c(t)$ are given by

$$m(t) = \cos(\omega_m t)$$

$$p(t) = \cos(\omega_m t - \pi/2) = \sin(\omega_m t)$$

$$c(t) = \cos(\omega_c t)$$

$$y_1(t) = \cos(\omega_m t) \cdot \cos(\omega_c t) = \frac{1}{2}[\cos(\omega_c - \omega_m)t + \cos(\omega_c + \omega_m)t]$$

$$y_2(t) = \sin(\omega_m t) \cdot \sin(\omega_c t) = \frac{1}{2}[\cos(\omega_c - \omega_m)t - \cos(\omega_c + \omega_m)t]$$

$$\therefore \quad s(t) = y_1(t) + y_2(t) = \cos(\omega_c - \omega_m)t \tag{2.8}$$

Equation (2.8) indicates that the lower sideband is selected.

Example 2.3
A carrier signal that has a frequency of 1.41 MHz is modulated by a signal that has a frequency range from 20 Hz to 10 kHz. Determine the frequency of the upper and lower sidebands.

Solution
The upper sideband USB frequency is given by

$$f_{USB} = f_c + f_m$$

where f_m is from 20 Hz to 10 kHz, thus, the f_{USB} frequency range is

from $(1.41 \times 10^6 + 20)$ Hz to $(1.41 \times 10^6 + 10 \times 10^3)$ Hz

and the f_{LSB} frequency range is

from $(1.41 \times 10^6 - 10 \times 10^3)$ Hz to $(1.41 \times 10^6 - 20)$ Hz

Figure 2.8 illustrates the upper and lower frequency bands of the frequency spectrum.

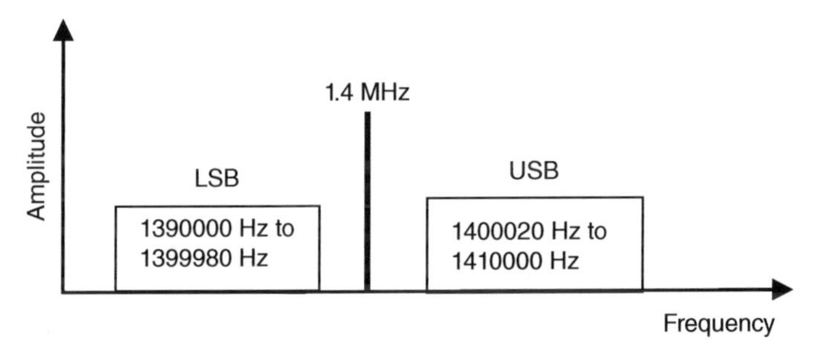

Figure 2.8 Solution of Example 2.3.

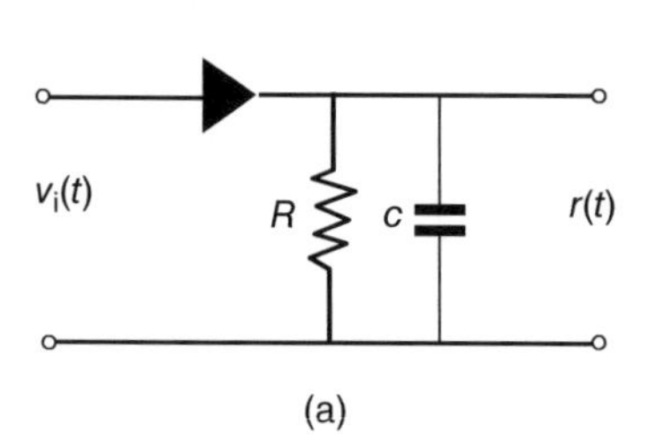

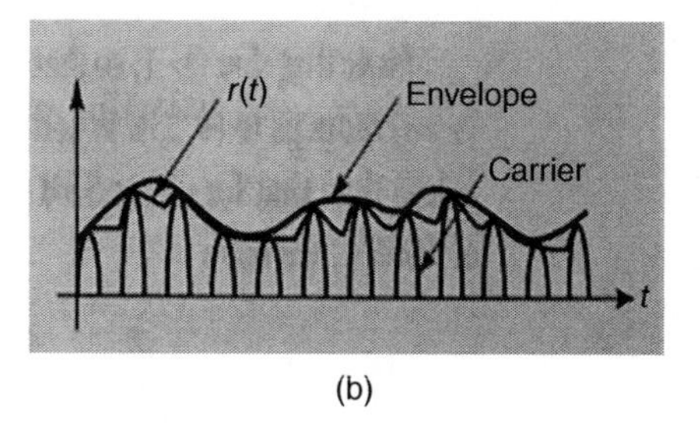

Figure 2.9 Envelope detection of an AM signal. (a) Envelope detector circuit. (b) Output waveform of a detector.

2.5 AM Demodulation

The purpose of an AM demodulator in a radio receiver is to recover the low-frequency baseband signal from the high-frequency carrier signal. The simplest and most widely used AM demodulator is the envelope detector method.

2.5.1 AM Envelope Detector

A typical AM envelope detector process is shown in Figure 2.9.

In Figure 2.9a, when the AM signal is applied to the diode, the capacitor will be charged during the positive half-cycle of the carrier. During the negative half-cycle, the diode does not conduct, and the capacitor begins to discharge through the resistor. Thus, the resulting output voltage $r(t)$ approximates the actual envelope of the baseband signal. The RC time constant should be large enough so that the capacitor voltage does not decay too quickly before the next carrier peak arrives but small enough so the output can track the envelope when it is decreasing. Figure 2.9b illustrates the output waveform of an envelope detector.

2.6 Frequency Modulation (FM)

FM [1–4] is an angle modulation type in which the frequency of the carrier signal changes in proportion to the amplitude of the baseband signal. Figure 2.10 illustrates the process of frequency modulation.

FM modulation is a good choice for mobile communications because the signal's amplitude does not change with the noise and interference level but at the expense of increased bandwidth. It is a compromise of increased spectrum for an improved SNR. Since the signal's amplitude does not change in FM, a nonlinear amplifier can be used, which results in a longer lifetime battery. However, FM has poor spectral efficiency compared to AM modulation or SSB modulation. The constant amplitude of the modulated carrier can be expressed as:

$$v_c(t) = V_c \cos\theta(t) \tag{2.9}$$

where V_c is the peak amplitude of the carrier, and $\theta(t)$ is its phase angle. The relation between the instantaneous angular frequency $\omega_i(t)$ and $\theta(t)$ is given by

$$\omega_i(t) = \frac{d\theta(t)}{dt} \tag{2.10}$$

The amount by which the instantaneous carrier differs from the reference (also called rest) frequency is referred to as the frequency deviation. The instantaneous angular frequency of FM signal

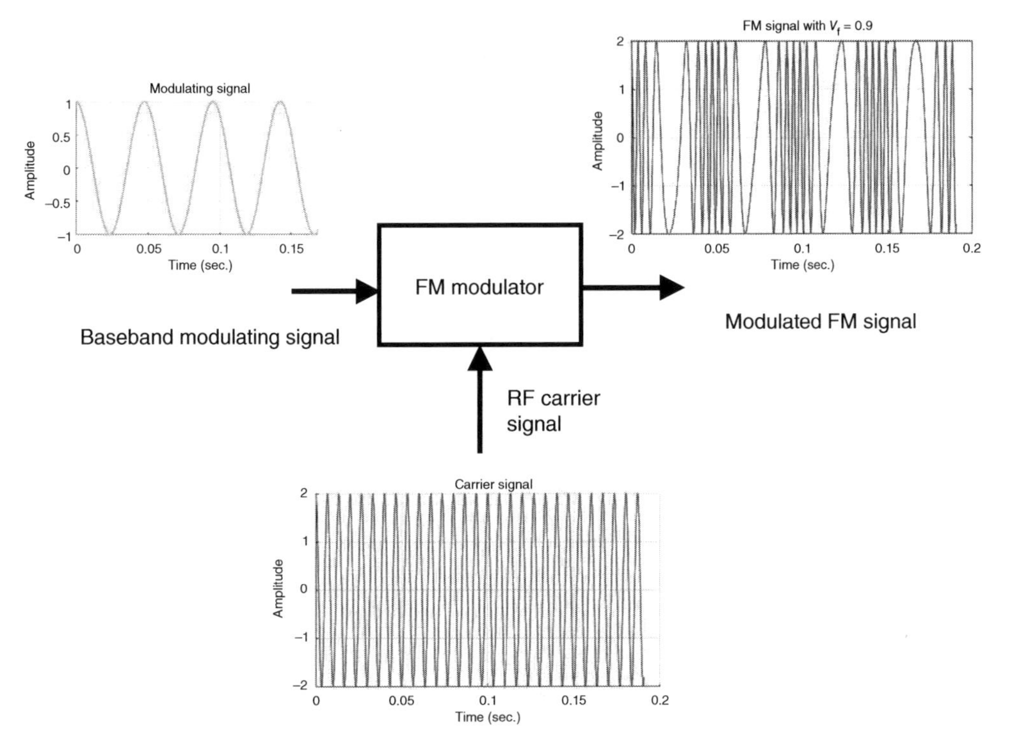

Figure 2.10 Illustration of FM modulation generation.

can be written as

$$\omega_i(t) = \omega_c + k_f v_m(t) \tag{2.11}$$

where ω_c is the reference carrier angular frequency, k_f is a constant that depends on the modulation circuit and has units of rad/sec, and $v_m(t)$ is the modulating signal.

$$\therefore\ \omega_i(t) = \omega_c + k_f V_m \cos \omega_m t \tag{2.12}$$

where $k_f V_m$ is the angular frequency deviation $\Delta\omega$. High-frequency deviation results in a wideband signal. The instantaneous angular frequency is expressed as

$$\omega_i(t) = \omega_c + \Delta\omega \cos \omega_m t \tag{2.13}$$

From Eq. (2.10), the corresponding instantaneous phase can be written as

$$\theta(t) = \int \omega_i(t)\,dt = \omega_c t + \frac{\Delta f}{f_m} \sin \omega_m t \tag{2.14}$$

$$\theta(t) = \omega_c t + M_f \sin \omega_m t \tag{2.15}$$

where $M_f = \Delta f / f_m$ is the maximum instantaneous frequency departure and is called the modulation index for FM. Substituting (2.15) in (2.9) gives the modulated FM signal, $v_{FM}(t)$,

$$v_{FM}(t) = V_c \cos(\omega_c t + M_f \sin \omega_m t) \tag{2.16}$$

2.6.1 Frequency Spectrum of FM Signal

The spectrum of an angle-modulated signal can be explained as follows:

From Euler's identity,

$$e^{j\theta} = \cos\theta + j\sin\theta,$$

and cos θ can be expressed as the real part of $e^{j\theta}$,

$$\therefore \quad \cos\,\theta = \text{Re}\,\{e^{j\theta}\}$$

Equation (2.16) can be written as

$$v_{\text{FM}}(t) = v_{\text{c}}\,\text{Re}\left\{e^{j\omega_{\text{c}}t} \times e^{jM_{\text{f}}\,\sin\,\omega_{\text{m}}t}\right\} \tag{2.17}$$

The term $e^{jM_{\text{f}}\,\sin\,\omega_{\text{m}}t}$ can be written as

$$e^{jM_{\text{f}}\,\sin\,\omega_{\text{m}}t} = \sum_{n=-\infty}^{n=\infty} J_{\text{n}}(M_{\text{f}})\,e^{jn\omega_{\text{m}}t} \tag{2.18}$$

where $J_n(M_{\text{f}})$ is the Bessel function of the first kind of order n with argument M_{f}.

$$\therefore V_{\text{FM}}(t) = V_{\text{c}} \sum_{n=-\infty}^{n=\infty} J_n(M_{\text{f}})\cos(\omega_{\text{c}} + \omega_{\text{m}})t \tag{2.19}$$

and

$$J_n(M_{\text{f}}) = J_{-n}(M_{\text{f}}) \qquad \text{for } n \text{ even}$$

$$J_n(M_{\text{f}}) = -J_{-n}(M_{\text{f}}) \quad \text{for } n \text{ odd}$$

$$\begin{aligned}\therefore\; v_{\text{FM}}(t) = v_{\text{c}}\{\, & J_{\text{o}}(M_{\text{f}})\cos\omega_{\text{c}}t - J_1(M_{\text{f}})\,[\cos(\omega_{\text{c}} + 2\omega_{\text{m}})t + \cos(\omega_{\text{c}} - 2\omega_{\text{m}})t] \\ & + J_2(M_{\text{f}})\,[\cos(\omega_{\text{c}} + 2\omega_{\text{m}})t + \cos(\omega_{\text{c}} - 2\omega_{\text{m}})t] \\ & - J_3(M_{\text{f}})\,[\cos(\omega_{\text{c}} + 3\omega_{\text{m}})t + \cos(\omega_{\text{c}} - 3\omega_{\text{m}})t] \ldots\ldots\}\end{aligned} \tag{2.20}$$

Equation (2.20) indicates that an FM waveform with a sinusoidal modulating signal has an infinite number of sidebands. However, the spectral components of higher-order sidebands become negligible and can be neglected. The term $v_{\text{c}}J_{\text{o}}(M_{\text{f}})$ is the peak amplitude of the carrier signal. The sidebands on either side of the carrier are separated by ω_{m} and are symmetrical in amplitude. Only sidebands with amplitudes greater than 1% of the unmodulated carrier are considered significant. As M_{f} increases, the number of significant sidebands increases, resulting in an increased bandwidth. Figure 2.11 shows the Bessel functions for different modulation indexes. From Figure 2.11, at modulation indexes of 2.4, 5.5, and 8.7, the carrier amplitude $J_{\text{o}}(M_{\text{f}})$ drops to zero. At those points, all the signal power is distributed through the sidebands.

The bandwidth of an FM signal is given by

$$\text{BW} \cong 2(\Delta f + f_{\text{m}}) = 2f_{\text{m}}\,(M_{\text{f}} + 1) \tag{2.21}$$

Equation (2.21) is referred to as Carson's rule. With $M_{\text{f}} = 0.25$, the FM signal has only a single pair of significant sidebands like those of AM modulation. This type of modulation is called narrowband FM (NBFM). FM signals can be classified as follows:

Narrowband FM, $M_{\text{f}} < 1$
Wideband FM, $1 < M_{\text{f}} < 10$
Ultra-wideband FM, $M_{\text{f}} > 10$

The primary objective of a NBFM system is to conserve spectrum. However, it is at the expense of the SNR. When the modulating signal is a pulse or binary wave train, the carrier is modulated by the fundamental wave and all its harmonics. Also, each harmonic produces multiple pairs of sidebands depending on the modulation index M_{f}.

The amplitudes of sidebands in an FM signal are $V_{\text{c}}\,J_{\text{o}}(M_{\text{f}})$, $V_{\text{c}}\,J_1(M_{\text{f}})$, $V_{\text{c}}\,J_2(M_{\text{f}})$, ……, as shown in Figure 2.12. The average power of an FM signal is given by

$$P = \frac{V_{\text{c}}^2}{2R} \tag{2.22}$$

where V_{c} is the amplitude of the carrier, and R is the load resistance.

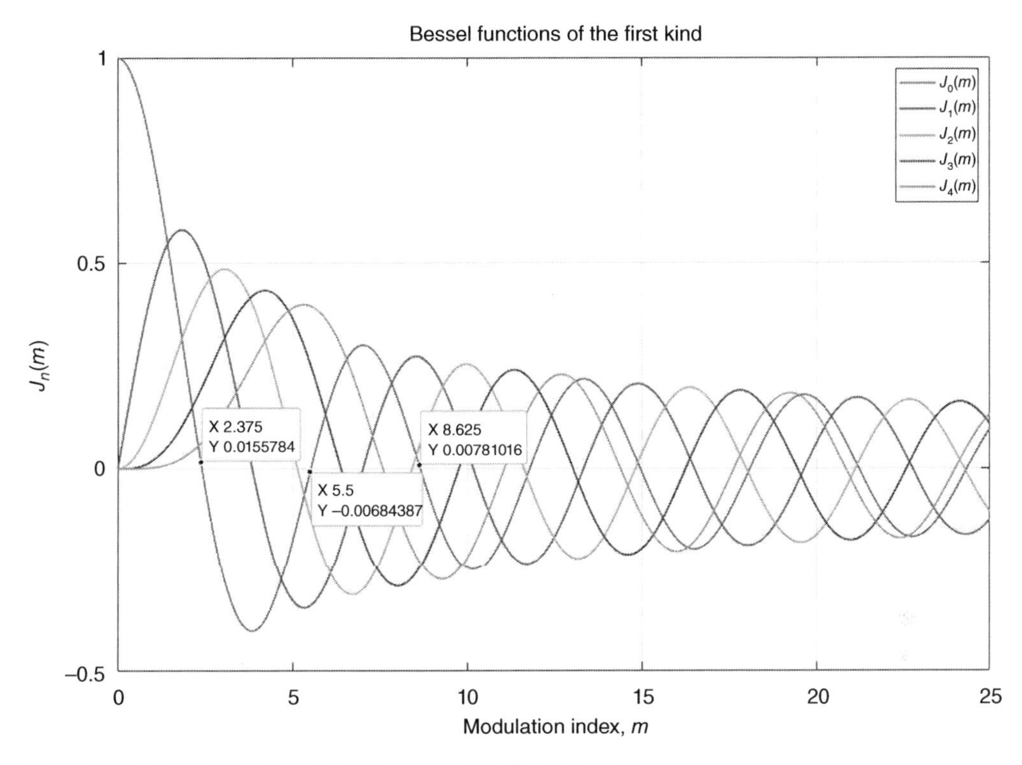

Figure 2.11 Bessel functions for different modulation indices M_f.

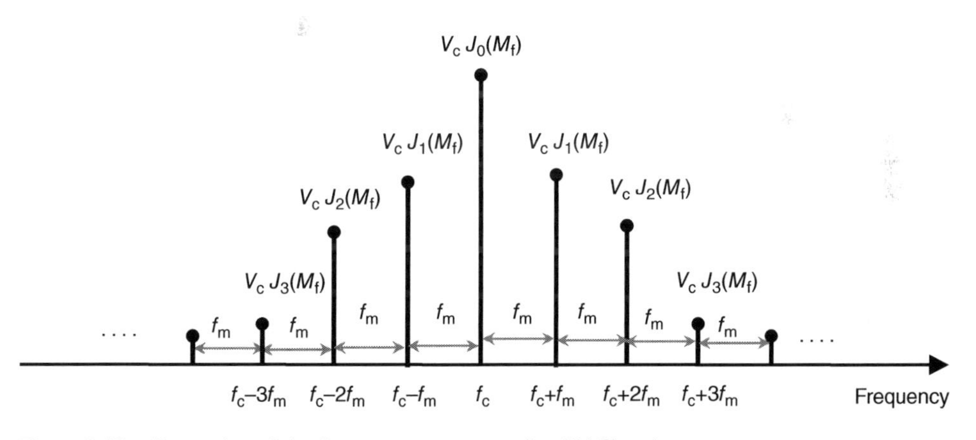

Figure 2.12 Illustration of the frequency spectrum of an FM Signal.

Example 2.4

Calculate the maximum bandwidth of an FM signal that has a deviation of 30 kHz and maximum modulating signal of 5 kHz.

Solution

The bandwidth (BW) of an FM signal can be determined using Carson's rule, thus,

$$\text{BW} = 2[\Delta f + f_m]$$

$$\text{BW} = 2 \times [30\ \text{KHz} + 5\ \text{kHz}] = 70\ \text{KHz}$$

Example 2.5

A 100 MHz carrier signal is frequency modulated with a 4 kHz sinusoid signal of 3 V_{peak}. If the sensitivity of the modulation is 2 kHz/V, determine the following:

(1) The maximum frequency deviation of the carrier
(2) The modulation index of the carrier
(3) The expression of the FM signal for a cosine carrier of 5 V_{peak}, when the modulating signal is also a cosine.

Solution

(1) $\Delta f = k_f V_m = 2\,\text{kHz/V} \times 3\,V_{peak} = 6\,\text{kHz}$
(2) $M_f = \Delta f / f_m = 6\,\text{kHz}/4\,\text{kHz} = 1.5$
(3) $v_{FM}(t) = 5\cos[(2\pi \times 10^8 t) + 1.5\sin(2\pi \times 4 \times 10^3)t]$

Example 2.6

Determine the spectrum of a 5 V_{rms} carrier when it is modulated such that the deviation equals 3 kHz when modulated by a 1 kHz signal.

Solution

$$M_f = \frac{\Delta f}{f_m} = \frac{3\,\text{kHz}}{1\,\text{kHz}} = 3$$

From Figure 2.13, the Bessel coefficients for $M_f = 3$ are:

$J_0(3) = -0.260; J_1(3) = 0.339; J_2(3) = 0.486; J_3(3) = 0.309;$
$J_4(3) = 0.132; J_5(3) = 0.043; J_6(3) = 0.011.$

Multiplying the Bessel coefficients by the carrier amplitude yields the spectrum shown in Figure 2.14

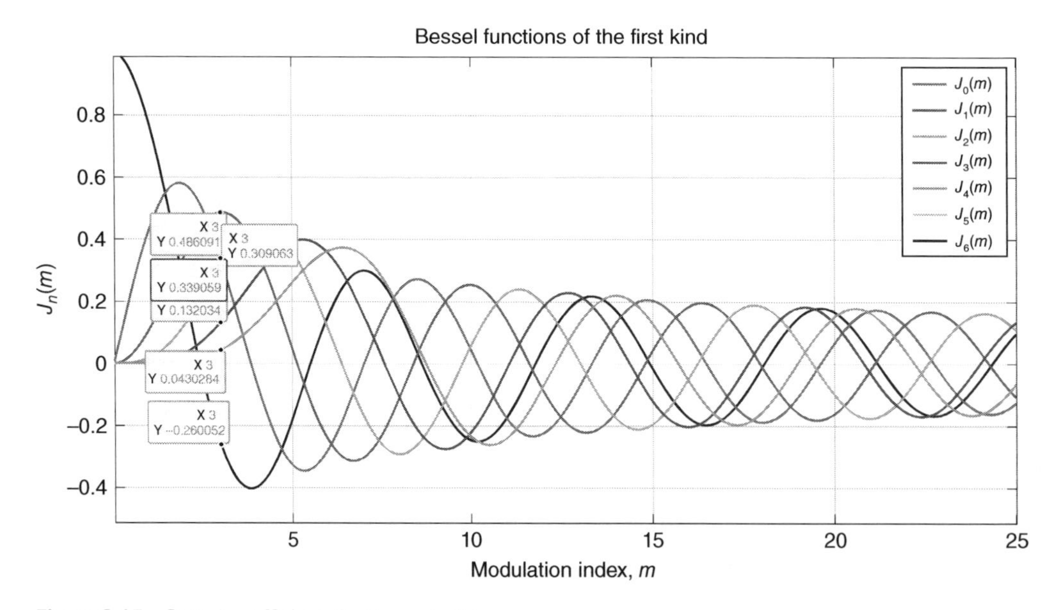

Figure 2.13 Bessel coefficient, for Example 2.6.

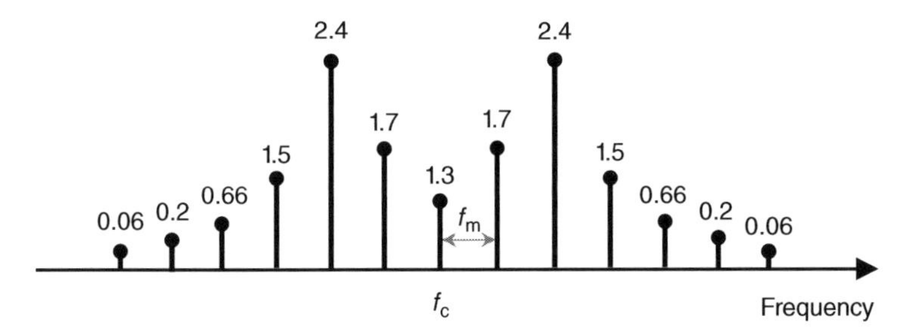

Figure 2.14 Spectrum of Example 2.6.

Example 2.7

From a spectrum analyzer's display, the carrier frequency of the displayed signal goes to zero when a 5 V_{rms} modulating signal is adjusted to 4.16 kHz. Determine the modulation sensitivity in Hz/V.

Solution

The first Bessel coefficient $J_0(M_f)$ goes to zero when the modulation index, M_f is 2.4

$$\therefore\ M_f = 2.405 = \frac{\Delta f}{f_m} = \frac{\Delta f}{4.16 \times 10^3};$$

therefore

$$\Delta f = 2.405 \times 4.16 \times 10^3 \approx 10\,\text{kHz}$$

The modulation sensitivity is given by

$$k_f = \frac{\Delta f}{V_m} = \frac{10 \times 10^3}{\sqrt{2} \times 5} = 1.414\,\text{kHz/V}$$

Example 2.8

Determine the approximate bandwidth required to transmit an FM signal having a carrier frequency of 100 MHz, a maximum modulating frequency of 5 kHz, and a modulation index of 5.

Solution

The bandwidth is given by

$$\begin{aligned}\text{BW} &\cong 2\,(\Delta f + f_m) = 2 f_m\,(M_f + 1)\\ &= 2 \times (5 \times 10^3)(5 + 1) = 60\,\text{kHz}\end{aligned}$$

The occupied bandwidth is from 99.97 to 100.03 MHz

Example 2.9

Determine the range of frequencies occupied by the following FM signal:

$$v_{FM}(\text{t}) = 5\cos\,[2\pi \times 10^8 t + 1.5\sin(8\pi \times 10^3 t)]$$

Solution

From the given FM equation, the modulation index $M_f = 1.5$, and

$$f_m = \frac{8\pi \times 10^3}{2\pi} = 4\,\text{kHz};$$

$$\therefore\ \text{BW} \cong 2 f_m\,(M_f + 1) = 2(4 \times 10^3)\,(1.5 + 1) = 20\,\text{kHz}$$

Hence, the occupied bandwidth is from (100 MHz − 10 kHz) to (100 MHz + 10 kHz), which means that 99.99 to 100.01 MHz.

Example 2.10

Determine the instantaneous frequency of a signal given by

$$e_c = E_c \cos(\omega_c t + \theta_o)$$

Solution

$$\theta(t) = \omega_c t + \theta_o, \text{ and } \quad \omega_i(t) = \frac{d\theta(t)}{dt}, \quad \text{thus} \quad \omega_i(t) = \omega_c(t)$$

$$\therefore \ f_i = f_c$$

Example 2.11

Determine the instantaneous angular frequency of a signal given by

$$\mathrm{e}_c = E_c \cos(\cos\ \omega_c t)$$

Solution

$$\theta(t) = \cos(\omega_c t),$$

and

$$\omega_i(t) = \frac{d\theta(t)}{dt} = -\ \omega_c \sin(\omega_c t)$$

2.7 Noise Suppression in FM Systems

Noise can impact FM signals, particularly the baseband modulating signals, because their higher frequency components have very low amplitudes. Thus, to maintain the SNR across the baseband, the higher frequency components of the baseband should be boosted (i.e., preemphasis) before passing through the modulator stage.

In an FM radio transmitter, the modulating signal passes through a simple RC high-pass filter with a time constant of 75 ms and a cut-off frequency of 2122 Hz. Thus, frequencies higher than 2122 Hz will be linearly enhanced by 6 dB/octave. In addition, the preemphasis circuit increases the level of high-frequency signals, so they become higher than the noise level.

In FM receivers, a simple RC low-pass filter with a time constant of 75 ms is used to return the frequency components of the baseband signal to their original levels. This RC circuit is called deemphasis and is placed after the FM demodulator. Figure 2.15 shows a frequency response of a combined preemphasis and deemphasis circuits.

2.8 FM Demodulation

The purpose of FM demodulation in FM receivers is to convert the frequency variation in the carrier signal back to a proportional voltage variation (i.e., baseband signal). Thus, the FM demodulator is a frequency-to-voltage converter. Figure 2.16 shows the input and output waveforms of an FM demodulator.

FM demodulators are also called detectors or discriminators. There are different FM demodulation techniques, which are frequency discrimination (e.g., *simple slope detection, balanced slope detector*) and phase discrimination (e.g., *Foster Seely detector, ratio detector, "PLL" detector*). The PLL FM demodulator is the most used one.

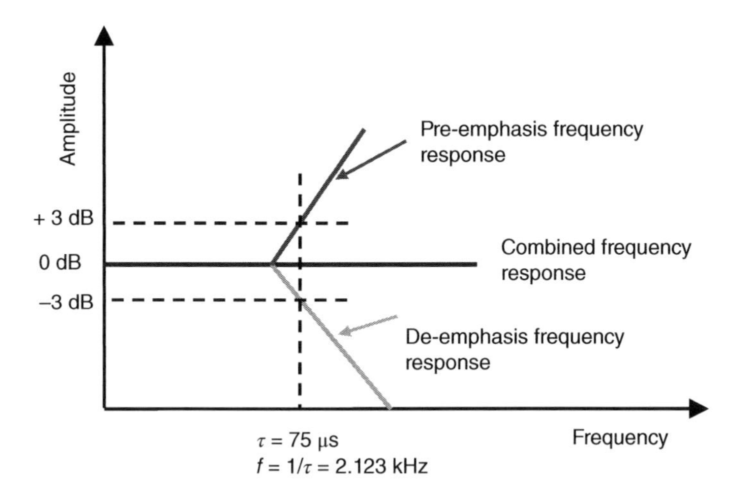

Figure 2.15 Frequency response of a combined preemphasis and deemphasis circuits.

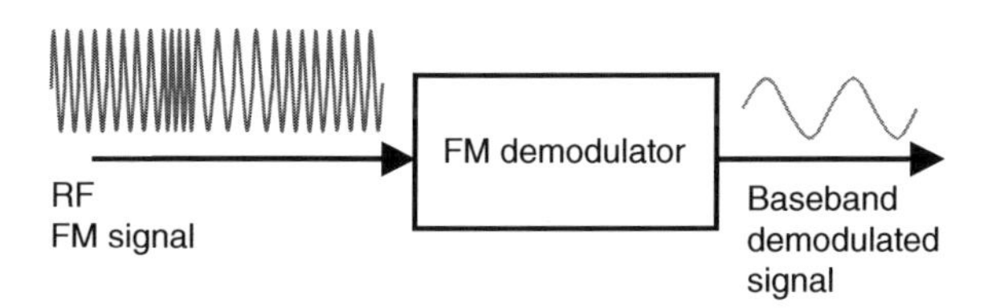

Figure 2.16 Input and output waveforms of an FM demodulator.

2.8.1 PLL FM Demodulator

A PLL is a frequency or phase feedback control circuit used in frequency demodulation and frequency synthesizers. Figure 2.17 shows the basic circuit of a PLL FM demodulator.

In Figure 2.17, the phase detector compares the FM-modulated signal with the voltage-controlled oscillator (VCO) signal. If the frequencies of the two signals (i.e., f_{VCO} and f_{FM}) are not the same, the phase detector produces an error signal which is filtered by a low-pass filter and then applied to the VCO. The dc voltage of the error signal changes the control voltage of the VCO, so the frequency of the VCO changes to match the frequency of the input FM signal. When the two frequencies are the same, the amplitude of the error signal is zero, and the VCO is locked to the input signal. However, once the frequency of the FM input signal changes again, a new dc voltage is produced at the output of the detector, and the VCO frequency changes to lock on the new input

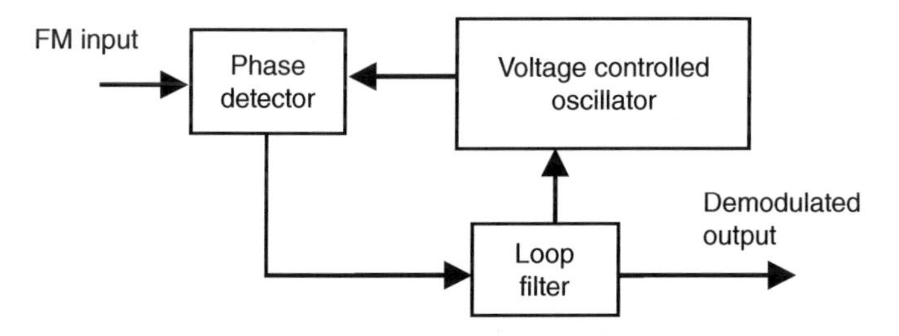

Figure 2.17 Basic building blocks of a PLL FM demodulator.

frequency. Thus, the PLL FM demodulator converts the frequency variations of the FM signal to voltage variations corresponding to the original baseband signal.

2.9 Phase Modulation (PM)

In PM systems [3], the phase angle $\theta(t)$ of the carrier signal varies in proportion to the change in the amplitude of the modulating signal. In PM, the phase angle of the carrier signal is expressed as

$$\theta(t) = \omega_c t + k_P\, v_m(t) \tag{2.23}$$

where k_P is a constant called ***phase sensitivity*** and has units of rad/volt and depends on the electrical parameters of the modulation circuit, and $v_m(t)$ is the modulating signal. The angle by which the phase differs from the reference value is referred to as the *phase departure*. The $k_P\, v_m(t)$ represents the phase departure, Eq. (2.23) can be written as

$$\theta(t) = \omega_c t + k_P\, V_m \cos \omega_m t \tag{2.24}$$

Substituting (2.24) in (2.9) gives an expression for a phase-modulated signal, $v_{PM}(t)$,

$$v_{PM}(t) = V_c \cos(\omega_c t + k_P\, V_m \cos \omega_m t) \tag{2.25}$$

$$v_{PM}(t) = V_c \cos(\omega_c t + \Delta\emptyset \cos \omega_m t).$$

$$v_{PM}(t) = V_c \cos(\omega_c t + M_P\, V_m \cos \omega_m t) \tag{2.26}$$

where $\Delta\emptyset = k_P V_m = M_P$, and M_P is referred to as a modulation index for PM (maximum phase deviation). Using Eq. (2.10), the instantaneous frequency deviation of a PM signal can be written as

$$\omega_i(t) = \frac{d}{dt}(\omega_c t + \Delta\emptyset \cos \omega_m t) = \omega_c - \omega_m \Delta\emptyset \sin \omega_m t \tag{2.27}$$

$$\therefore\ f_i(t) = f_c - f_m \Delta\emptyset \sin \omega_m t \tag{2.28}$$

Equation (2.28) indicates that the instantaneous frequency is influenced by the phase deviation. Thus, any variation in the phase will result in a variation in the frequency. Figure 2.18 shows a phase-modulated signal, and Figure 2.19 shows PM signals with different phase modulation indices.

Example 2.12

Determine the instantaneous phase of a signal that has a frequency of $f_i = \cos(\omega_c t)$, and $\theta_o = 0$

Solution

$$\omega_i = 2\pi f_i = 2\pi \cos(\omega_c t), \quad \text{and} \quad \omega_i(t) = \frac{d\theta(t)}{dt}$$

$$\therefore\ \theta_i(t) = \int_0^t 2\pi \cos(\omega_c t)\, dt$$

$$\theta_i(t) = \frac{2\pi}{\omega_c} \sin(\omega_c t)$$

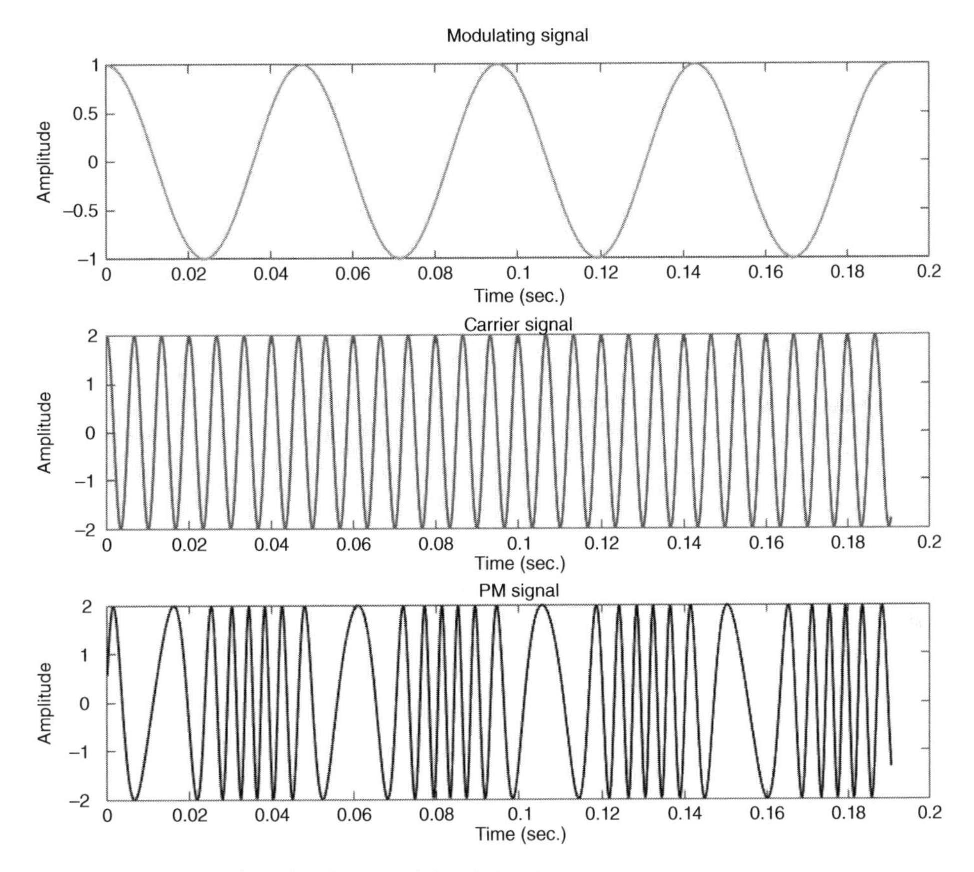

Figure 2.18 Illustration of a phase-modulated signal.

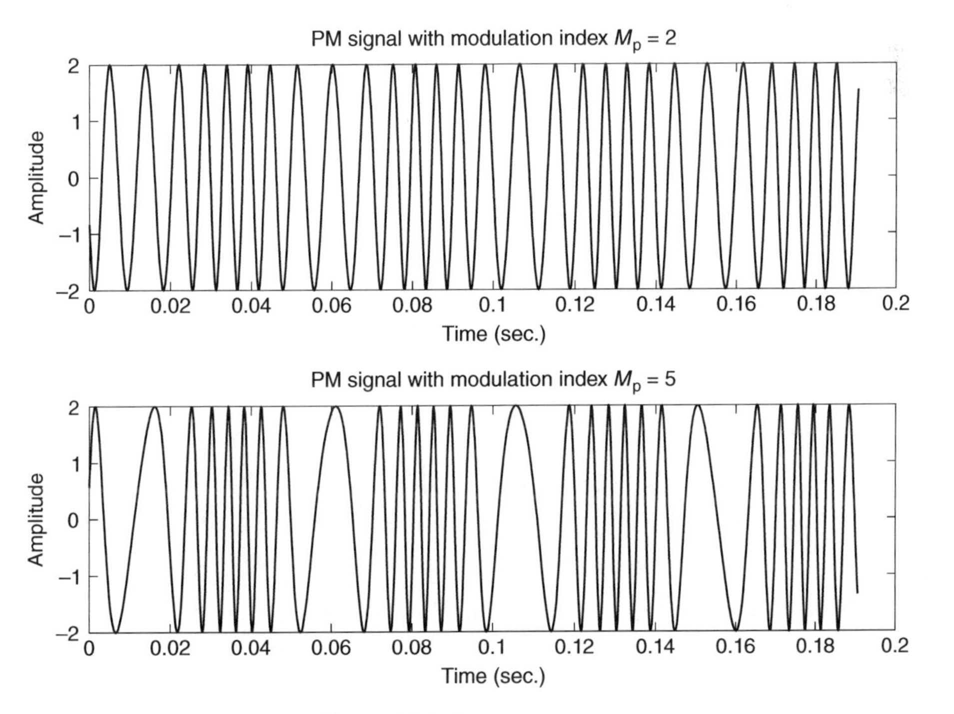

Figure 2.19 PM signals with different M_p indices.

Review Questions

2.1 Draw a block diagram of an analog communication system and explain the functionality of each block.

2.2 What is the minimum bandwidth of an AM signal that can be transmitted and still carry the necessary information?

2.3 What are the advantages of SSB transmission over the conventional AM?

2.4 For an AM signal that has $V_{max(p-p)}$ of 5.9 volts and $V_{min(p-p)}$ of 1.2 volts, determine the modulation index and calculate the signal's V_{max} and $V_{min.}$

2.5 For an AM broadcast station that is transmitting on a frequency of 980 MHz and allowed to transmit modulating frequency up to 5 kHz, determine the upper and lower sidebands and the occupied bandwidth of the transmitted signal.

2.6 An AM transmitter has a carrier power of 30 W. If the percentage modulation is 85%, calculate the total power and the power in one sideband.

2.7 For an FM transmitter that operates on frequency of 915 MHz and maximum FM deviation $\pm$ 12.5 KHz, determine the maximum and minimum frequencies that occur during the modulation process.

2.8 Determine the deviation ratio if the maximum deviation of an FM signal is 25 KHz and the maximum modulating frequency is 1.5 kHz.

2.9 Calculate the maximum modulating frequency that can be used to achieve a modulation index of 2.2 with a deviation of 7.48 kHz.

2.10 Using the graph of Bessel function in Figure 2.11, determine the carrier and the first four sidebands of an FM signal with a modulation index of 4.

2.11 What is the primary advantage of FM over AM?

2.12 Explain the process of preemphasis, and why it is used in FM?

2.13 Describe the process of deemphasis in FM radio receivers.

2.14 Calculate the modulation index if a 162 MHz carrier is deviated by a 12 kHz due to a 2 kHz modulating signal.

2.15 Determine the relative amplitudes of the fourth pair of sidebands for an FM signal with modulation index of 8.

2.16 Determine the modulation index at which the amplitude of the first pair of sidebands goes to zero.

References

1 Bhagwandas Pannalal Lathi, Zhi Ding, *Modern Digital and Analog Communication Systems*, Oxford University Press, 2018.
2 Rodger E. Ziemer, William H. Tranter, *Principles of Communications: Systems, Modulation, and Noise*, 7th ed., Wiley, 2015.
3 Leon W. Couch, *Digital and Analog Communication Systems*, 8th ed., Pearson, 2013.
4 Masoud Salehi, John G. Proakis, *Fundamentals of Communication Systems*, 2nd ed., Pearson, 2013.
5 Andreas F. Molisch, *Wireless Communications*, Wiley, 2010.
6 Michael Moher, Simon Haykin, *Communication Systems*, 5th ed., Wiley, 2009.

Suggested Readings

Ahmed M. Alaa, "Narrowband and wideband frequency modulation spectral characterization and bandwidth calculation," *IEEE Potentials*, Volume: 42, Issue: 1, 2023.

Andria Nicolaou, Antonis Kakas et. al., "An explainable artificial intelligence model in the assessment of brain MRI lesions in multiple sclerosis using amplitude modulation – frequency modulation multi-scale feature sets," *24th International Conference on Digital Signal Processing (DSP)*, 2023.

Chenxia Liu, Tao Liu, Tianwei Jiang, Song Yu, "Stable 2.4 GHz radio frequency transmission based on phase modulation," *Joint Conference of the European Frequency and Time Forum and IEEE International Frequency Control Symposium (EFTF/IFCS)*, 2022.

Dinh Le, Ashik Amin, Tahmid Ibne Mannan, Seungdeog Choi, "Sinusoidal frequency modulation carrier wave topology," *IEEE Energy Conversion Congress and Exposition (ECCE)*, 2022.

Gregor Lasser, Connor Nogales et al., "Wideband phase modulator MMIC for K-band supply-modulated power amplifier linearization," *51st European Microwave Conference (EuMC)*, 2022.

Ivan Horbatyi, Ivan Tsymbaliuk, "Neural network based approach for demodulation of signals with amplitude modulation of many components," *IEEE 16th International Conference on Advanced Trends in Radioelectronics Telecommunications and Computer Engineering (TCSET)*, 2022.

Lyubomir B. Laskov, Veska M. Georgieva, "Analysis of amplitude modulation and demodulation in MATLAB simulink environment," *56th International Scientific Conference on Information, Communication and Energy Systems and Technologies (ICEST)*, 2021.

Marc Bauduin, André Bourdoux, "Pi/K phase modulation for MIMO digitally modulated radars," *IEEE Radar Conference (RadarConf22)*, 2022.

Michael Nickerson, Bowen Song et al., "Broadband and low residual amplitude modulation phase modulator arrays for optical beamsteering applications," *Conference on Lasers and Electro-Optics (CLEO)*, IEEE, 2022.

Qianwei Zeng, Peng Yang et al., "Design of phase modulation antenna array with stable overall efficiencies," *IEEE Antennas and Wireless Propagation Letters*, Volume: 21, Issue: 2, 2022.

Qianwei Zeng, Peng Yang et al., "Phase modulation technique for harmonic beamforming in time-modulated arrays," *IEEE Transactions on Antennas and Propagation*, Volume: 70, Issue: 3, 2022.

Radu Gabriel Bozomitu, Ştefan Corneliu Stoica, "A robust radiocommunication system for FM transmission based on software defined radio technology," *IEEE 28th International Symposium for Design and Technology in Electronic Packaging (SIITME)*, 2022.

Xi Chen, Qiao Xiang, "Learning from FM communications: toward accurate, efficient, all-terrain vehicle localization," *IEEE/ACM Transactions on Networking*, Volume: 31, Issue: 1, 2023.

3

Digital Communication Systems

This chapter provides the necessary background for understanding the performance of digital communication systems. It covers an overview of digital communication fundamentals, analog-to-digital and digital-to-analog converters, digital modulation including binary phase shift keying (BPSK), frequency-shift keying (FSK), phase-shift keying (PSK), quadrature amplitude modulation (QAM), spectral efficiency and noise, frequency hopping spread spectrum, direct sequence spread spectrum, and orthogonal frequency division multiplexing (OFDM). Examples and review questions are provided to help the readers understand the covered topics.

3.1 Chapter Objectives

On reading this chapter, the reader will be able to:

- Draw a block diagram of a digital communication system and explain the functionality of each block.
- Explain the difference between non-return to zero (NRZ) and return to zero (RTZ) signals.
- Draw a block diagram of an analog-to-digital converter (ADC) and explain its operation.
- Explain what the Nyquist frequency is.
- List the key specifications of ADC and DAC devices.
- Write an expression for the dynamic range of an ADC.
- Draw a block diagram of a frequency shift keying (FSK) transmitter and explain the basic concept of FSK modulation.
- Illustrate the constellation of a BPSK and an 8-PSK modulation.
- Explain the basic concept of a differential phase-shift keying (DPSK) modulator.
- Draw a block diagram for a 16-quadrature amplitude modulation (QAM) modulator and illustrate its constellation diagram.
- Draw a block diagram of a 16-QAM demodulator and explain its operation.
- Explain what is meant by spectral efficiency.
- Explain the basic principle of frequency-hopping spread spectrum and what is the processing gain?
- Draw a block diagram of a frequency-hopping spread spectrum demodulator.
- Draw a block diagram of a direct sequence spread spectrum (DSSS) and explain its basic concept.

Essentials of RF Front-end Design and Testing: A Practical Guide for Wireless Systems, First Edition. Ibrahim A. Haroun.

– Explain the advantages of the orthogonal frequency division multiplexing (OFDM), technique, and draw block diagrams for the baseband OFDM transmitter and receiver.

3.2 Overview of Digital Communication

Digital communication [1–3] is a process in which the information (i.e., a physical variable such as voice, video image, data, temperature, speed, etc.) is sent from the source to the destination in a digital format. If the information is an analog signal, it gets converted to a digital signal to enable digital processing (i.e., source encoding, channel encoding, and modulation) in the transmitter. At the receiver side, the received digitally modulated signal gets converted back to the physical variable of the information that was sent. Figure 3.1 shows a simplified block diagram of a basic digital communication system.

In Figure 3.1, the information source in the transmitter is a transducer that converts a physical variable to a time-varying electrical signal (e.g., a microphone that converts sound to an electrical signal). The source encoder converts the output of the transducer to a series of binary digits (i.e., 1s and 0s). This binary digital signal is then applied to a channel encoder, which adds redundant bits to the data to enable the receiver to correct any errors during the transmission. The output of the channel encoder is connected to a modulator, which modulates the RF carrier according to the change in the digital sequence (i.e., converts the digital sequence into an analog electrical signal). The output signal from the modulator is then applied to the transmitter's RF front-end to get amplified to the required transmit power level. Finally, the output signal from the transmitter is applied to an antenna, which converts the RF signal to electromagnetic waves that propagate in free space at the speed of light.

In the receiver, the demodulator inverses the modulation process (i.e., converts the RF signal to binary data). The received RF signal gets down-converted and demodulated, and the demodulator's output is then applied to the channel decoder. The decoder removes the redundant bits that were added to the original data. The output of the channel decoder is then connected to the source decoder to convert the binary sequence to a time-varying electrical signal. Finally, the signal from the source decoder gets converted to the original physical variable by the destination transducer (e.g., a loudspeaker converts the electrical signal into sounds).

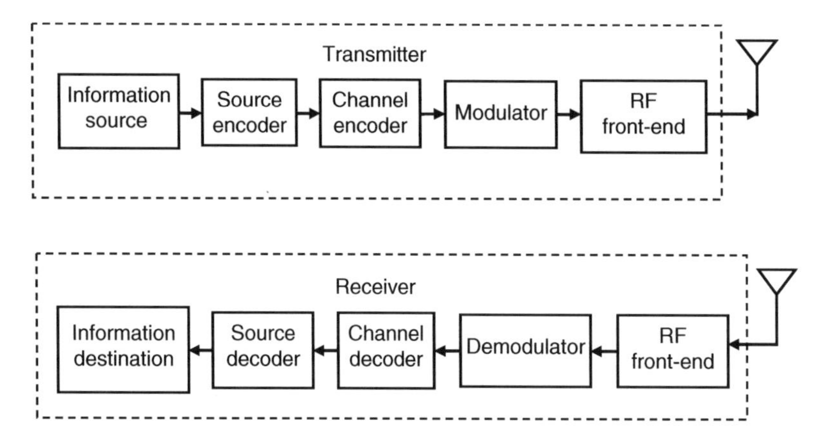

Figure 3.1 Block diagram of a basic digital communication system.

3.3 Types of Digital Signals

Whether digital signals are transmitted by baseband or broadband, they get encoded before the transmission. The following are the primary encoding signals used in digital communication systems.

3.3.1 Non-Return to Zero (NRZ) Signal

In non-return to zero (NRZ) encoding signals, the signal remains at the binary level for the entire bit time. Therefore, consecutive binary 1s do not return to zero during an interval. Figure 3.2 shows NRZ and return to zero signals.

The disadvantage of the NRZ is that the rise time and fall time of the signal introduce a dc component, which could make a binary "0" to be interpreted as a binary "1" and causes bit-error.

3.3.2 Return to Zero (RTZ) Signal

The return to zero (RTZ) signal reduces the dc component but creates problems as the frequency increases. Figure 3.2b shows an RTZ signal.

3.3.3 Non-Return to Zero (Bipolar)

In NRZ (bipolar) encoding signals, consecutive 1s and 0s do not return to zero. Figure 3.3a shows a NRZ bipolar signal.

3.3.4 Return to Zero (Bipolar)

In RTZ (bipolar) signals, consecutive 1s and 0s return to zero, which gives more reduction in the dc component. Figure 3.3b shows an RTZ bipolar signal.

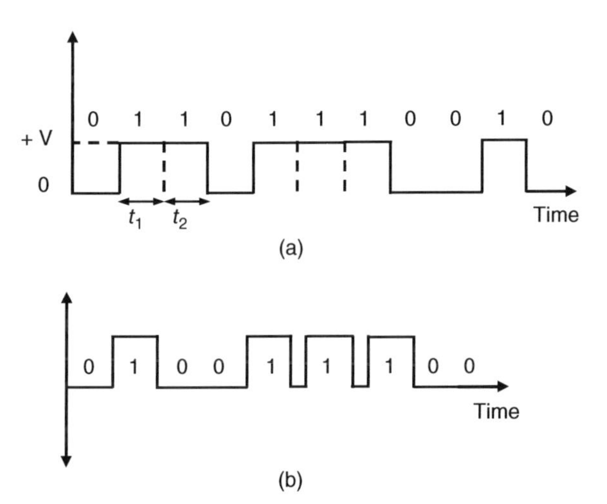

Figure 3.2 (a) NRZ signals and (b) RTZ signals.

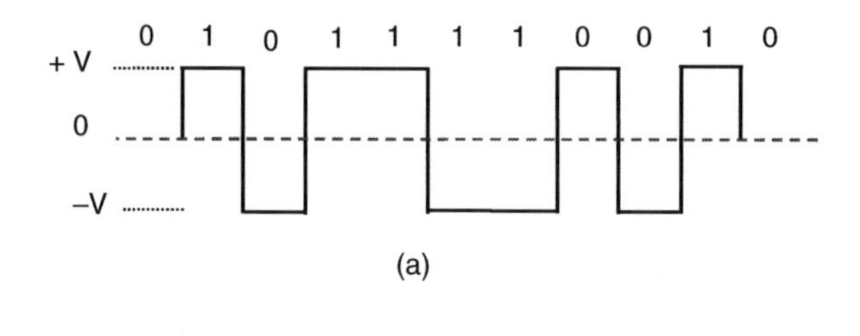

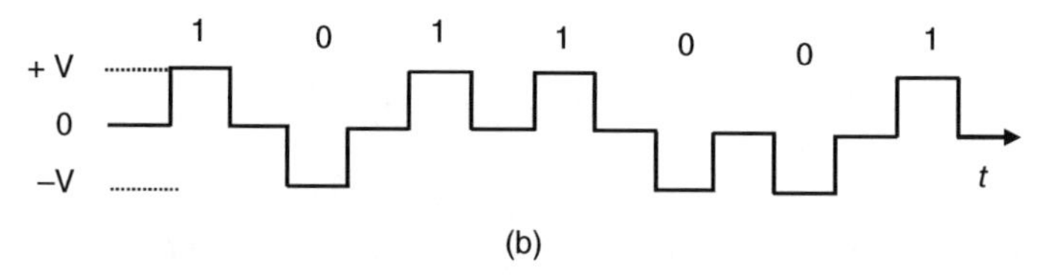

Figure 3.3 (a) NRZ bipolar signal and (b) RTZ bipolar signal.

3.4 Data Conversion

In digital communication radio systems, the process of converting analog signals into digital signals is done by A/D (analog-to-digital) converters, and the process of converting digital signals back to their equivalent analog is performed by digital-to-analog (D/A) converters.

3.4.1 Analog-to-Digital Conversion

Analog-to-digital converters are transducers that convert time-varying electrical signals to a binary sequence. The A/D conversion is the process of sampling the analog signals at regular time intervals, as shown in Figure 3.4.

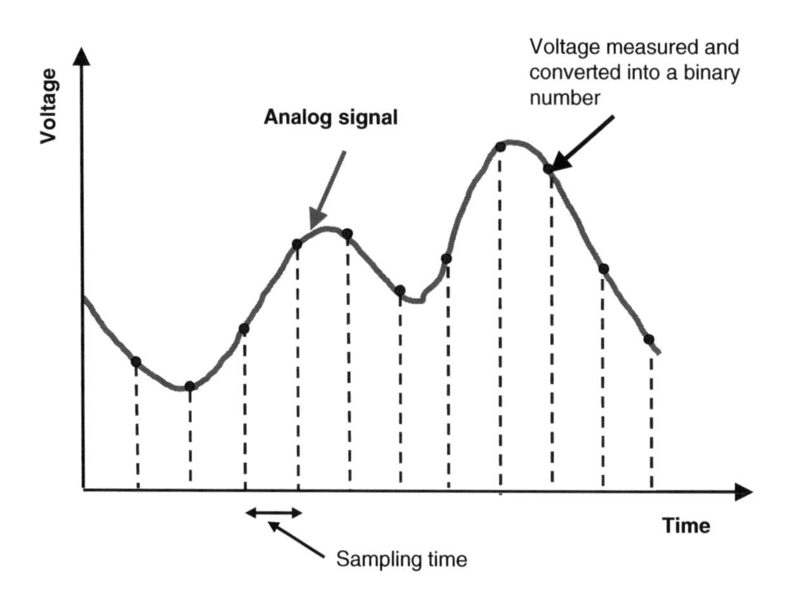

Figure 3.4 Sampling of an analog signal.

The instantaneous values of the measured sample get converted into a series of discrete binary numbers representing the samples. The sampling frequency, f_s, is a critical parameter that impacts the signal bandwidth. The frequency f_s is given by

$$f_s \geq 2f_{max} \tag{3.1}$$

where f_{max} is the maximum high-frequency content of the signal. The minimum sampling frequency is called the Nyquist frequency. In practice, the sampling rate is 2.5 to 3 times higher than the Nyquist minim frequency. A sampling rate less than the Nyquist rate causes an undesirable effect called ***aliasing*** (aliasing is a new signal produced near the original signal). Aliasing can be eliminated by a low-pass antialiasing filter in front of the ADC to cut off signals that are more than half the sampling rate. The number of levels in the sampled signal is 2^N, where N is the number of bits, and the number of increments is $2^N - 1$. There is some error associated with the conversion process, and it is referred to as a quantization error. However, this error can be reduced by having smaller step increments (i.e., more bits). The greater the number of bits, the greater the number of increments over the range and the smaller the quantization error. Assume using 10 bits ADC (i.e., $2^{10} = 1024$ voltage levels) and the input range is 0 to 5 volts; this gives a voltage step that is $5/1023 = 4.887 \times 10^{-3} = 4.887$ mV. Thus, the maximum error that can occur is less than 5 mV. A sample and hold circuit are usually used with ADC to hold the sample until the next sample arrives. The number of samples is called the sample rate and is measured in Hz. The resolution (i.e., step-size) of ADC is given by

$$\text{Resolution} = \frac{\text{Full scale range (FSR)}}{2^N - 1} = \frac{V_{max} - V_{min}}{2^N - 1} \tag{3.2}$$

Figure 3.5 shows a diagram of the analog to digital conversion process.

The key specifications of an ADC are as follows:

- Speed
- resolution
- dynamic range
- signal-to-noise ratio
- number of bits
- spurious free dynamic range

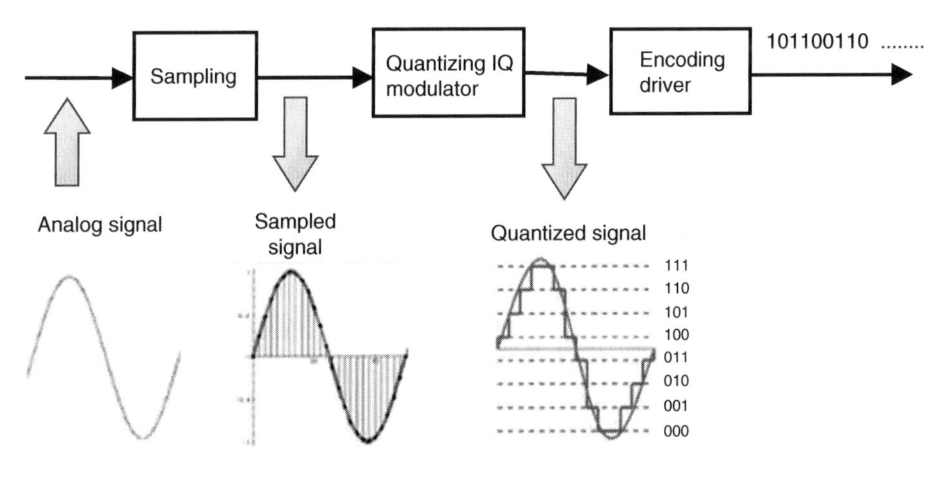

Figure 3.5 Analog to digital conversion process.

The dynamic range (DR) of an ADC is expressed in dB as

$$\mathrm{DR} = 20\log(2^N - 1)$$

The signal-to-noise ratio (SNR) impacts the performance of an ADC, it is the ratio of the actual input voltage to the total noise in the system (i.e., clock noise, power supply ripple, quantization noise, and external signal coupling). A *spurious free dynamic range* (SFDR) is defined as the ratio of the rms signal voltage to the voltage value of the highest spur (spur is any unwanted signal that results from intermodulation distortion) expressed in dB.

Example 3.1
For an ADC that uses 14-bits, the voltage range is −6 to +6 volts. Determine the following:

(a) the number of discrete levels (i.e., binary codes) that are represented
(b) the number of voltage increments used to divide the voltage range
(c) the resolution of digitization

Solution

(a) The number of discrete levels is

$$2^N = 2^{14} = 16384$$

(b) The number of voltage increments is

$$2^N - 1 = 2^{14} - 1 = 16384 - 1 = 16383$$

(c) The resolution is

$$\text{Resolution} = \frac{\text{Fullscalerange(FSR)}}{2^N - 1} = \frac{V_{\max} - V_{\min}}{2^N - 1} = \frac{6-(-6)}{16383} = 7.32 \times 10^{-4} = 0.732\ \text{mV}$$

Example 3.2
Determine the dynamic rang in dB for a 12-bit ADC converter.

Solution
The dynamic range, DR, is calculated as

$$\mathrm{DR} = 20\ \log(2^N - 1) = 20\ \log(2^{12} - 1) = 72.24\ \mathrm{dB}$$

3.4.2 Digital-to-Analog Conversion

Digital-to-analog converters are transducers that convert a given input digital word to a proportional output analog voltage, which represents a physical variable. Figure 3.6 shows a representation of an 8-bit D/A converter.

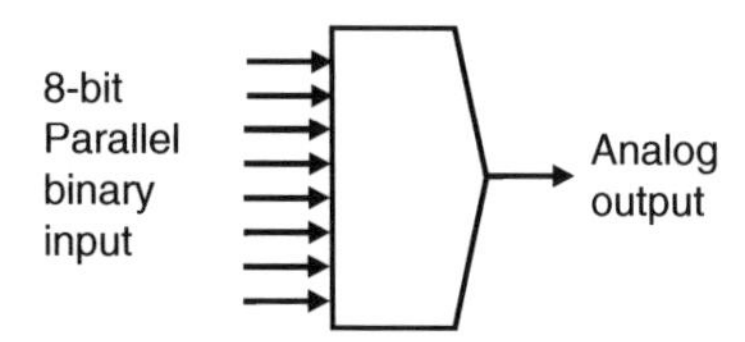

Figure 3.6 8-bit D/A converter.

The dynamic range of an analog-to-digital converter is given by

$$DR = 20 \log(2^N - 1) \quad (3.3)$$

The key specifications of D/A converters are speed, resolution, error, and settling time. ***Speed*** is the fastest rate at which the D/A converter can produce output steps (i.e., the number of samples per second). ***Resolution*** is the smallest increment voltage that the D/A converter produces over its voltage range, and it is related to the number of input bits. ***Error*** is a percentage of the full-scale voltage, typically +/− 0.1%. As an example, for an 8-bit D/A converter with 10-volt reference, the error is $10 \times 0.001 = 0.01$ V, or 10 mV.

Settling time is the time it takes for the voltage at the output of the D/A converter to settle to within +/− 1/2 LSB (least significant bit).

3.5 Digital Modulation

In digital modulation systems [4–7], the modulating signal is of a digital nature (i.e., binary), and the carrier signal is of an analog nature. Changing (switching) the carrier's parameters (amplitude, frequency, or phase) by the modulating signal results in a digitally modulated signal. The advantage of digital modulation includes noise immunity, ease of multiplexing, and ease of processing. The main types of digital modulation include FSK, PSK, QAM, andOFDM [8–11]. QAM modulation combines amplitude and PSK modulation, supporting a very high data rate in narrow bandwidths. On the other hand, OFDM operates over a very wide bandwidth and can achieve very high rates in a noisy communication channel.

3.5.1 Frequency-Shift Keying (FSK)

FSK is the simplest modulation type, also called binary FSK. It transmits digital information by switching the carrier signal between two sine wave frequencies, in contrast to the FM modulation where the carrier is continuously varied with the analog input signal. The two frequencies represent binary 0s and 1s, a binary 1 is referred to as mark f_m, and a binary 0 is referred to as space f_s. Figure 3.7 shows an FSK signal.

The frequency shift between f_m and f_s is called deviation Δf,

$$\Delta f = f_m - f_s \quad (3.4)$$

the modulation index, m, of FSK modulation is given by

$$m = \Delta f \, T \quad (3.5)$$

where T is the bit time ($T = 1$/data rate). The FSK signal is expressed as

$$v(t) = A \sin(\omega_c \pm \Delta\omega)t \quad (3.6)$$

where "A" is the carrier's amplitude, and ω_c is the carrier's radian frequency. In FSK, the signal amplitude does not change and that simplifies the design of the RF amplifiers, and it does not require high linearity amplifiers. Each symbol (i.e., mark or space) is one bit, and the output frequency changes each time the binary input changes. Thus, the output bid rate is equal to the input bit rate. However, if the mark and space frequencies are generated from two different oscillators, they will not be phase coherent, which causes abrupt changes during the 0-to-1 and 1-to-0 transition. The glitches or phase discontinuities produce more harmonics and wider bandwidth,

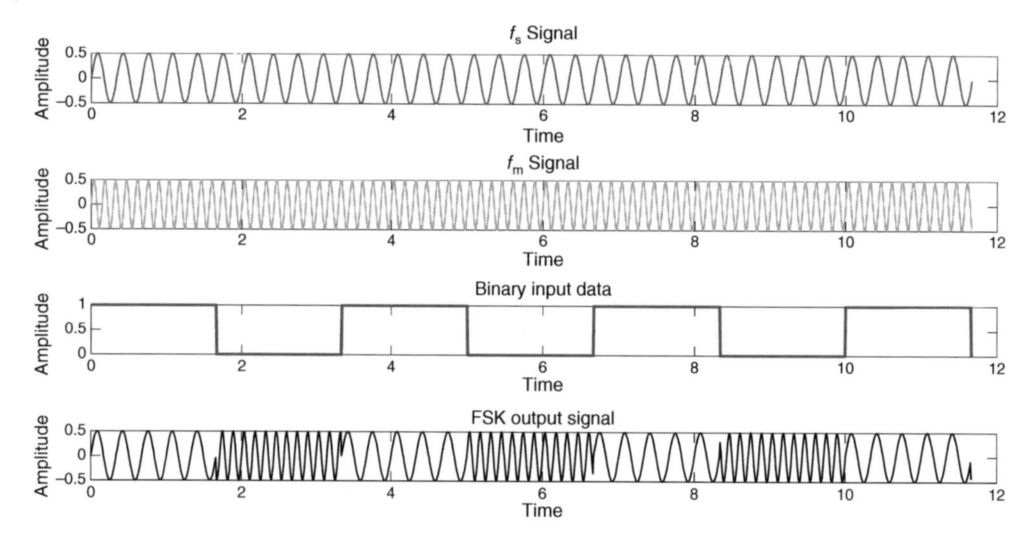

Figure 3.7 Illustration of FSK signal.

making the demodulation difficult and increasing the bit-error rate (BER). This problem can be removed if the periods of the sine-wave signal cross zero at the mark-to-space and space-to-mark transition; this modulation is called continuous-phase frequency-shift keying (CPFSK), also called coherent FSK. Figure 3.8 illustrates the generation of an FSK signal.

An improved variant of CPFSK is called minimum shift keying (MSK), where the mark and space frequencies are some integers multiple of the bit clock frequency. The MSK ensures that the signals are synchronized, and that there are no phase discontinuities. This type of modulation is spectral efficient. The bandwidth of a MFSK is given by

$$\mathrm{BW} = 2\,(f_\mathrm{d} - f_\mathrm{b}) \tag{3.7}$$

where f_d is the frequency deviation, and f_b is the bit rate. The bandwidth can be reduced by pre-filtering the binary modulating signal to remove some harmonics and lengthening the rise/fall time. This type of modulation is called Gaussian MSK (GMSK). Multilevel FSK allows more bits per symbol, including 4 FSK, 8 FSK, and 16 FSK. An FSK that uses more than two frequencies is called "M-ary" FSK.

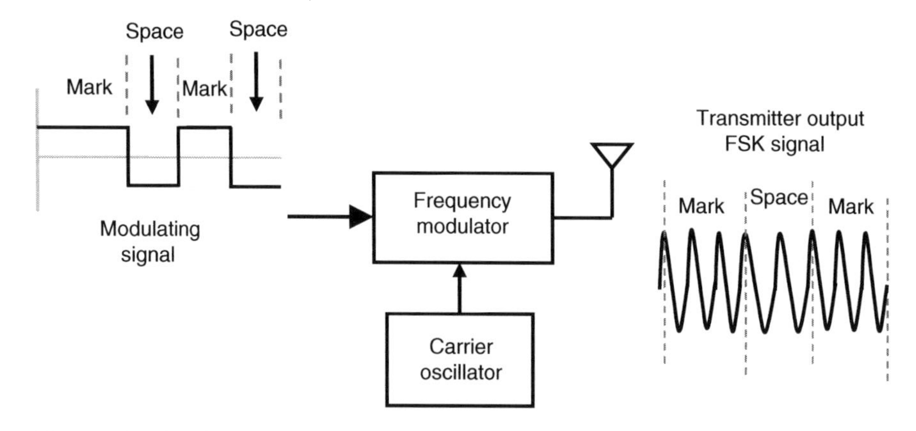

Figure 3.8 Simplified block diagram of an FSK transmitter.

3.5.2 Phase-Shift Keying (PSK)

PSK is an efficient modulation technique. In a PSK scheme, the binary signal to be transmitted causes the phase of the sine-wave carrier to shift between two phases of 0 or 180 degrees. The simplest form of PSK is the binary PSK (BPSK). Figure 3.9 illustrates the constellation of a BPSK signal, and Figure 3.10 shows the output of a BPSK modulator.

A block diagram of a BPSK modulator is shown in Figure 3.11.

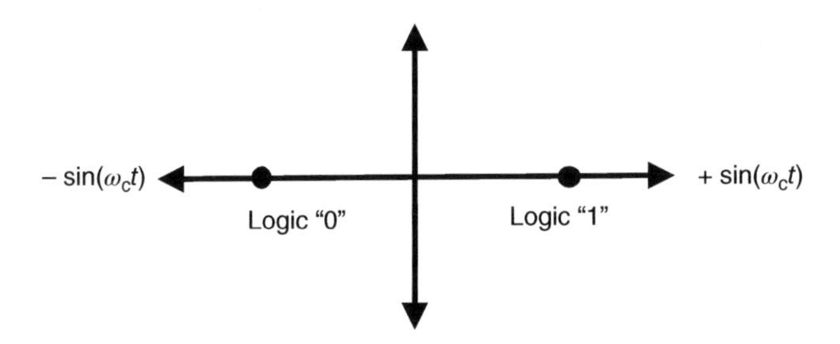

Figure 3.9 Illustration of a BPSK constellation diagram.

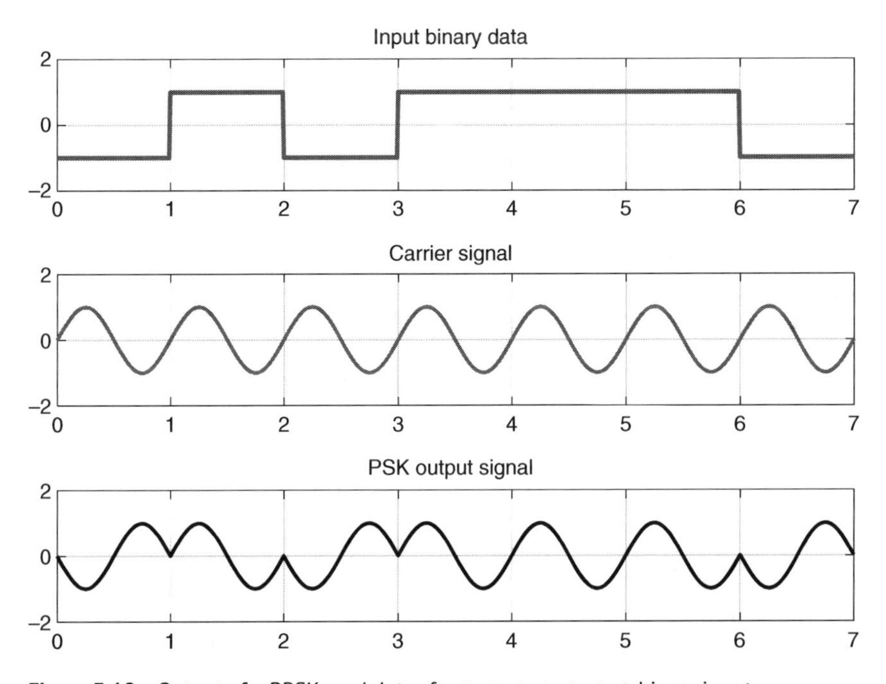

Figure 3.10 Output of a BPSK modulator for `0 1 0 1 1 1 0` binary input.

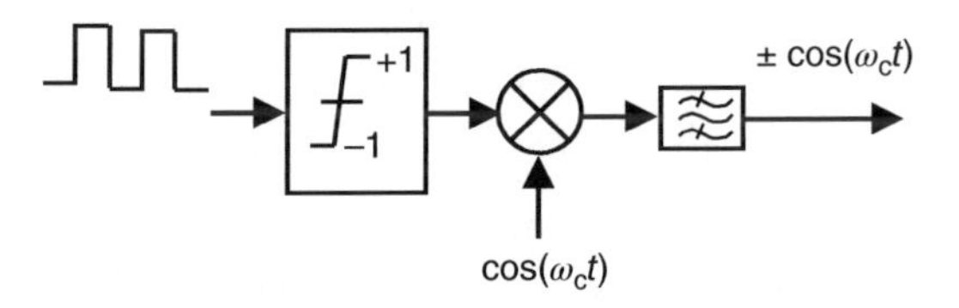

Figure 3.11 Block diagram of a BPSK modulator.

Figure 3.11 shows that when the signal at the input of the balanced mixer is high, the carrier is multiplied by +1, and the mixer's output is identical to the carrier. Conversely, when the signal at the mixer's input is low, the carrier signal is multiplied by –1, which means the signal is inverted (i.e., 180 phase shift). Therefore, the balanced mixer acts as a phase-reversing switch. The BPSK signal is expressed as

$$v(t) = A\cos(\omega_c t + \emptyset) \tag{3.8}$$

where "A" is the carrier's amplitude and ϕ is either 0 or 180 degrees; 180 degrees is equivalent to multiplication by –1, so a PSK signal can be written as

$$v_{\text{PSK}}(t) = \pm A\ \cos(\omega_c t) \tag{3.9}$$

Since there are only two states in BPSK modulation, the symbol rate equals the bit rate. A constellation diagram is often used to represent phase changes as shown in Figure 3.12. Figure 3.12 shows a constellation diagram of 8 PSK modulation.

In a constellation diagram, the phase is defined as an angle, and the amplitude is the distance from the constellation's origin to the symbol's location. Because the PSK signal has a constant amplitude, all the symbols lie on one circle. Like other modulation techniques, the spectral efficiency can be increased by increasing the number of states, a 4-PSK has four states, and each one is separated by 90 degrees. In 4 PSK, each symbol has two bits (00, 01, 11, and 10). Thus, the bit rate is twice the symbol rate. A 4-PSK is also called quadrature PSK (QPSK). The mapping of the states to bits is flexible, but it should be the same in both the transmitter and the receiver. The spectral efficiency can be increased by using higher order PSK such as 8-PSK, 8-PSK has eight possible states, and each symbol has 3 bits (2^3). The state transitions should avoid crossing the constellation's origin. It becomes problematic when the carrier's amplitude goes to zero temporarily because this causes the signal's peak-to-average power ratio (PAPR) to be high. The PAPR complicates the selection and design of the transmitter's power amplifier. The problem of transitions through the origin can be avoided by using variants of PSK such as offset quadrature phase-shift keying (OQPSK) and differential phase-shift keying (DPSK).

DPSK modulation is a noncoherent type (i.e., it does not need a synchronous carrier for demodulation). In DPSK modulation, the input bitstream is modified to a new bitstream such that the next

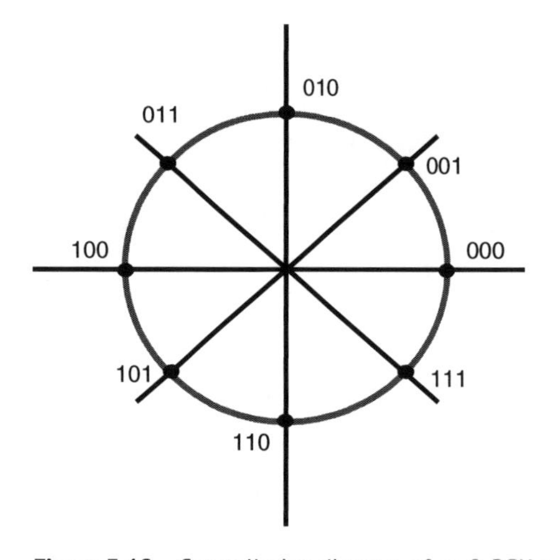

Figure 3.12 Constellation diagram of an 8-PSK modulation.

bit depends on the last bit. The DPSK is done by applying the input bitstream to an XNOR gate, and the output of the XNOR is applied to a 1-bit delay circuit before being applied back to the input as shown in Figure 3.13. Figure 3.13 shows a block diagram of a DPSK modulator, and Table 3.1 illustrates the phase transitions of a DPSK modulator.

The output of the XNOR is zero whenever the input bit and the last bit are not the same. Thus, a transition can be represented by a "0" symbol, and no transition is represented by "1." The bandwidth of a DPSK signal is $1/T_b$, where T_b is the bit period.

3.5.3 Quadrature Amplitude Modulation (QAM)

The QAM, is one of the most popular modulation techniques because of its enhanced spectral efficiency over limited channel bandwidths. This modulation technique uses both amplitude and phase modulation of the carrier, and it uses two orthogonal carries sin $(\omega_c t)$ and cos $(\omega_c t)$ signals. Figure 3.14 shows a block diagram of a 16-QAM modulator.

Common QAM variants are 16-QAM, 64-QAM, 256-QAM, and 1024-QAM. Figure 3.15 shows a constellation diagram of the 16-QAM and 64-QAM signals.

The max distance from the constellation's origin to the symbol's position determines the required signal power; the minimum distance between the symbols determines the noise immunity. Hence, higher order QAM schemes could have a higher bit error rate if the radio channel condition is noisy. However, QAM systems dynamically adapt modulation order based on the channel condition. If the channel condition is good, higher order modulation is used, and if the channel condition is poor, a lower order modulation is used. Because QAM modulation involves amplitude modulation, the

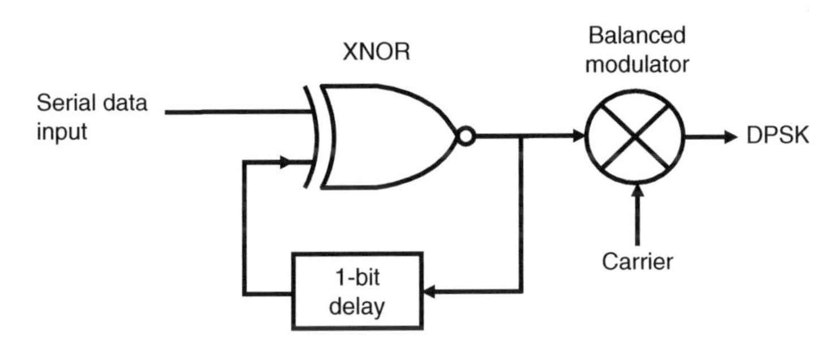

Figure 3.13 DPSK modulator.

Table 3.1 Transitions of DPSK modulator.

Input $b(t)$		1	1	0	0	1	0
Previous bit $d(t\text{-}T)$		1	1	1	0	1	1
Delayed bit $d(t)$	1	1	1	0	1	1	0
Phase @ DPSK output		0	0	π	0	0	π

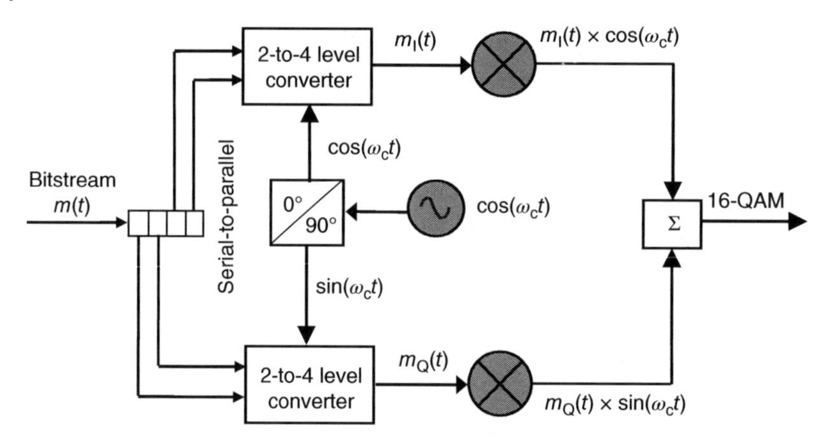

Figure 3.14 Block diagram of a 16-QAM modulator.

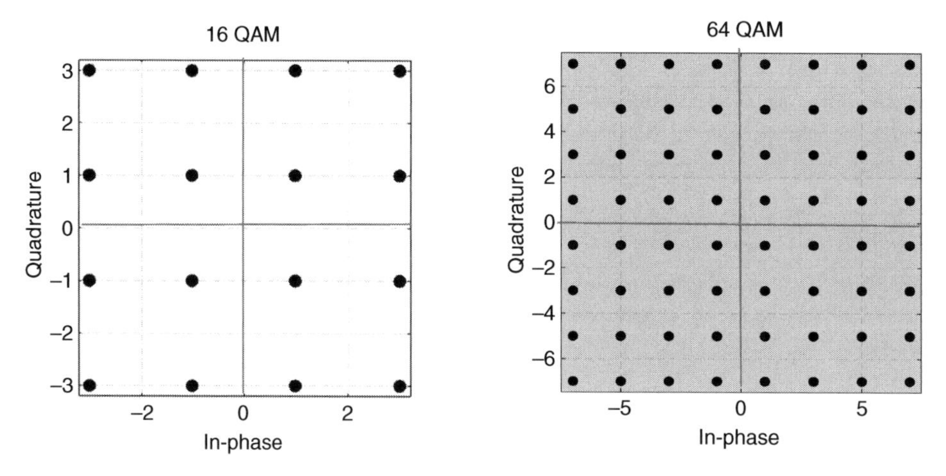

Figure 3.15 Constellation diagrams of a 16 QAM and 64 QAM.

linearity of the transmitter's power amplifier is a key system specification for QAM-type transmitters. The mathematical expression of a QAM signal is given by

$$y_{QAM}(t) = m_I(t)\cos\omega_c t + m_Q(t)\sin\omega_c t \tag{3.10}$$

where $m_I(t)$ and $m_Q(t)$ are the amplitudes of the in-phase and quadrature signals. In Figure 3.15, each point on the constellation diagram represents one of the 16 phase-amplitude positions that represents a 4-bit binary symbol (i.e., $2^4 = 16$ states). Figure 3.16 shows a block diagram of a 16-QAM demodulator.

In Figure 3.16, when the received QAM signal is mixed with $\cos(\omega_c t)$ in the I (In-phase) branch, the signal at the output of the mixer is given by

$$m_I(t)\cos^2(\omega_c t) + m_Q(t)\sin(\omega_c t) \times \cos(\omega_c t) \tag{3.11}$$

The 2nd term of Eq. (3.11) is zero due to the multiplication of $\sin(\omega_c t)$ and $\cos(\omega_c t)$. The first term of (3.11) can be expressed as $m_I(t)/2 + \cos(2\omega_c t)/2$. Using a low-pass filter at the mixer's output removes the carrier frequency component (i.e., $2\omega_c t$) and leaves the baseband signal $m_I(t)/2$. Similarly, the mixer's output of the quadrature branch is $m_Q(t)/2$ combining the baseband signals from the two filters provides the original message signal m(t).

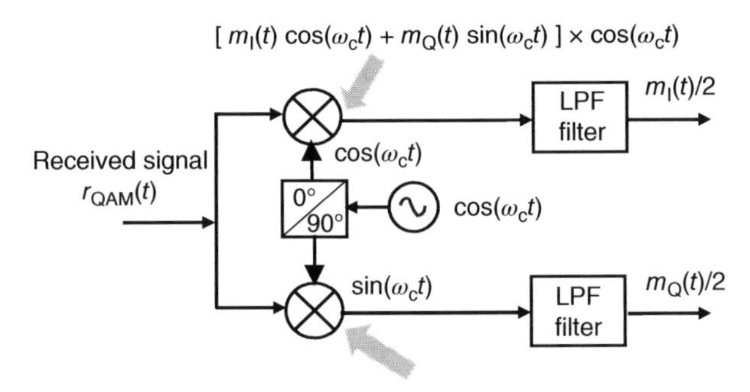

Figure 3.16 Block diagram of a 16-QAM demodulator.

3.6 Spectral Efficiency and Noise

Spectral efficiency is a measure of how fast information can be transmitted in a limited bandwidth. It is expressed as bps/Hz, and it depends on the modulation scheme. Table 3.2 lists the spectral efficiency for different modulation schemes.

Higher order modulation schemes have higher spectral efficiency. However, the spectral efficiency depends on the noise in the transmit channel. The greater the noise the smaller the signal-to-noise and the higher the probability of error in the receiver. The probability of error, P_e, is estimated as a function of E_b/N_o, where E_b is the energy per bit transmitted (energy = power × time), and N_o is noise power density in watts per hertz, N_o is given by

$$N_o = KT = N/B \tag{3.12}$$

where,

K = Boltzman's constant 1.38×10^{-23}
T = Temperature in Kelvins
B = Bandwidth in Hz
N = Power of the carrier signal, C.

Thus,

$$\frac{E_b}{N_o} = \frac{P_{tb}}{KT} \tag{3.13}$$

Table 3.2 Spectral efficiency vs. modulation scheme.

Modulation	Spectral efficiency
BPSK	1
QPSK	2
8-PSK	3
16-QAM	4
256-QAM	8

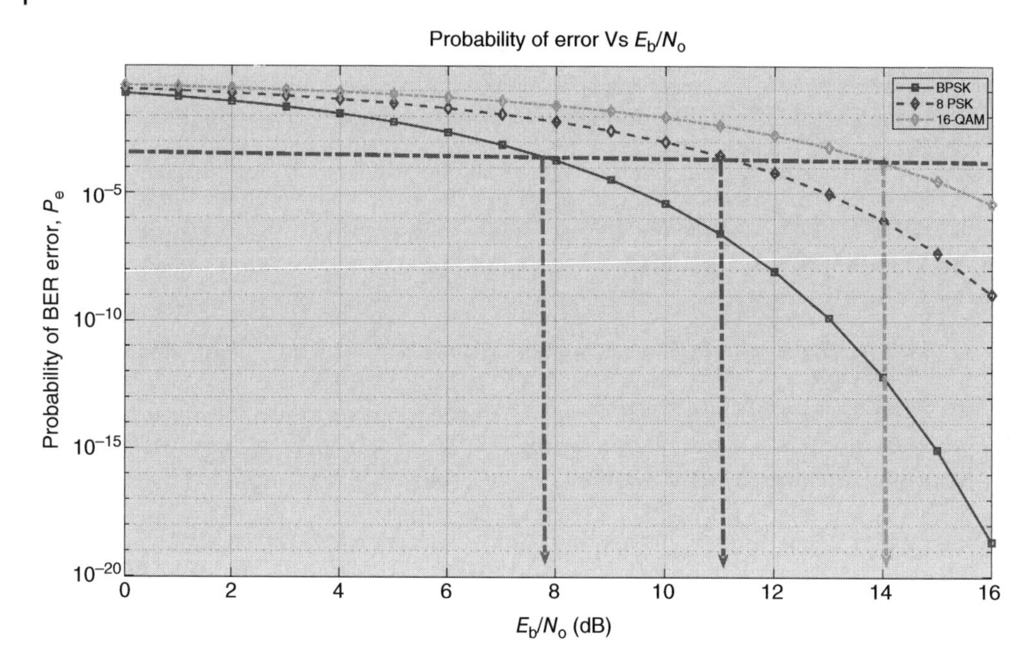

Figure 3.17 Probability of error vs. E_b/E_o for different modulation schemes.

where t_b is the bit time, and E_b/N_o can be expressed as

$$\frac{E_b}{N_o} = \left(\frac{C}{N}\right)\left(\frac{B}{f_b}\right) \tag{3.14}$$

where f_b is the bit rate ($f_b = 1/t_b$). Figure 3.17 shows simulation results for the probability of vs. E_b/N_o for different modulation schemes.

3.7 Wideband Modulation

Wideband modulation is a technique that spreads the transmitted signal over a wide bandwidth. This technique makes the signal immune to interference and jamming. It is also a multiplexing technique because it allows two or more signals to use the same bandwidth without interference. Therefore, it is a spectrum-efficient modulation type. The two most widely used wideband modulation techniques are spread-spectrum (SS) and OFDM [8–11]. Spread spectrum is used in cellular phones and referred to as code-division multiple access (CDMA) technology; OFDM is used in wireless technologies including 4G long term evolution (LTE) and 5G cellular phone systems.

3.7.1 Frequency-Hopping Spread Spectrum

In frequency-hopping spread spectrum (FHSS) systems, the frequency of the carrier signal is changed according to a predetermined sequence called pseudorandom noise (PN). The PN rate is higher than the rate of the serial binary data that modulates the carrier. Figure 3.18 shows a block diagram of an FHSS transmitter.

In Figure 3.18, the binary data d(t) to be transmitted is applied to an FSK modulator that is driven by the carrier signal, $A\cos(\omega_c t)$. The FSK modulator produces a modulated signal $s_d(t)$ that is

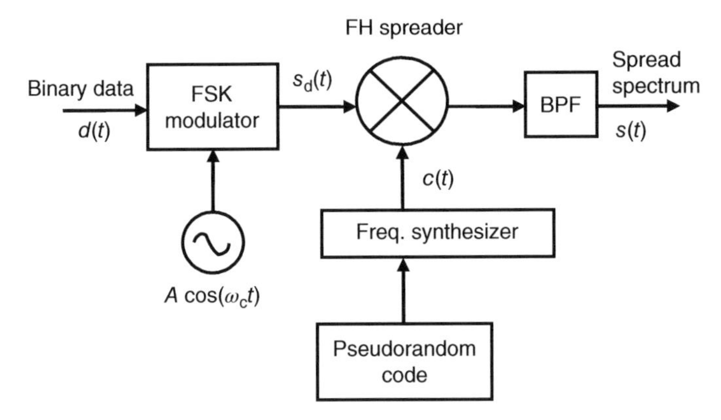

Figure 3.18 Block diagram of a frequency-hopping spread spectrum transmitter.

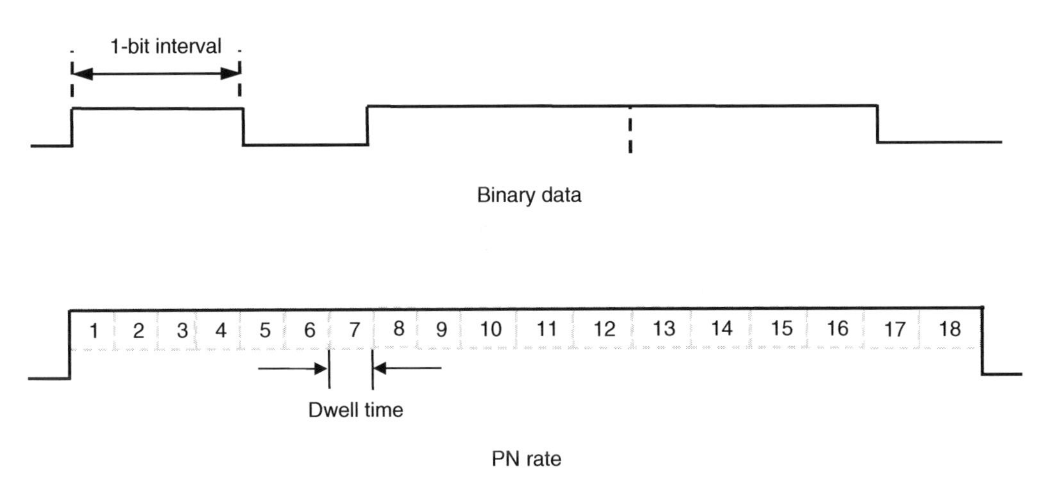

Figure 3.19 Serial binary data stream and a PN code rate.

applied to the FH spreader and mixed with the frequency synthesizer's signal, $c(t)$. The synthesizer is driven by the PN signal, which is a serial pattern of binary 0s and 1s. Figure 3.19 shows an example of a binary data stream and a PN code rate.

In Figure 3.19, the data bit interval is four times the dwell period (the time that the synthesizer remains on a single frequency). In this example, the carrier frequency changes for each dwell interval. Hence, the carrier frequency jumps fast all over the frequency band. The dwell time can be as short as 10 ms. The randomness of the PN code makes the signal at the output of the mixer appear as background noise and avoids the problem of failing communication at a particular frequency because of fading, interference, or jamming. Figure 3.20 shows a pseudorandom frequency-hope pattern.

There are two kinds of frequency hopping; these are slow frequency hopping (SFH) and fast frequency hopping (FFH). In SFH, the transmitter remains on one frequency band during the transmission of multiple data bits, $T_d > T_b$. In FFH, the transmitter hops in many frequency bands during the transmission of a single bit, $T_d < T_b$. Figure 3.21 shows a block diagram of a frequency-hopping spread spectrum receiver.

In Figure 3.21, the frequency synthesizer that drives the de-spreader must have the same pseudorandom code as the one generated in the transmitter.

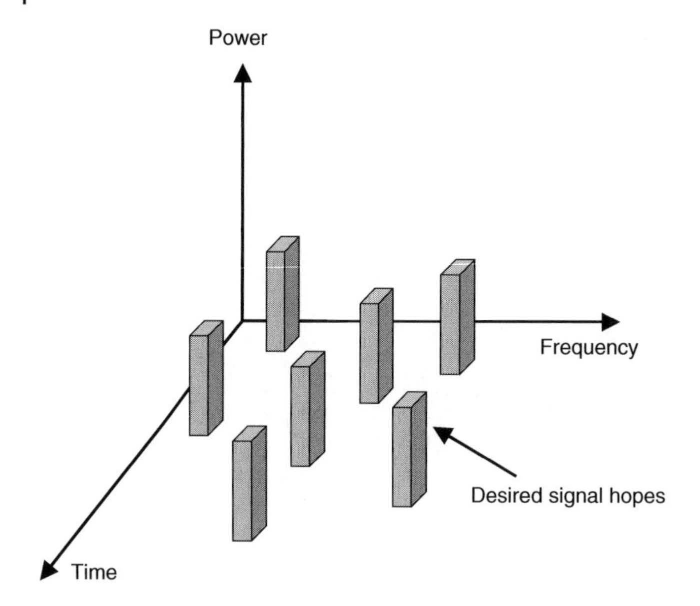

Figure 3.20 Pseudorandom frequency-hope pattern.

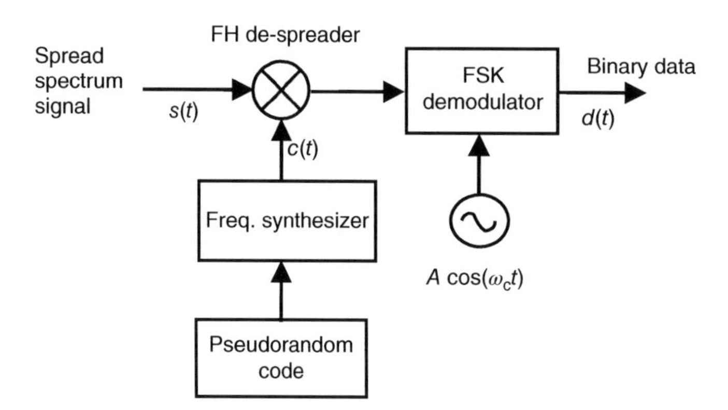

Figure 3.21 Block diagram of a frequency-hopping spread spectrum receiver.

3.7.2 Direct-Sequence Spread Spectrum

The direct-sequence spread spectrum (DSSS) is a modulation technique that makes the transmitted signal wider in bandwidth than the information bandwidth. In a DSSS system, the binary data is multiplied by the high-rate PN code sequence to acquire the spreading. Figure 3.22 shows a block diagram of a DSSS transmitter.

In Figure 3.22, the one-bit time of the PN code is called a chip, and the PN code rate is called the ***chip rate***. The carrier signal phase is switched between 0° and 180° by the 1s and 0s of the binary data. The modulated signal $S_d(t)$ is then mixed with the PN code $c(t)$, and the obtained signal $s(t)$ at the mixer's output is a spread spectrum signal. The ratio of the bandwidth of the spread signal to the unspread signal is called the ***processing gain*** and is given by

$$G_p = \frac{\mathrm{BW}}{f_b} = \frac{1/T_c}{1/T_b} = \frac{T_b}{T_c} \tag{3.15}$$

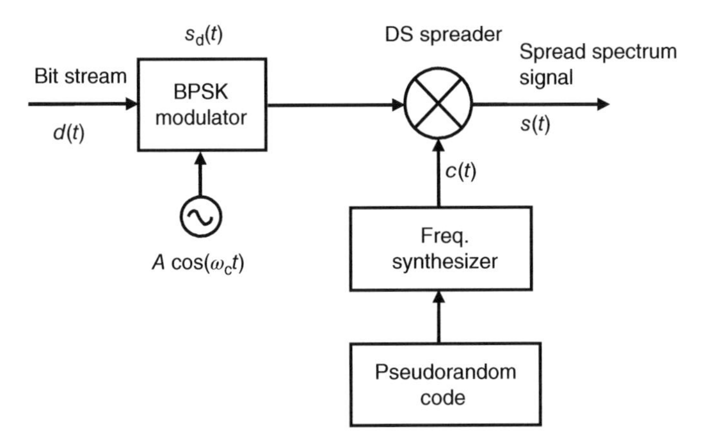

Figure 3.22 Block diagram of a DSSS transmitter.

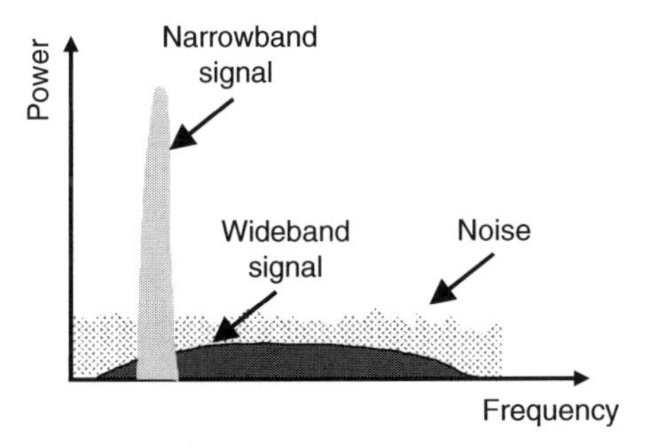

Figure 3.23 Narrowband and spread spectrum signals.

where BW is the bandwidth of the spread signal and f_b is the data rate. The DSSS signal $s(t)$ is given by

$$s(t) = d(t) \times A\cos(\omega_c t) \times c(t) \tag{3.16}$$

where $d(t)$ is the binary data signal, $c(t)$ is the PN code, and $A\cos(\omega_c t)$ is the carrier signal ("A" is the carrier amplitude). Figure 3.23 shows narrowband and spread spectrum signals, and Figure 3.24 shows a block diagram of a DSSS receiver.

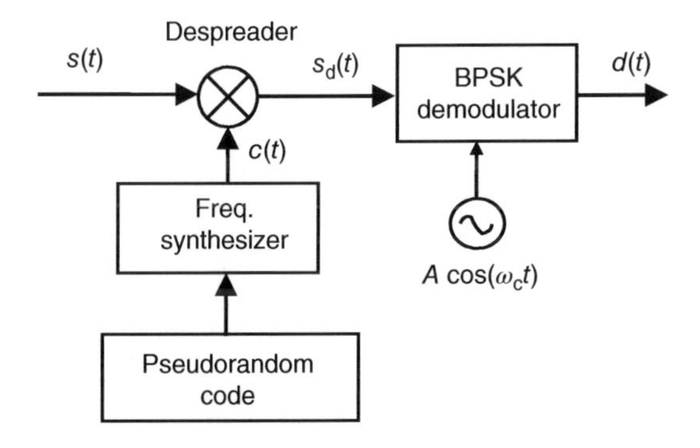

Figure 3.24 Block diagram of a DSSS receiver.

In Figure 3.24, the incoming spread spectrum signal $s(t)$ is multiplied by the PN signal $c(t)$ and the produced signal $s_\text{d}(t)$ is given by

$$s_\text{d}(t) = \text{d}(t) \cdot A\cos(\omega_\text{c}t) \cdot c(t) \times c(t)$$

and

$$c(t) \times c(t) = 1$$

$$\therefore \quad s_\text{d}(t) = \text{d}(t) \times A\cos(\omega_\text{c}t) \tag{3.17}$$

from Eq. (3.17), by passing the signal $s_\text{d}(t)$ through a BPSK demodulator, the original binary data $\text{d}(t)$ can be recovered.

3.7.3 Orthogonal Frequency Division Multiplexing (OFDM)

OFDM [8–11] is a modulation and multiplexing technique that converts high-speed serial data to parallel lower speed subcarriers. Each subcarrier is modulated using a different modulation scheme (QPSK, QAM, etc.). OFDM systems are used in many wireless technologies, including fourth-generation (4G), LTE and 5G cellular systems. The advantages of OFDM over single-carrier schemes include coping with severe channel conditions, robustness against intersymbol interference (ISI) and fading caused by multipath propagation, and higher spectral efficiency. Figure 3.25 illustrates the basic concept of an analog OFDM transmitter.

In Figure 3.25, the transmitter requires multiple local oscillators, each of which must have small phase noise and small frequency drift. Digital OFDM is used to overcome the limitation of analog OFDM. Figure 3.26 shows block diagrams of a digital OFDM transmitter and an OFDM receiver.

In a digital OFDM system, the channel subcarriers overlap without interfering because they are orthogonal to each other (i.e., the peak of one subcarrier falls on NULLs of the other sub-carriers), as shown in Figure 3.27b.

In an OFDM transmitter, a digital signal processing technique called ***Inverse Fast Fourier Transform*** **(IFFT)** is implemented to convert the parallel signals to orthogonal signals in the time domain. The data from the IFFT is then applied to a D/A converter, and the output of the D/A converter is a baseband OFDM.

At the OFDM receiver side, the baseband OFDM signal is converted from analog to digital and then applied to the FFT block. The FFT converts the digitized waveforms from samples in time to samples in frequency, the output of the FFT is then applied to the demodulator to extract the original binary data.

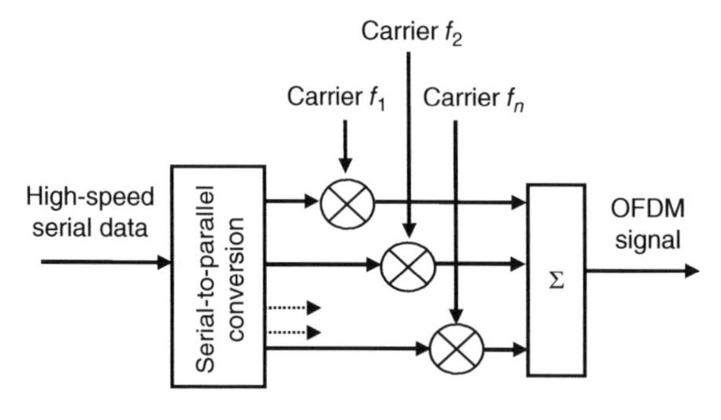

Figure 3.25 Block diagram of an analog OFDM transmitter.

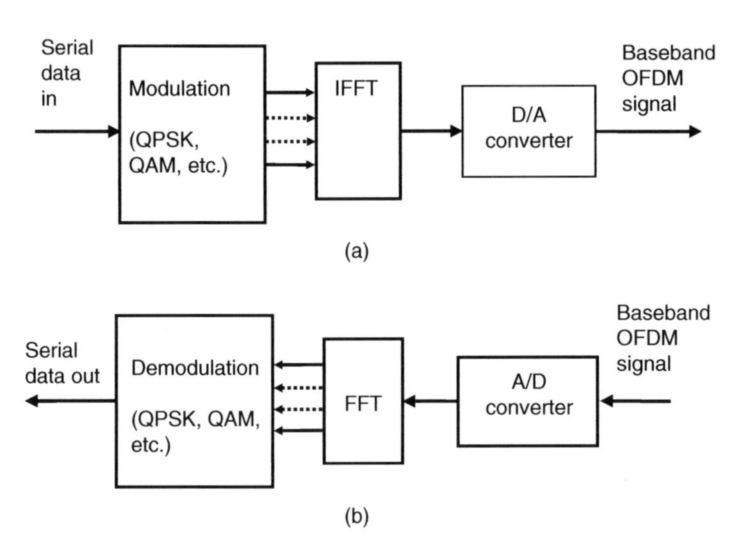

Figure 3.26 Block diagrams of a digital OFDM transmitter (a) and OFDM receiver (b).

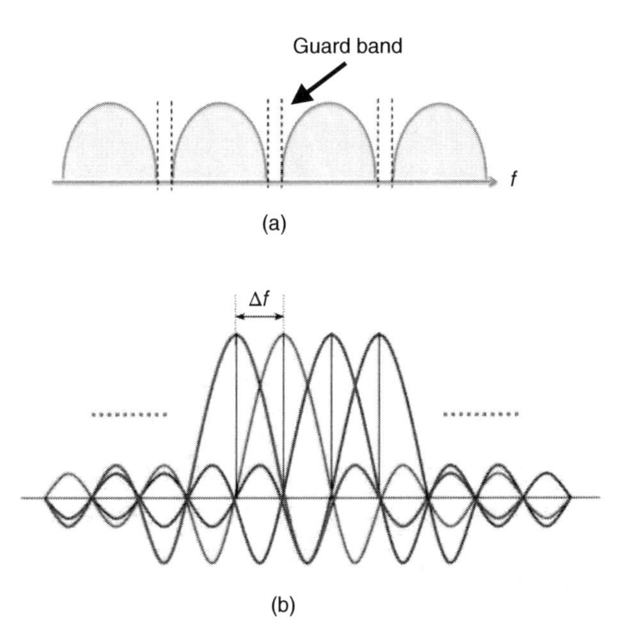

Figure 3.27 Representation of (a) frequency division multiplexing and (b) OFDM signals.

The main disadvantage of OFDM is its large PAPR, which requires a high linearity power amplifier in the transmitter. Also, to maintain orthogonality in an OFDM system, the frequency of the oscillators in both the transmitter and the receiver needs to have small phase noise and small frequency drift. However, the advantages of an OFDM system outweigh the disadvantages.

Review Questions

3.1 Explain the advantages of digital communications over analog communications and draw a block diagram of a digital modulation system.

3.2 Compare and contrast NRZ and return to zero bipolar digital signals.

3.3 Draw a block diagram of a digital-to-analog converter and explain the function of each block.

3.4 Explain and illustrate the process of digital-to-analog conversion.

3.5 List the key specifications of analog-to-digital and digital-to-analog converters.

3.6 For a voltage range of −5 to +5 volts and a 14-bit ADC, determine the following:
(a) the number of voltage increments used to divide the voltage range
(b) the resolution of digitization

3.7 Determine the dynamic range in dB for a 14-bit ADC converter.

3.8 Explain and illustrate the process of frequency shift keying.

3.9 For an FSK modem with $f_{\mathrm{m}} = 1.2$ kHz and $f_{\mathrm{s}} = 1.8$ kHz, determine the modulation index if the bit rate is 1200 bps.

3.10 Draw a block diagram of a binary phase shift keying (BPSK) modulator and illustrate its operation.

3.11 Draw the constellation diagram of an 8-PSK modulation.

3.12 Draw a block diagram of a differential PSK modulator and explain its operation.

3.13 Draw a block diagram of a 16-QAM modulator and explain its operation.

3.14 Compare and contrast a 16-QAM and 64-QAM modulation schemes.

3.15 Draw a block diagram of a 16-QAM demodulator and describe its demodulation process.

3.16 List the spectral efficiency for QPSK, 16-QAM, and 256-QAM modulation schemes.

3.17 Draw a block diagram of a frequency-hopping spread spectrum transmitter and illustrate its operation.

3.18 Explain the basic concept of DSSS and draw a block diagram for a DSSS transmitter.

3.19 Explain and illustrate the basic concept of frequency multiplexing and what is the primary advantage of multiplexing?

3.20 Compare and contrast the FDM and OFDM multiplexing techniques.

References

1. Bernard Sklar, Fredric Harris, *Digital Communications: Fundamentals and Applications*, 3rd ed., Pearson, 2020.
2. Leon W. Couch, *Digital and Analog Communication Systems*, 8th ed., Pearson, 2013.
3. John G. Proakis, Masoud Salehi, *Digital Communication*, 5th ed., McGraw-Hill, 2008.
4. John G. Proakis, Gerhard Bauch, Masoud Salehi, *Modern Communication Systems Using MATLAB*, Cengage Learning, 2013.
5. John G. Proakis, Masoud Salehi, *Fundamentals of Communication Systems*, Pearson Australia Pty Limited, 2015.
6. Rodger E. Ziemer, William H. Tranter, *Principles of Communications: Systems, Modulation, and Noise*, 7th ed., Wiley, 2015.
7. Michael Moher, Simon Haykin, *Communication Systems*, Wiley, 2009.
8. Hermann Rohling, *OFDM Concepts for Future Communication Systems*, Springer, 2011.
9. Y. J. Liu, *Introduction to OFDM Receiver Design and Simulation*, Artech House Publishers, 2019.
10. CIhan Tepedelenlioglu, Adrash B. Narasimhamurthy, *OFDM Systems for Wireless Communications*, Springer International Publishing, 2010.
11. Samuel C. Yang, *OFDMA System Analysis and Design*, Artech House, 2010.

Suggested Readings

Huajun Chen, Ning Ran, Lina Yuan, "BER research of OFDM systems under QPSK and 16-QAM modulation," *IEEE International Conference on Advances in Electrical Engineering and Computer Applications (AEECA)*, 2022.

Carl D'heer, Patrick Reynaert, "A 135 GHz 32 Gb/s direct-digital modulation 16-QAM transmitter in 28 nm CMOS," *IEEE 48th European Solid State Circuits Conference (ESSCIRC)*, 2022.

Puneeth Kumar DN, M. N. Eshwarappa, "Analysis of PAPR and BER in MIMO OFDM 5G systems with different modulation and transform techniques," *IEEE 2nd International Conference on Mobile Networks and Wireless Communications (ICMNWC)*, 2022.

Wei-Cheng Huang, Chih-Chieh Chiang, Yuan-Pei Wang, Juinn-Horng Deng, "Design of a new high-order QAM FM communication transceiver and verification of mmWave software radio platform," *IEEE VTS Asia Pacific Wireless Communications Symposium (APWCS)*, 2022.

Yuhao Lian, Xinyue Bi, Peisen Zhao, Yifan Jiang, "A research on modulation and detection in OFDM-IM system," *3rd Asia-Pacific Conference on Communications Technology and Computer Science (ACCTCS)*, 2023.

Tonghe Liu, "A review on the 5G enhanced OFDM modulation technique," *3rd Asia-Pacific Conference on Communications Technology and Computer Science (ACCTCS)*, 2023.

Muhammad Fahad Munir, Abdul Basit, Wasim Khan, et al., "Frequency quadrature amplitude modulation based scheme for dual function radar and communication systems," *International Conference on Engineering and Emerging Technologies (ICEET)*, 2022.

Sam Razavian, Sidharth Thomas et al., "A 0.4 THz efficient OOK/FSK wireless transmitter enabling 3 Gbps at 20 meters," *IEEE BiCMOS and Compound Semiconductor Integrated Circuits and Technology Symposium (BCICTS)*, 2022.

R. Swathika, S. M. Dilip Kumar, "Analysis of BER performance over AWGN and Rayleigh channels using FSK and PSK modulation schemes in LoRa based IoT networks," *International Conference on Intelligent and Innovative Technologies in Computing, Electrical and Electronics (IITCEE),* 2023.

T. Sairam Vamsi, Sudheer Kumar Terlapu, M. Vamshi Krishna, "Investigation of channel estimation techniques using OFDM with BPSK QPSK and QAM modulations," *International Conference on Computing, Communication and Power Technology (IC3P)*, 2022.

4

High-Frequency Transmission Lines

This chapter provides a solid background for understanding the high-frequency transmission lines that are used in wireless systems, and their impact on the system's performance. It covers transmission lines terminated in matched and mismatched loads, the input impedance of a transmission line of any load, quarter-wave transmission line impedance matching, lossy transmission lines, and transmission line parameters including reflection coefficient and standing wave ratio. Furthermore, the chapter discusses several types of high-frequency transmission lines including coaxial cables, microstrip, coplanar waveguide (CPW), and strip lines. The Smith chart and its use for solving transmission line problems, scattering parameters of two-port networks, single- and double-stubs matching techniques, and chain parameters are also discussed. Throughout the chapter, sufficient examples are provided to help solve transmission line problems both analytically and using the Smith chart. Review questions, list of references, and suggested readings that present the recent research on high-frequency transmission lines are provided at the end of the chapter.

4.1 Chapter Objectives

On working through this chapter, the reader will be able to:

- Calculate the transmission line parameters such as characteristic impedance, voltage-standing wave ratio (VSWR), reflection coefficient, attenuation factor, and complex propagation constant.
- Compare and contrast microstrip, coplanar waveguide (CPW), and coaxial transmission lines.
- Calculate the characteristic impedance of a quarter-wavelength line to match different impedances.
- Use the Smith chart to calculate VSWR, reflection coefficient, and input impedance of transmission lines.
- Use the Smith chart to determine the length of a transmission line to have a specific input impedance for a given load and characteristic impedances.
- Derive the S-matrix of a two-port T-section network.
- Explain the use of a vector network analyzer.
- Calculate the ABCD parameters for a two-port T-section and a π-section.
- Calculate the lengths of open- and short-circuit stubs to achieve a perfect matching.
- Derive the scattering matrix of a two-port network consisting of a π-section circuit.

Essentials of RF Front-end Design and Testing: A Practical Guide for Wireless Systems, First Edition. Ibrahim A. Haroun.

4.2 RF Transmission Lines Overview

Transmission lines [1–3] are critical in radio systems because they carry the electrical signal from component to component and influence the system's performance. They are used in wireless communication systems to connect the transmitter's output signal to an antenna; therefore, they could significantly impact the transmit power, which determines the communication range. The insertion loss of a transmission line is a key design parameter and should be kept low, so it does not affect the system's link budget. Also, the received signal at the receiver is connected from the antenna to the receiver's input by a transmission line. Thus, the transmission line's insertion loss affects the signal level at the receiver's input and could degrade the system's ability to detect weak signals if the line's insertion loss is high.

There are various types of high-frequency transmission lines, but the common types are the two-wire lines, coaxial lines, and planar lines including microstrip, stripline, and coplanar waveguide. The simplest form of a transmission line is the two-wire line, which consists of two identical conductors separated by a dielectric material. Figure 4.1 shows the common types of transmission lines. The two-wire lines are commonly used with TV receiver antennas. Figure 4.1b shows a coaxial transmission line that consists of two cylindrical conductors sharing the same axis (coaxial) and separated by a dielectric material. There are several coaxial transmission lines, including flexible, semi-rigid, and rigid cables. Figure 4.1c shows a planner microstrip transmission line [2–4] that consists of a flat strip of conducting material over a ground plane and is separated by a dielectric substrate. Planar transmission lines are easily fabricated by printed circuit technology and are convenient to use with surface mount devices and components.

High-frequency signals propagate along a transmission line as transverse electromagnetic (TEM) waves [5]. In a TEM propagation, both the electric E and magnetic H fields are perpendicular to each other at all points along the transmission line. Also, they are perpendicular to the direction of propagation. Figure 4.2 shows cross-sectional views of the E and H fields surrounding parallel two-wire line, coaxial transmission line, and microstrip line. Transmission lines can be

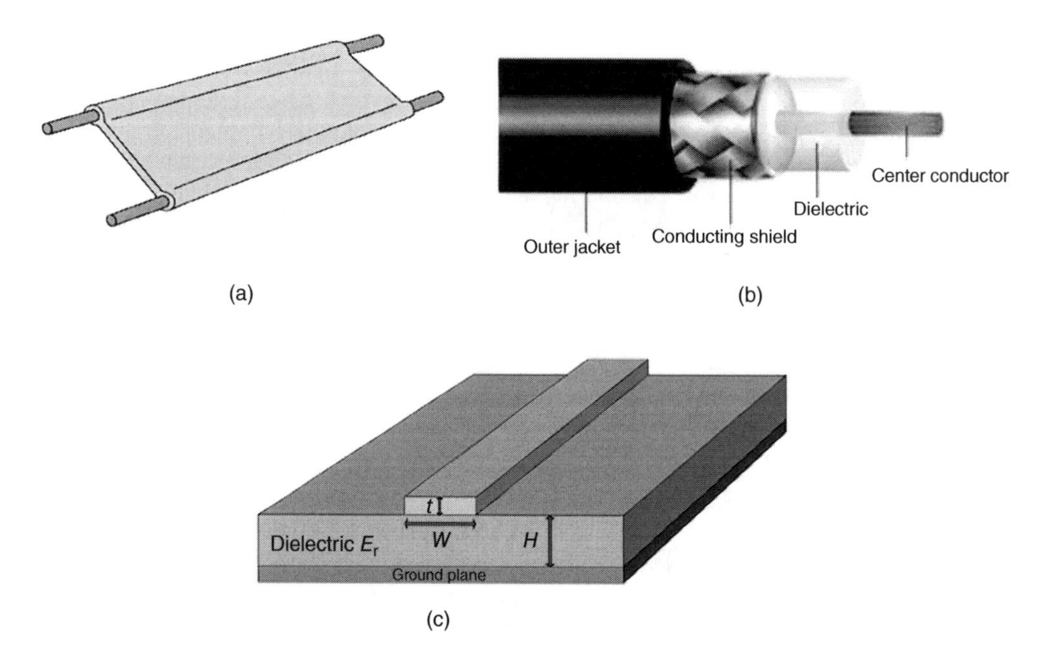

Figure 4.1 (a) Two-wire transmission line, (b) coaxial transmission line, and (c) microstrip transmission line.

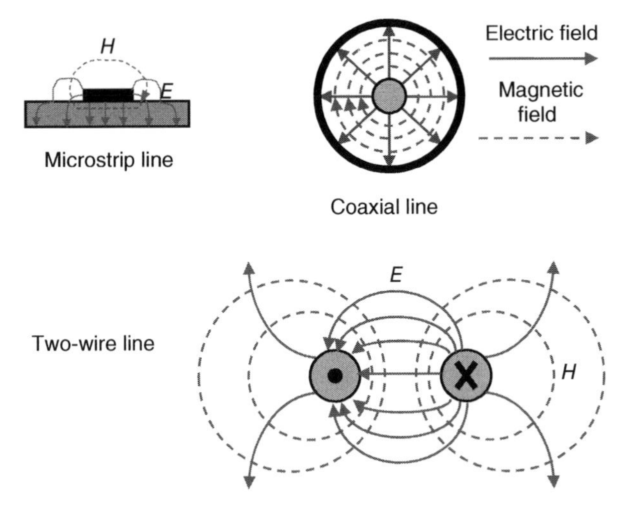

Figure 4.2 Cross-sectional views of the E and the H fields surrounding parallel two-wire line, coaxial transmission line, and microstrip line.

designed and simulated using RF/microwave software design tools such as Keysight Advanced Design System (ADS) [6] and Ansys high-frequency structure simulator (HFSS) [7]. Also, there are online free tools for calculating the parameters of microstrip, stripline, and CPW lines.

4.3 Transmission Line Analysis

The analysis of transmission lines is required to determine their design parameters such as the characteristic impedance and the complex propagation constant. A transmission line is considered as a segments of lengths dz with a constant cross section along its length and can be modeled by the circuit shown in Figure 4.3. In this circuit model approach, the transmission line is analyzed in terms of voltage, current, and impedance.

The parameters of a transmission line section dz are

R = Resistance per unit length [Ω/m]
L = Inductance per unit length [Henry/m]
G = Conductance per unit length [℧/m]
C = Capacitance per unit length [Farad/m]

Applying both Kirchhoff's voltage and current laws to the circuit shown in Figure 4.3 and taking the limits as Δz goes to 0, results in two equations referred to as the telegrapher's equations of a transmission line and are given by

$$\frac{\mathrm{d}V(z)}{\mathrm{d}Z} = -(R + j\omega L)I(z) \tag{4.1}$$

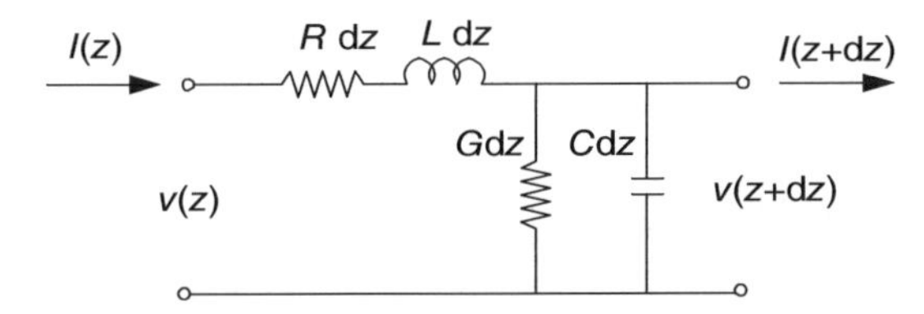

Figure 4.3 Circuit model of a transmission line section of length dz.

$$\frac{dI(z)}{dZ} = -(G + j\omega C)V(z) \tag{4.2}$$

Taking the derivative of Eqs. (4.1) and (4.2) results in the wave equations of $V(z)$ and $I(z)$, which are given by

$$\frac{\partial^2 V}{\partial Z^2} - \gamma^2 V = 0 \tag{4.3}$$

$$\frac{\partial^2 I}{\partial Z^2} - \gamma^2 I = 0 \tag{4.4}$$

and

$$\gamma = \alpha + j\beta = \sqrt{(R + j\omega L)(G + j\omega C)} \tag{4.5}$$

where γ is the complex propagation constant with units m^{-1}, α is the attenuation coefficient with units of Neper per meter (Np/m, *one neper*= 8.68 dB), and β is the phase constant with units of radians per meter (rad/m). The solutions of (4.3) and (4.4) are given by

$$V(z) = V_o^+ e^{-\gamma Z} + V_o^- e^{\gamma Z} \tag{4.6}$$

$$I(z) = I_o^+ e^{-\gamma Z} + I_o^- e^{\gamma Z} \tag{4.7}$$

The first terms of Eqs. (4.6) and (4.7) represent waves propagating in the positive z-direction (+), while the second terms of these equations represent waves propagating in the negative z-direction (−). Substituting Eq. (4.6) in Eq. (4.1) results in

$$I(z) = \frac{\gamma}{R + j\omega L}\left[V_o^+ e^{-\gamma Z} - V_o^- e^{\gamma Z}\right] \tag{4.8}$$

From Eqs. (4.8) and (4.7), $I_o^+(z)$ and $I_o^-(z)$ can be written as

$$I_o^+(z) = \frac{\gamma}{R + j\omega L}\left(V_o^+\right) \quad \text{and} \quad I_o^-(z) = \frac{\gamma}{R + j\omega L}\left(-V_o^-\right) \tag{4.9}$$

Thus, the characteristic impedance z_o is defined as

$$z_o = \frac{V_o^+}{I_o^+} = \frac{-V_o^-}{I_o^-} = \frac{(R + j\omega L)}{\gamma}$$

$$\therefore\ z_o = \sqrt{\frac{R + j\omega L}{G + j\omega C}}\ \Omega \tag{4.10}$$

and Eq. (4.7) can be written as

$$I(z) = \frac{V_o^+}{z_o} e^{-\gamma Z} - \frac{V_o^-}{z_o} e^{\gamma Z} \tag{4.11}$$

For low-loss transmission lines, $\alpha \ll \beta$, and the characteristic impedance can be written as

$$\therefore\ z_o = \sqrt{\frac{L}{C}}\ \Omega \tag{4.12}$$

The phase constant β and the phase velocity v_p are given by

$$\beta = \omega\sqrt{LC} \tag{4.13}$$

$$v_p = \frac{\omega}{\beta} = \frac{1}{\sqrt{LC}}, \qquad \beta = \frac{2\pi}{\lambda} \tag{4.14}$$

The phase velocity depends on the properties of the dielectric material between the conductors of the transmission line. These properties are permittivity, ε_o, which describes the energy storage associated with the electric field, and permeability, μ_o, which describes the energy storage associated with the magnetic field. Thus,

$$v_p = \frac{1}{\sqrt{\mu\varepsilon}} \tag{4.15}$$

In free space, the permittivity is ε_o, and the permeability is μ_o, and they are given by

$$\varepsilon_o = 8.845 \times 10^{-12}\ \text{F/m}, \qquad \mu_o = 4\pi \times 10^{-7}\ \text{H/m} \tag{4.16}$$

Therefore, in free space, the phase velocity is the same as the speed of light and expressed as

$$v_p = c = \frac{1}{\sqrt{\mu_o\varepsilon_o}} = 3 \times 10^8\ \text{m/s} \tag{4.17}$$

and

$$\lambda_o = \frac{c}{f} \tag{4.18}$$

where λ_o is the wavelength in free space, and λ_g is the guided wavelength on a transmission line. The relative permittivity ε_r (also known as dielectric constant) and the relative permeability μ_r are defined as

$$\varepsilon_r = \frac{\varepsilon}{\varepsilon_o}, \qquad \mu_r = \frac{\mu}{\mu_o}, \qquad \lambda_g = \frac{\lambda_o}{\varepsilon_r} \tag{4.19}$$

Most materials have $\mu_r = 1$. The parameters Z_o, α, and β are the key design parameters of high-frequency, microwave, and mmWave transmission lines. The electrical signals that propagate in transmission lines experience some attenuation due to lossy substrate and imperfect conductors (i.e., $\sigma \neq \infty$). At high frequencies, the skin-effect δ must be considered when the metal loss is considered, δ is given by

$$\delta = \frac{1}{\sqrt{\pi f \mu \sigma}} \tag{4.20}$$

The skin depth δ is defined as the distance from the medium surface to where the magnitude of the field of a wave traveling in the medium is reduced to $1/e$ (i.e., 37%, $e = 2.718$) relative to the wave's magnitude at the medium's surface. The conductor loss can also be described in terms of its surface resistance R_s which is given by

$$R_s = \frac{1}{\sigma\delta} \tag{4.21}$$

Substituting Eq. (4.20) in Eq. (4.21), the surface resistance can be expressed as

$$R_s = \sqrt{\frac{\pi f \mu}{\sigma}} \tag{4.22}$$

Example 4.1

For a transmission line that has an inductance of 1.119 μH/m and a capacitance of 12.3 pF/m, determine the line's characteristic impedance.

Solution

$$Z_o = \sqrt{\frac{L}{C}}$$

$$Z_o = \sqrt{\frac{1.119 \times 10^{-6}}{12.3 \times 10^{-12}}} = 301.6\,\Omega$$

Example 4.2

For a transmission line that has the following characteristics:

$Z_o = 52\,\Omega$, $C = 96.8$ pF/m, and attenuation of 15.4 dB/100 m at 400 MHz, determine the attenuation and phase constant per meter of the line, and the phase velocity.

Solution

$$\alpha = (15.4/100) \times (1/8.86) = 0.0174\,\text{Np/m}$$

$$\beta = \omega Z_o C = 2\pi(400 \times 10^6) \times 52(96.8 \times 10^{-12}) = 12.65\,\text{rad/m}$$

$$v_p = \frac{\omega}{\beta} = \frac{2\pi(400 \times 10^6)}{12.65} = 1.9868 \times 10^8\,\text{m/s}$$

Example 4.3

For a copper wire that has a resistivity of 1.725×10^{-8} Ω/m, determine the skin depth when the frequency of operation is 6 GHz.

Solution

The skin depth is given by

$$\delta = \frac{1}{\sqrt{\pi f \mu \sigma}}$$

$$\delta = \sqrt{\frac{1.725 \times 10^{-8}}{\pi(6 \times 10^9)(4\pi \times 10^{-7})}} = 8.53 \times 10^{-7}\,\text{m} = 0.853\,\mu\text{m}$$

Example 4.4

For a coaxial cable that has a relative permittivity of 20 and operating at frequency of 1.9 GHz, determine the length of a quarter-wavelength line.

Solution

The guided wavelength λ_g is

$$\lambda_g = \frac{\lambda_o}{\varepsilon_r} = \frac{3 \times 10^8/1.9 \times 10^9}{20} = 0.035\,\text{m}.$$

$$\therefore \quad \frac{\lambda_g}{4} = 8.83\,\text{mm}$$

Example 4.5

For a transmission line that has a 10 cm length and a phase constant of 3 rad/m, determine the electrical length of the line.

Solution
The phase constant β is

$$\beta = \frac{2\pi}{\lambda}, \quad \therefore \quad \lambda = \frac{2\pi}{3} = 2.094 \text{ m}$$

and

$$\text{Length } l = \frac{0.1}{2.094}\lambda = 0.0477\lambda$$

Example 4.6
For a transmission that has a resistance of 100 Ω/m, inductance of 80 nH/m, conductance of 1.6 S/m, and capacitance of 200 pF/m, calculate the attenuation factor, phase constant, phase velocity, and the characteristic impedance for a traveling wave at 2 GHz on the line.

Solution
The propagation constant γ is

$$\gamma = \sqrt{(R + j\omega L) \times (G + j\omega C)}$$
$$= \sqrt{(100 + j(2\pi \times 2 \times 10^9) \times 80 \times 10^{-9}) \cdot (1.6 + (2\pi \times 2 \times 10^9) \times 200 \times 10^{-12})}$$

$$\therefore \text{ Attenuation} = \text{Re}(\gamma) = 17.94 \text{ Np/m},$$

and

$$\text{Phase constant} = \text{Im}(\gamma) = 51.85 \text{ rad/m}$$

The phase velocity v_p is

$$v_p = \frac{\omega}{\beta} = \frac{2\pi(200 \times 10^9)}{51.85} = 2.42 \times 10^8 \text{ m/s}$$

The characteristic impedance Z_o is

$$Z_0 = \sqrt{\frac{(R + j\omega L)}{(G + j\omega C)}} \;\Omega = \sqrt{\frac{(100 + j(2\pi \times 2 \times 10^9) \times 80 \times 10^{-9}}{(1.6 + (2\pi \times 2 \times 10^9) \times 200 \times 10^{-12})}} = 17.9 + j4.27 \;\Omega$$

4.4 Reflection Due to Impedance Mismatch

When a transmission line is connected to a load (or source) different from its characteristic impedance (i.e., mismatched), reflections of the signals on the line will occur. As a result, the incident power will be reflected toward the source. The reflected power could change or impair the system's performance. The amount of reflection that is caused by a mismatched load is expressed in terms of a reflection coefficient, Γ, the reflection coefficient at the load is given by

$$\Gamma_L = \frac{Z_L - Z_o}{Z_L + Z_o} = \frac{V_o^-}{V_o^+} \tag{4.23}$$

where Z_L and Z_o are the load and characteristic impedances of the transmission line, respectively. V_o^+ and V_o^- are the incident and reflected voltage waves. $|\Gamma_L|$ varies between 0 and 1, and it is 0 for a perfectly matched line.

Example 4.7

For a transmission with a characteristic impedance of 50 Ω and terminated in a shunt capacitor of 10 pF and a resistor of 60 Ω, determine the load impedance and the load reflection coefficient if the operating frequency is 5 GHz.

Solution

$$\Gamma = \frac{Z_L - Z_o}{Z_L + Z_o}, \quad Y_L = \frac{1}{R} + j\omega C = \frac{1}{60} j(2\pi \times 5 \times 10^9) \cdot (10 \times 10^{-12}) = 0.0167 + j0.32$$

$$\therefore \; Z_L = \frac{1}{Y_L} = (0.1684 - j3.174)\,\Omega$$

$$\Gamma = \frac{(0.1684 - j3.174) - 50}{(0.1684 - j3.174) + 50} = 0.993\angle 7.3°$$

Example 4.8

A 30-watt radio transmitter is connected to an antenna that has an input impedance $Z_A = 80 + j40$ through a 50 Ω cable. How much power is delivered to the antenna?

Solution

The reflection coefficient at the antenna input is given by

$$\Gamma = \frac{Z_{ANT} - Z_o}{Z_{ANT} + Z_o} = \frac{(80 + j40) - 50}{(80 + j40) + 50} = 0.297 + j0.216$$

The power delivered to the antenna is given by

$$P_{ant} = P_{inc}(1 - |\Gamma|^2) = 30 \times 0.8649 = 25.945 \text{ W}$$

4.5 Voltage Standing Wave Ratio VSWR

The *Voltage standing wave ratio*, VSWR, is a measure of the mismatch in a transmission system. When a transmission line is mismatched, the voltage along the line will vary periodically with a period of one-half of the wavelength λ of the operating frequency. The VSWR is given by

$$\text{VSWR} = \frac{V_{max}}{V_{min}} = \frac{1 + |\Gamma|}{1 - |\Gamma|} \tag{4.24}$$

where V_{max} and V_{min} are the maximum and minimum voltages along the transmission line, a VSWR of 1 indicates a perfect match. The mismatch can also be quantified as a return loss *RL*, which is given by

$$RL = -20 \log |\Gamma| \tag{4.25}$$

for a perfectly matched line, $\Gamma = 0$.

Example 4.9

For a 50 Ω lossless transmission line that is terminated with an impedance Z_L, determine the following:

(a) VSWR on the line if $Z_L = 50 - j50\,\Omega$
(b) VSWR on the line if $Z_L = 50\,\Omega$

Solution

$$\text{(a)}\ \Gamma = \frac{Z_L - Z_o}{Z_L + Z_o} = \frac{(50 - j50) - 50}{(50 - j50) + 50} = 0.447\angle 116.57^\circ$$

$$\text{VSWR} = \frac{1 + |\Gamma|}{1 - |\Gamma|} = \frac{1 + 0.447}{1 - 0.447} = 2.62$$

$$\text{(b)}\ \Gamma = \frac{(50) - 50}{(50) + 50} = 0$$

A reflection coefficient of 0 means the transmission line is perfectly matched.

$$\text{VSWR} = \frac{1 + 0}{1 - 0} = 1$$

Example 4.10

For a voltage standing wave caused by a mismatched load that has a maximum value of 50 V and a minimum value of 30 V, determine the SWR (i.e., VSWR in dB)

Solution

$$\text{VSWR} = \frac{V_{max}}{V_{min}} = \frac{50}{30} = 1.67$$

$$\therefore\ \text{VSWR(dB)} = 20\log_{10}(1.67) = 4.4\ \text{dB}$$

Example 4.11

Determine the maximum impedance that can occur on a transmission line with a VSWR of 4 and characteristic impedance of $Z_o = 75\,\Omega$.

Solution

The maximum impedance can be expressed as

$$Z_{max}(\Omega) = (\text{VSWR})\,Z_{max} = 4 \times 75 = 300\ \Omega$$

Example 4.12

For a transmission line with characteristic impedance of $Z_o = 100\ \Omega$ and minimum impedance of 45 Ω, determine the return loss.

Solution

The minimum impedance can be expressed as

$$Z_{min}\,(\Omega) = \frac{Z_o}{\text{VSWR}}$$

$$\therefore\ \text{VSWR} = \frac{Z_o}{Z_{min}\,(\Omega)} = \frac{100}{45} = 2.22$$

and

$$\rho = \frac{\text{VSWR} - 1}{\text{VSWR} + 1} = 0.379$$

$$\text{Return loss} = 20\log_{10}(0.379) = 8.43\ \text{dB}$$

4.6 Input Impedance of a Transmission Line

The input impedance of a transmission line varies with the position on the line because of the forward and backward traveling waves. Figure 4.4 represents the input impedance of a transmission line at any position *d*.

For a lossless (i.e., $\alpha = 0$) transmission line, the input impedance at any position *d* is defined as

$$Z(d) = \frac{V(d)}{I(d)} = Z_o \frac{e^{j\beta d} + \Gamma_o e^{-j\beta d}}{e^{j\beta d} - \Gamma_o e^{-j\beta d}} \tag{4.26}$$

At $d = 0$, Γ_o is given by

$$\Gamma_o = \frac{Z_L - Z_o}{Z_L + Z_o} \tag{4.27}$$

Substituting (4.27) in (4.26) gives

$$Z_{in}(l) = Z_o \frac{(Z_L + Z_o)\, e^{j\beta l} + (Z_L - Z_o)e^{-j\beta l}}{(Z_L + Z_o)e^{j\beta l} - (Z_L - Z_o)e^{-j\beta l}} = Z_o \frac{Z_L(e^{j\beta l} + e^{-j\beta l}) + Z_o(e^{j\beta l} - e^{-j\beta l})}{Z_o(e^{j\beta l} + e^{-j\beta l}) + Z_L(e^{j\beta l} - e^{-j\beta l})} \tag{4.28}$$

$$\cos(\beta d) = \frac{(e^{j\beta l} + e^{-j\beta l})}{2}, \qquad \sin(\beta d) = \frac{(e^{j\beta d} - e^{-j\beta l})}{2j}$$

$$\therefore\ Z_{in} = Z_o \frac{Z_L + jZ_o \tan(\beta l)}{Z_o + jZ_L \tan(\beta l)} \tag{4.29}$$

If the load is a short or open circuit, the input impedance becomes pure imaginary (i.e., no real part exists), such a line is called a stub. With $Z_L = 0$ and the line's length less than a quarter-wavelength at the operating frequency, the input impedance of the line becomes inductive, and Eq. (4.29) can be written as

$$Z_{in} = Z_o \frac{jZ_o \tan(\beta l)}{Z_o} = jZ_o \tan(\beta l) \tag{4.30}$$

From Eq. (4.30), a short-circuited line with length of $l < \lambda_g/4$ looks like an inductor with inductance L_s

$$\therefore\ \omega L_s = Z_o \tan(\beta l), \qquad \text{and} \qquad L_s = \frac{Z_o}{\omega} \tan\left(\frac{2\pi}{\lambda_g} l\right). \tag{4.31}$$

From Eq. (4.31), for large values of inductance L_s, sections of transmission lines of high characteristic impedance are required and can be implemented using microstrip lines with narrow strips.

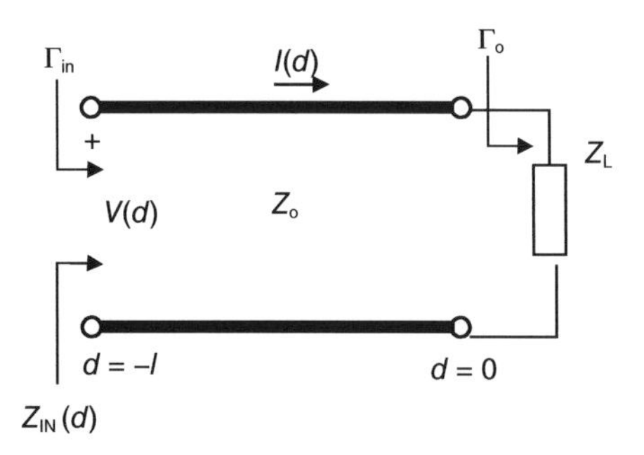

Figure 4.4 Input impedance of the transmission line at any position *d*.

A transmission line with *open-circuit load* has an input impedance given by

$$Z_{in} = -jZ_o \cot(\beta l) \tag{4.32}$$

For a line with length of $l < \lambda_g/4$, an open-circuited section of the line can be implemented as a capacitor C_o, thus

$$\frac{1}{\omega C_o} = Z_o \cot(\beta l), \quad \text{and} \quad C_o = \frac{1}{Z_o}\frac{\cot(\beta l)}{\omega} \tag{4.33}$$

Equation (4.33) indicates that the capacitance of an open-stub C_o, is inversely proportional to the characteristic impedance Z_o. Open stubs are used as on-board matching elements. The reflection coefficient looking into a loaded transmission line is expressed as

$$\Gamma_{in} = \frac{V^- e^{-j\beta\ell}}{V^+ e^{+j\beta\ell}} = \Gamma_L e^{-j2\beta l}$$

Example 4.13
Determine the input impedance at a point one-eight wavelength in front of a transmission line that has a characteristic impedance of 50 Ω and terminated in an open load.

Solution
The input impedance of a transmission line terminated in an open load is given by

$$Z_{in} = -jZ_o \cot(\beta l) = -j50 \cot(\pi/4) = -j50\ \Omega$$

The input impedance is a capacitive reactance.

Example 4.14
Determine the input impedance at 1.5 wavelengths in front of a transmission line with a characteristic impedance of 50 Ω and terminated in an 85 Ω load.

Solution
Because the load is repeated every one-half wavelength on the transmission line and the line is three half-wavelengths long, thus the input impedance is 85 Ω.

Example 4.15
Determine the input impedance at a point of 0.45 wavelengths on a transmission line that is terminated in a shorted load if the line has a characteristic impedance of 75 Ω.

Solution

$$Z_{in} = Z_o \frac{Z_{L+} jZ_o \tan(\beta l)}{Z_{o+} jZ_L \tan(\beta l)}$$

$$\beta l = \frac{2\pi}{\lambda}(0.45\,\lambda) = 2.827 \text{ rad} = 161.38°$$

$$\therefore\ Z_{in} = 75\frac{0 + j75\tan(161.38°)}{75 + j(0)\tan(161.38°)} = -j24.37\ \Omega$$

The input impedance is a capacitive reactance.

Example 4.16
For the transmission line shown in Figure 4.5, determine the load reflection coefficient and the input impedance.

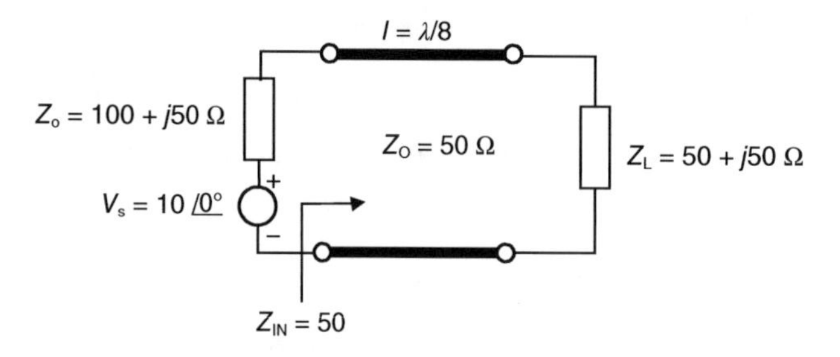

Figure 4.5 Transmission line circuit for Example 4.16.

Solution

The load reflection coefficient Γ is

$$\Gamma = \frac{Z_L - Z_o}{Z_L + Z_o} = \frac{(50 + j50) - 50}{(50 + j50) + 50} = 0.4472\angle 63.43°$$

The input impedance is determined as follows:

$$\beta l = \frac{2\pi}{\lambda} \times \frac{\lambda}{8} = \frac{\pi}{4} = 45°$$

$$Z_{in} = Z_o \frac{Z_L + jZ_o \tan(\beta l)}{Z_o + jZ_L \tan(\beta l)} = 50 \frac{(50 + j50) + j50 \tan(45°)}{50 + j(50 + j50) \tan(45°)} = 100 - j50\ \Omega$$

Example 4.17

For the transmission shown in Figure 4.6, determine the reflection coefficient at the load, the VSWR, the return loss, and the input impedance.

Solution

The reflection coefficient is given by

$$\Gamma_o = \frac{Z_L - Z_o}{Z_L + Z_o} = \frac{(130 + j90) - 50}{(130 + j90) + 50} = 0.598\angle 21.8$$

$$\text{VSWR} = \frac{1 + |\Gamma|}{1 - |\Gamma|} = \frac{1 + 0.598}{1 - 0.598} = 3.98$$

$$\text{Return loss} = 20 \log_{10}|\Gamma| = -4.46\ \text{dB}$$

$$\beta l = \frac{2\pi}{\lambda}(0.3\lambda) = 108^o$$

$$Z_{in} = Z_o \frac{Z_{L+} jZ_o \tan(\beta l)}{Z_{o+} jZ_L \tan(\beta l)} = 50 \frac{(130 + j90) + j50 \tan(108°)}{50 + j(50 + j50) \tan(108°)} = 12.75 + j5.83\ \Omega$$

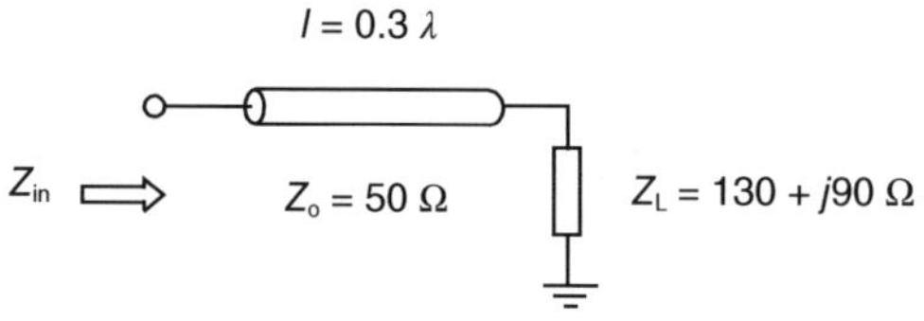

l = 0.3 λ

Zin ⟹ Zₒ = 50 Ω ZL = 130 + j90 Ω

Figure 4.6 Example 4.17.

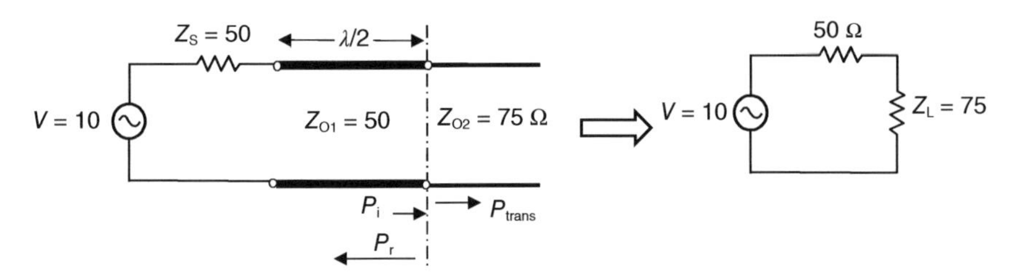

Figure 4.7 Example 4.18.

Example 4.18

For the 50 Ω transmission line shown in Figure 4.7, calculate the incident power, the reflected power, and the power transmitted into the line when that has a characteristic impedance of 75 Ω.

Solution

The incident power at the interface between the two transmission lines is given by

$$P_{inc} = (I)^2 \times Z = \left(\frac{V/\sqrt{2}}{Z_s + Z_o}\right)^2 Z_o = \frac{1}{2}\left(\frac{10}{50+50}\right)^2 (50) = 0.25\ \text{W}$$

The power reflected at the interface is given by

$$P_{ref} = P_{inc} \times |\Gamma|^2 = P_{inc} \times |(Z_L - Z_o)/(Z_L + Z_o)|^2$$

$$\therefore\ P_{ref} = 0.25 \times \left|\frac{75-50}{75+50}\right|^2 = 0.01\ \text{W}$$

The transmitted power at the interface is given by

$$P_{trans} = P_{inc} - P_{ref} = 0.25 - 0.01 = 0.24\ \text{W}$$

4.7 Quarter-Wave Transformer

For a transmission line of length l that is a quarter wavelength ($\lambda/4$) long of the frequency of transmission, Eq. (4.29) can be written as

$$Z_i = \frac{Z_o^2}{Z_L} \tag{4.34}$$

Eqn. (4.34) indicates that, to match a load impedance Z_L to an input impedance Z_i, the characteristic impedance of the quarter-wavelength section to be used as impedance matcher must have an impedance given by

$$Z_o = \sqrt{Z_i \cdot Z_L} \tag{4.35}$$

A transmission line of length $\lambda/4$ acts as a step up or step-down transformer, depending on whether Z_L is greater or less than Z_o.

$Z_L > Z_o$ provides step-down transformation
$Z_L < Z_o$ provides step-up transformation

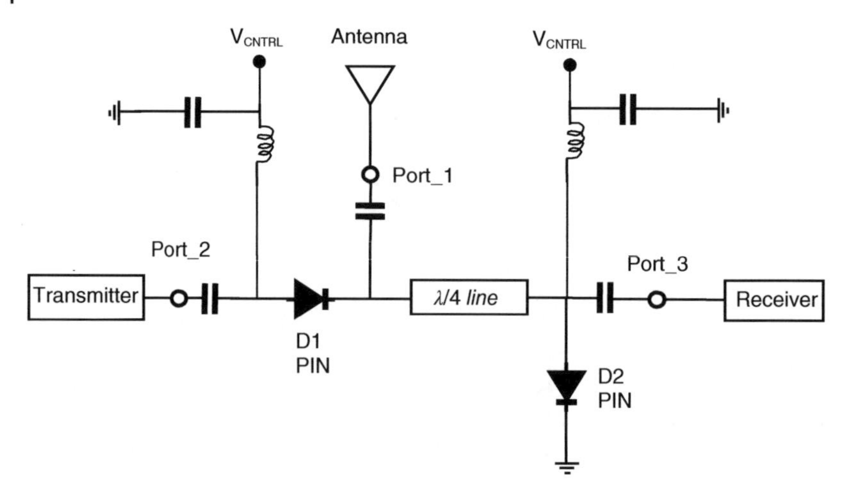

Figure 4.8 Transmit receive switch using quarter-wavelength lines as impedance transformers.

This type of matching is only useful over a small range of frequencies. As the frequency changes the length of the matching section will no longer be a quarter wavelength. Quarter-wave matching is beneficial for situations where the impedances are resistive. Figure 4.8 illustrates the concept of using a quarter-wavelength transmission line in a transmit receive switch.

In Figure 4.8, when the diodes D1 and D2 are turned on by the control voltage (V_{CNTRL}), the RF signal passes from the transmitter to the antenna port_1 but will be blocked by the quarter-wavelength transmission line section that is shortened through D2. The resonant quarter-wave line acts as an open circuit to the RF signal. When the PIN diodes D1 and D2 are turned off, both D1 and D2 become open circuits allowing the RF signal to pass from the antenna port_1 to the receiver port_3, whereas the RF signal is blocked from entering the transmitter port_2.

Example 4.19

Determine the characteristic impedance of a quarter-wavelength transmission line to match a 75-Ω load line with a 50-Ω source impedance.

Solution

The characteristic impedance of a quarter-wavelength transmission line is

$$Z_o = \sqrt{Z_i \cdot Z_L} = \sqrt{75 \times 50} = 61.23\ \Omega$$

4.8 Planar Transmission Lines in Radio Systems

Planar transmission lines are indispensable in any RF, microwave, and mmWave (millimeter wave) radio transceiver front-ends. They are used as filters, matching elements, phase shifters, antenna feed networks, power splitters/combiners, couplers, balanced mixers, balanced amplifiers, and interconnect lines. Therefore, they influence the system's performance. Figure 4.9 shows various RF circuits using microstrip transmission line elements.

In Figure 4.9a, the amplifier's input and output impedance matching elements are implemented using microstrip lines. Figure 4.9b shows a quadrature branch-line coupler where the coupler's

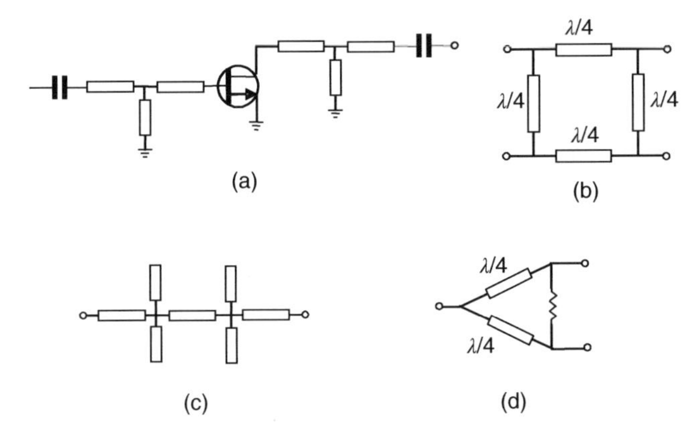

Figure 4.9 Transmission line elements in RF circuits: (a) low-noise amplifier, (b) branch-line coupler, (c) low-pass filter, and (d) Wilkinson power splitter.

branches are implemented with quarter-wave microstrip lines. Figure 4.9c shows a low-pass filter, where the filter's series elements are implemented with high impedance microstrip lines, while the shunt elements are implemented with microstrip open-stubs. Figure 4.9d shows a Wilkinson power splitter whose branches are implemented with quarter-wave microstrip lines. Microstrip lines are also used in phased array antennas and antenna distribution networks.

4.8.1 Planar Transmission Line Types

There are different planar transmission lines, and each type has its advantages and disadvantages. The most suitable type to achieve optimal performance depends on the required application. Figure 4.10 shows the most used RF and mmWave planar transmission lines.

Figures 4.10a–d show the cross section of a microstrip line, CPW (coplanar waveguide) [4], conductor-backed CPW, and stripline, respectively. The fundamental mode of propagation of these transmission lines is frequently referred to as quasi-TEM because of its similarity with pure TEM modes. Figure 4.11 shows a microstrip structure and its electric and magnetic fields.

In microstrip structures, losses occur in the dielectric and in the conductors, in addition to radiation losses. The characteristic impedance of a microstrip line is governed by the conductor line

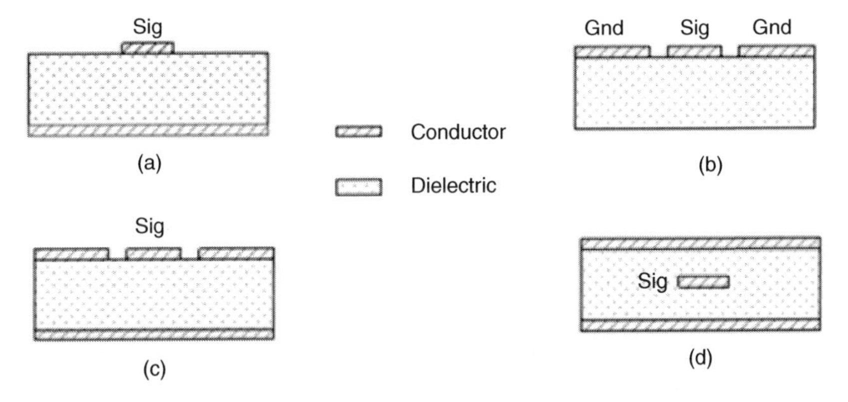

Figure 4.10 Cross-section of RF/mmWave planar transmission lines: (a) microstrip, (b) coplanar waveguide, (c) conductor-backed CPW, and (d) strip line.

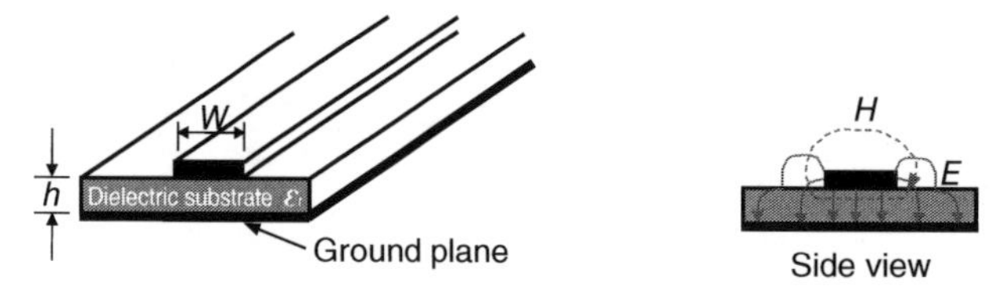

Figure 4.11 Microstrip line structure and its E (electric) and H (magnetic) fields.

width, w, the thickness of the dielectric substrate, h, and the permittivity of the dielectric, ε. Low impedance microstrip lines are wide, while high impedance microstrip lines are narrow.

For $w/h \leq 1$, the characteristic impedance of a microstrip line is given by

$$Z_o = \frac{60}{\sqrt{\acute{\varepsilon}_r}} \ln\left(\frac{8h}{W} + \frac{W}{4h}\right) \tag{4.36}$$

where $\acute{\varepsilon}_r$ is the effective relative permittivity that takes into account the combined substrate-air dielectric, and it is given by

$$\acute{\varepsilon}_r = \frac{\varepsilon_r + 1}{2} + \frac{\varepsilon_r - 1}{2}\left[\left(1 + \frac{12h}{W}\right)^{1/2} + 0.04\left(1 - \frac{h}{W}\right)^2\right] \tag{4.37}$$

For $w/h \geq 1$, the characteristic impedance is given by

$$Z_o = \frac{120\pi/\sqrt{\acute{\varepsilon}_r}}{\frac{w}{h} + 0.667 \ln\left(\frac{h}{W} + 1.444\right) + 1.393} \tag{4.38}$$

where

$$\acute{\varepsilon}_r = \frac{\varepsilon_r + 1}{2} + \frac{\varepsilon_r - 1}{2}\left(1 + \frac{12\text{h}}{\text{W}}\right)^{-1/2}$$

The limited impedance range of microstrip and its dispersion phenomenon (frequency dependent line parameters) makes the CPW a better choice for high-frequency operations. Furthermore, in microstrip lines, the height of the substrate is fixed by the fabrication process; therefore, the width of the signal conductor is the only design parameter that can be used to control the line's impedance. Thus, microstrip lines do not provide design flexibility to achieve high impedances. Meanwhile, CPW transmission lines provide design flexibility because the impedance can be controlled not only by the width of the signal conductor but also by the gap between the signal conductor and the ground-plane strips. In addition, since both the signal-conductor (Sig) and the ground-plane (Gnd) strips are placed on the same metal layer, no vias will be required to connect shunt components to the ground plane. In addition, better isolation can be obtained with CPW lines because the ground-plane strips act as a shield.

4.9 Smith Chart

Smith chart is a polar graph that is used for solving transmission line problems and the design of impedance matching networks. Figure 4.12 shows a Smith chart, the radially scaled parameters are printed below the polar portion of the chart.

In Figure 4.12, the horizontal centerline, ***Scale A***, represents either resistance or conductance. The circles on the chart represent circles of constant resistance or conductance. The curved lines

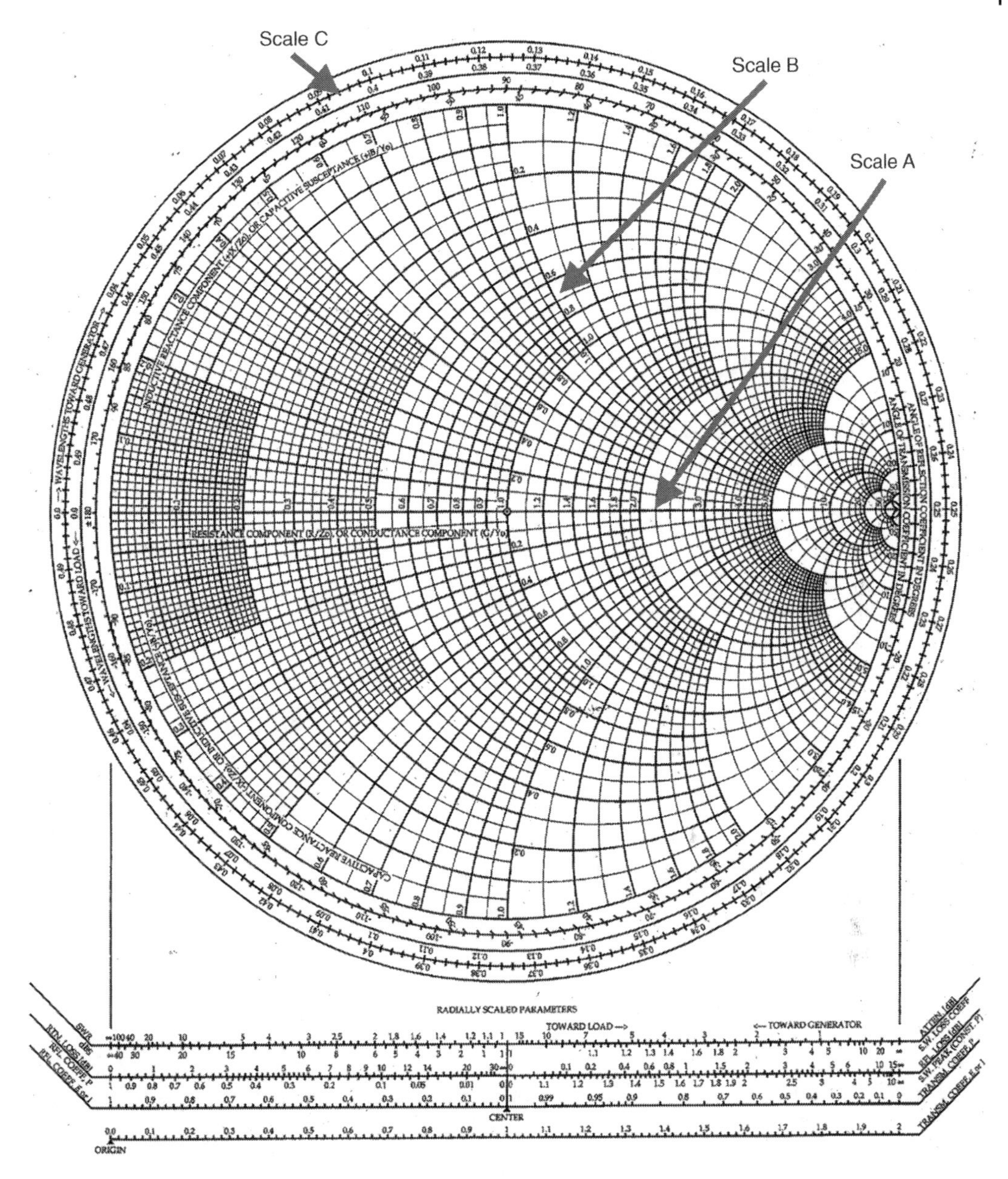

Figure 4.12 Smith chart.

from the outer circle that terminate on the centerline on the right side are the lines of constant reactance or susceptance. ***Scale B*** represents either reactance or susceptance that is located on the circumference of the circle. ***Scale C*** is the outermost circular scale and represents distance in normalized wavelengths. The length along a transmission line is represented on the chart by the outer wavelength scale. Overall, there are three scales on the outer perimeter of the chart, namely, wavelength toward the generator, wavelength toward the load, and reflection coefficient angle in degrees as shown in Figure 4.13. The direction toward the load is counterclockwise, whereas the direction toward the generator is clockwise.

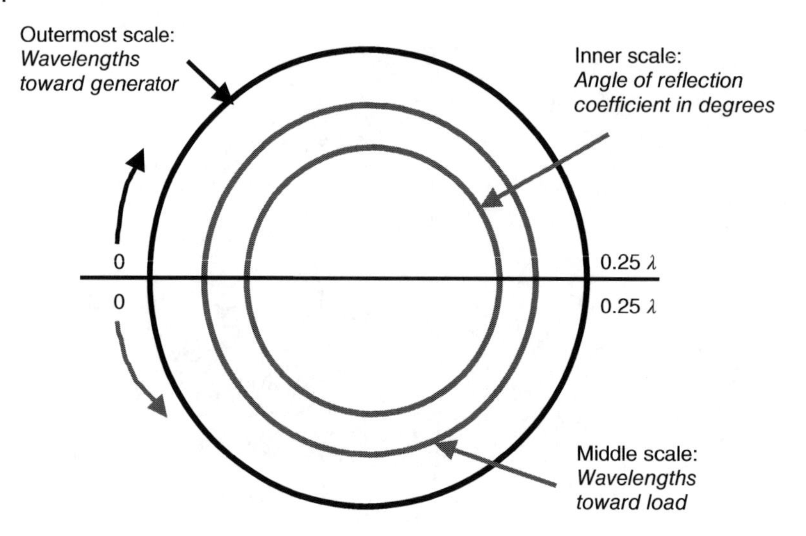

Figure 4.13 Perimeter scale of Smith chart.

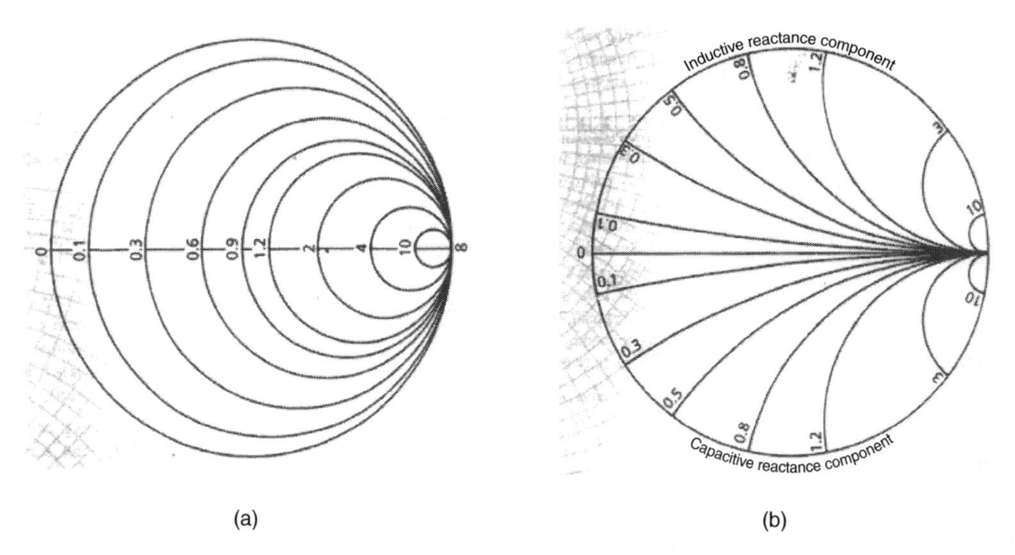

Figure 4.14 Constant reactance and constant resistance circles. (a) Constant resistance circle. (b) Constant reactance circle.

Each point on a constant resistance circle has the same resistance as other points on the circle, Figure 4.14a shows constant resistance circles. Each point on the reactance curve has the same reactance as any other point on that curve. All curves above the centerline of the chart represent inductive reactance ($+jX$), whereas all curves below the centerline represent capacitive reactance ($-jX$). Figure 4.14b shows the constant reactance circles.

Any point on the Smith chart represents an impedance, $Z = R \pm jX$; the location of a normalized impedance of $z = (1 + j1)$ on the Smith chart is the intersection of $r = 1$ circle and $x = 1$ curve as shown in Figure 4.15. The Smith chart can be used to convert any impedance, Z, to an admittance, Y, and vice versa. The conversion can be done by drawing a line from the location of the normalized impedance through the center of the chart to intersect with the VSWR circle; the intersection point is the location of the normalized admittance. If both the impedance and

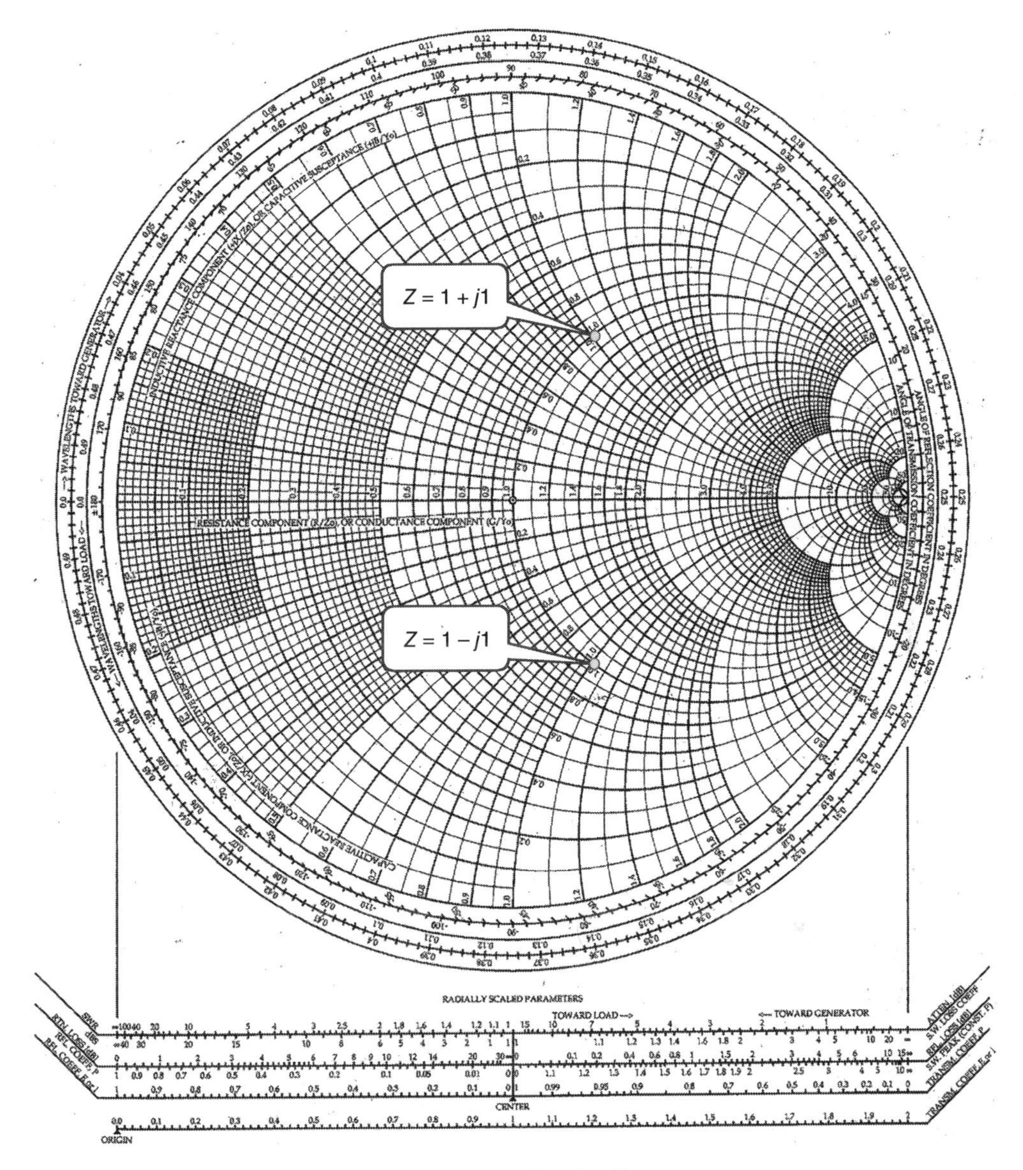

Figure 4.15 Illustration of normalized impedance of $z = 1 + j1$.

the admittance charts are plotted together, one on the other, the new chart is called the *immittance chart*. Figure 4.16 shows an immittance chart, in Figure 4.16, the series inductors and capacitors are presented on the impedance chart, whereas parallel inductors and capacitors are presented on the admittance chart.

The Smith chart is normalized with respect to the characteristic impedance, admittance, or wavelength as shown in Figure 4.17. The normalized chart is useful for transmission lines of any characteristic impedance or frequency. One revolution of the chart represents half-wavelength of the transmission line. For values greater than half-wavelength, subtract multiples of half-wavelength until a value that is less than half-wavelength is obtained.

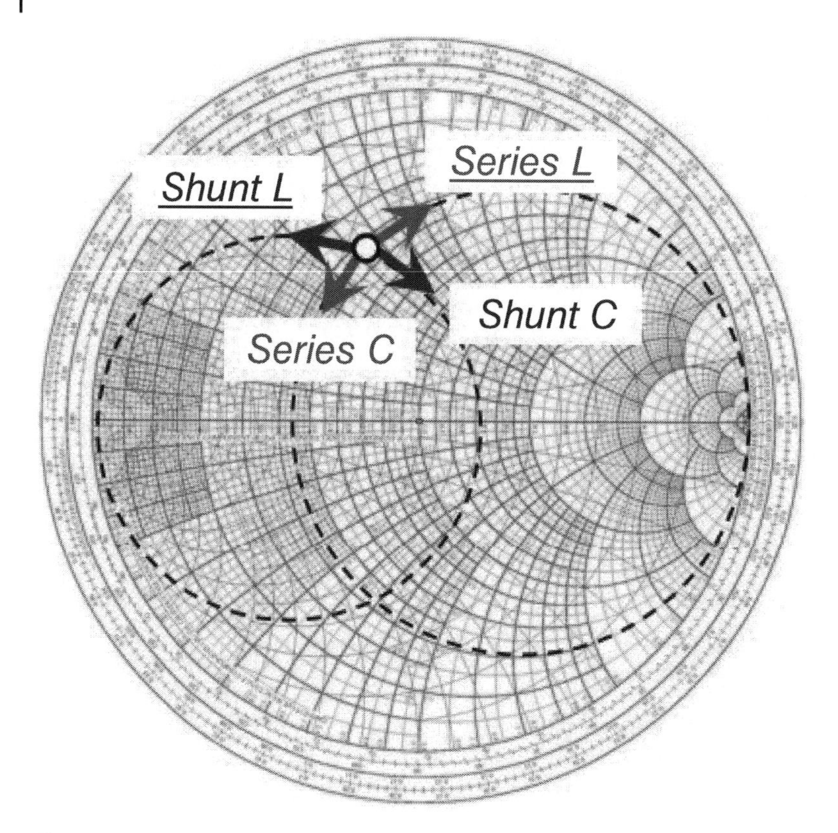

Figure 4.16 Immittance chart showing shunt and series inductances and capacitance.

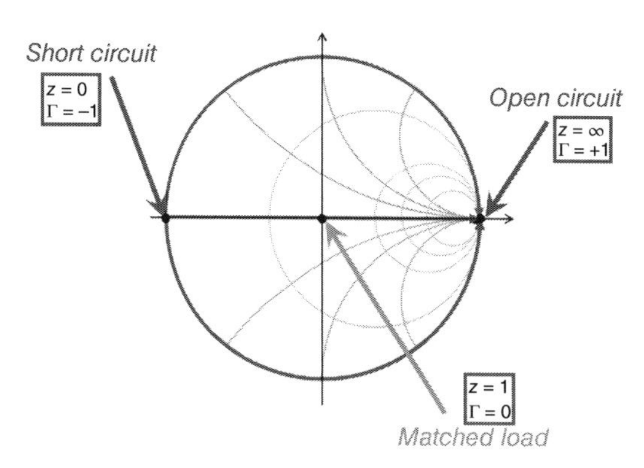

Figure 4.17 Illustration of short, matched, and open impedances.

Example 4.20

For a transmission line that has a characteristic impedance of $Z_o = 50\,\Omega$ and terminated in an impedance of $Z_L = 100 - j50\,\Omega$, determine the reflection coefficient Γ using Smith chart.

Solution

Normalize the load impedance Z_L, with respect to the characteristic impedance Z_o by dividing Z_L by Z_o.

$$z_n = Z_L/Z_o = 2 - j1$$

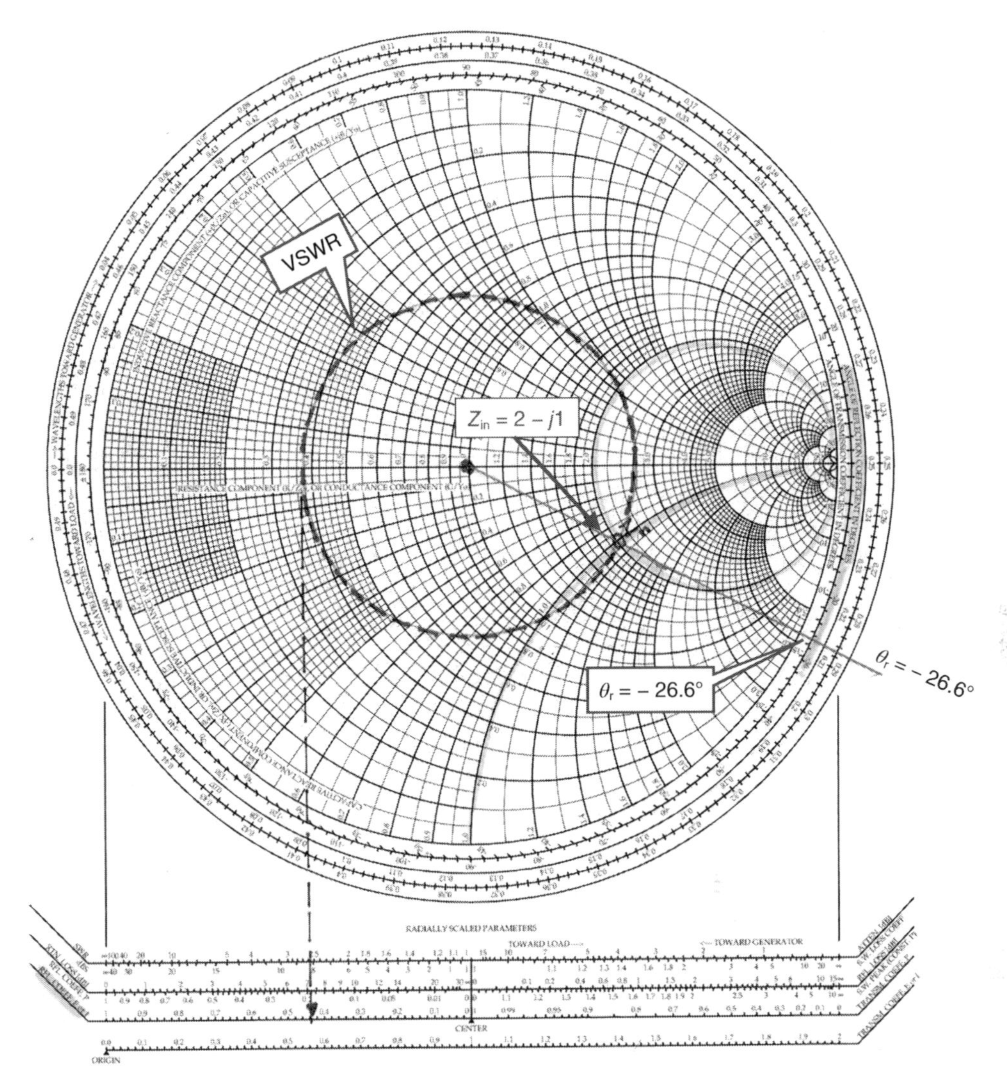

Figure 4.18 Solution of Example 4.20.

Locate the normalized impedance z_n on the chart (i.e., intersection of the curved line, $x = -j1$, and the constant resistance circle, $r = 2$, as shown in Figure 4.18).

Draw a line from the center of the chart through the load (Z_L/Z_o) to the Inner circle scale of "*Angle of reflection coefficient in degrees.*" The angle between the line drawn and the horizontal line on the chart gives the angle of $\Gamma(\theta_r)$, which is −26.6 degrees as shown in Figure 4.18.

From the center of the chart, draw a VSWR circle with radius equal to the length from the center to the location of z_n on the chart.

From the intersection of the VSWR circle with the horizontal resistance line to the left side of the circle's center, draw a vertical line to the *radial scaled parameter* portion of the chart. Read the value of $|\Gamma|$ from the radial scaled parameters of the chart; the value is 0.44

The following equations can be used to validate the results obtained from the Smith chart.

$$\Gamma = \frac{Z_L - Z_o}{Z_L + Z_o}, \quad \theta_r = \text{angle}(\Gamma), \quad |\Gamma| = \text{abs}\,(Z_L - Z_o)/\text{abs}(Z_L + Z_o)$$

Figure 4.18 illustrates the solution of this Example.

The admittance Y is the reciprocal of the impedance Z and given by

$$Y = \frac{1}{R + jx} = G + jB \tag{4.39}$$

where B is the susceptance, and G is the conductance. The normalized admittance is $y = Y/Y_o$, where Y_o is characteristic admittance.

Example 4.21

Using the Smith chart, determine the VSWR of a transmission line that has a characteristic impedance of 75 Ω and terminated in an impedance of $Z_L = 50 - j100\,\Omega$.

Solution

Normalize the load impedance Z_L by dividing it by the characteristic impedance Z_o, then locate the normalized impedance on the Smith chart (intersection of the resistive circle and the reactive curve)

$$\therefore\ z_n = \frac{(50 - j100)}{75} = 0.667 - j1.33$$

From the center of the chart, draw a VSWR circle with radius equal to the length from the center of the chart to the location of Z_L/Z_o.

Draw a vertical line from the intersection of the VSWR circle with the horizontal resistance line on the left of the circle's center to the radially scaled parameters, then read the value of the VSWR, which is 4.5. Figure 4.19 shows the solution of this Example.

Example 4.22

Using the Smith chart, determine the admittance Y_L of an impedance $Z_L = 50 + j100\ \Omega$ connected to a transmission line, which has a characteristic impedance of is $Z_o = 50\ \Omega$.

Solution

$$z_n = \frac{(50 + j100)}{50} = 1 + j2\,\Omega$$

Locate the normalized impedance (Z_L/Z_o) on the chart and draw a VSWR circle through the Z_L/Z_o location.

Extend a line from the (Z_L/Z_o) location through the center of the chart to the intersection with the VSWR circle on the opposite side (i.e., Y_L/Y_o) as shown in Figure 4.20. The value of the normalized admittance from the Smith chart is $Y_L/Y_o = 0.2 - j0.4$. Thus,

$$Y_L = (0.2 - j0.4) \times Y_o = 0.2 - j0.4 \times 1/Z_o$$
$$= 0.004 - j0.008$$

The solution of this Example is shown in Figure 4.20.

Example 4.23

For a transmission line that has a characteristic impedance $Z_o = 50$ and terminated in a shorted load, determine the line's length that is needed to have an input impedance of $j75$ at operating frequency of 7 GHz.

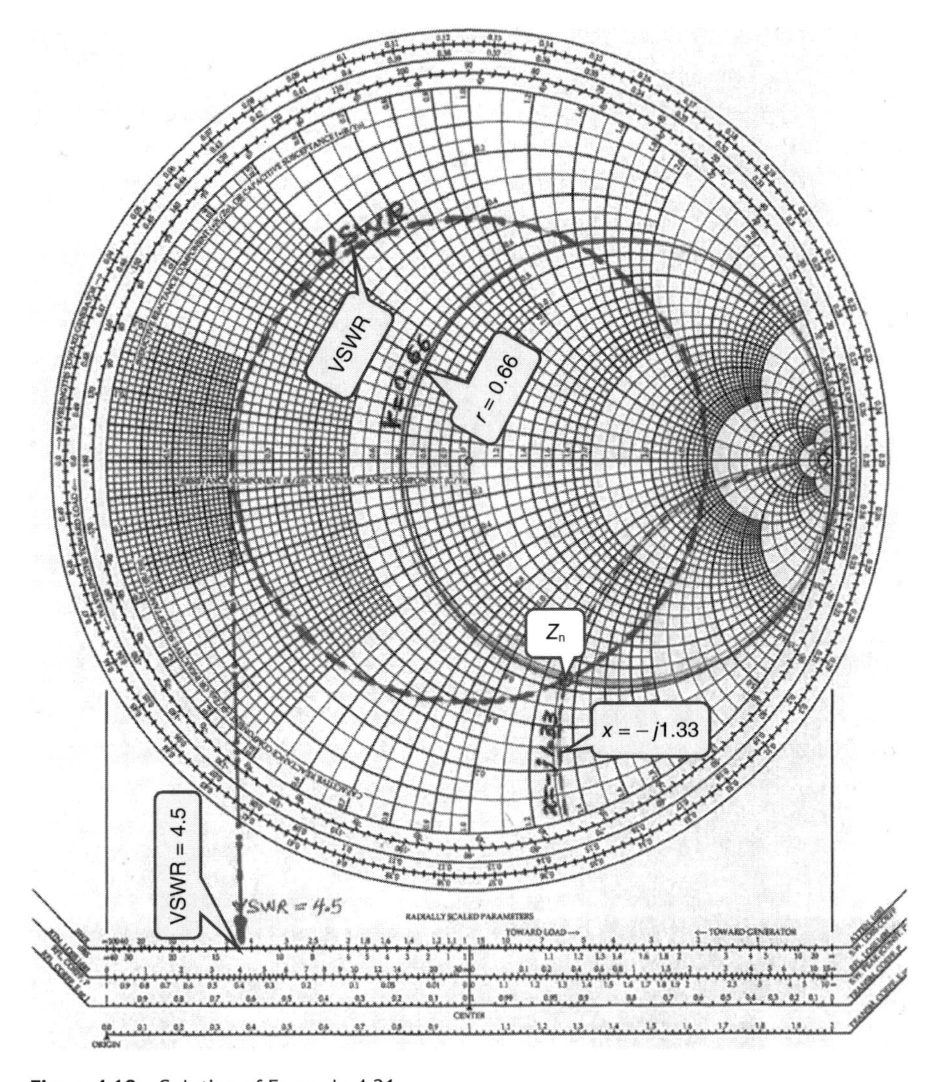

Figure 4.19 Solution of Example 4.21.

Solution

Calculate the normalized impedance as follows:

$$z_{\mathrm{n}} = \frac{(j75)}{50} = j1.5$$

Locate the normalized impedance z_{n} on the chart and draw a line from the center of the chart through the location of z_{n} to the scale of "*wavelength toward generator*."

From the Smith chart graph, the distance from the shorted load to the required input of $j1.5$ is $0.156\ \lambda$, and λ is estimated as follows

$$\lambda = \frac{3 \times 10^{10}}{7 \times 10^{9}} = 4.286\ \mathrm{cm}$$

The length of the transmission line is

$$l = 0.156\ \lambda = 0.156 \times 4.286 = 0.6686\ \mathrm{cm}$$

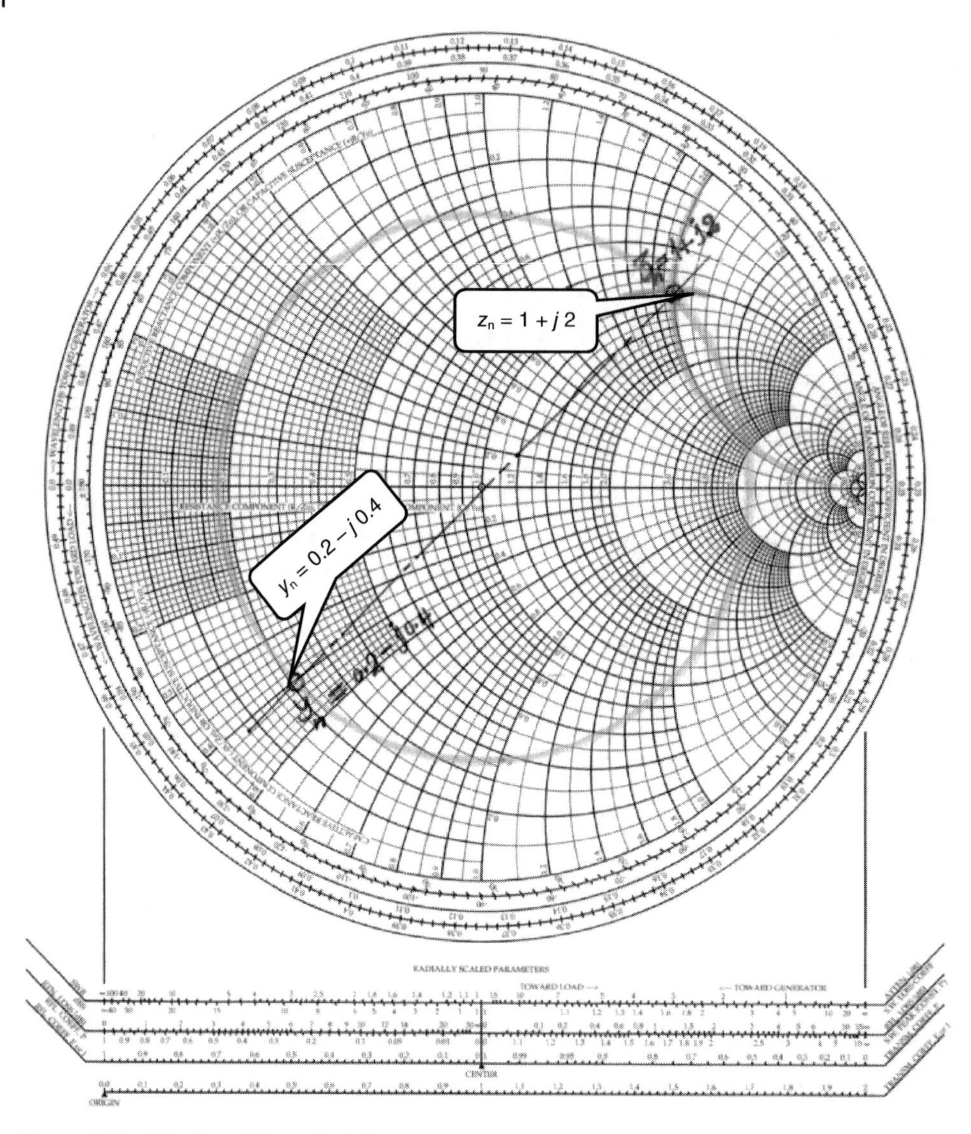

Figure 4.20 Solution of Example 4.22.

Figure 4.21 illustrates the solution of this Example.

Example 4.24

Using the Smith chart, determine the input impedance of a 15 cm transmission line that has a characteristic impedance of $Z_0 = 50$ and terminated in an open load at the operating frequency of 400 MHz.

Solution

The wavelength is

$$\lambda = \frac{3 \times 10^{10}}{400 \times 10^{6}} = 75\,\text{cm}$$

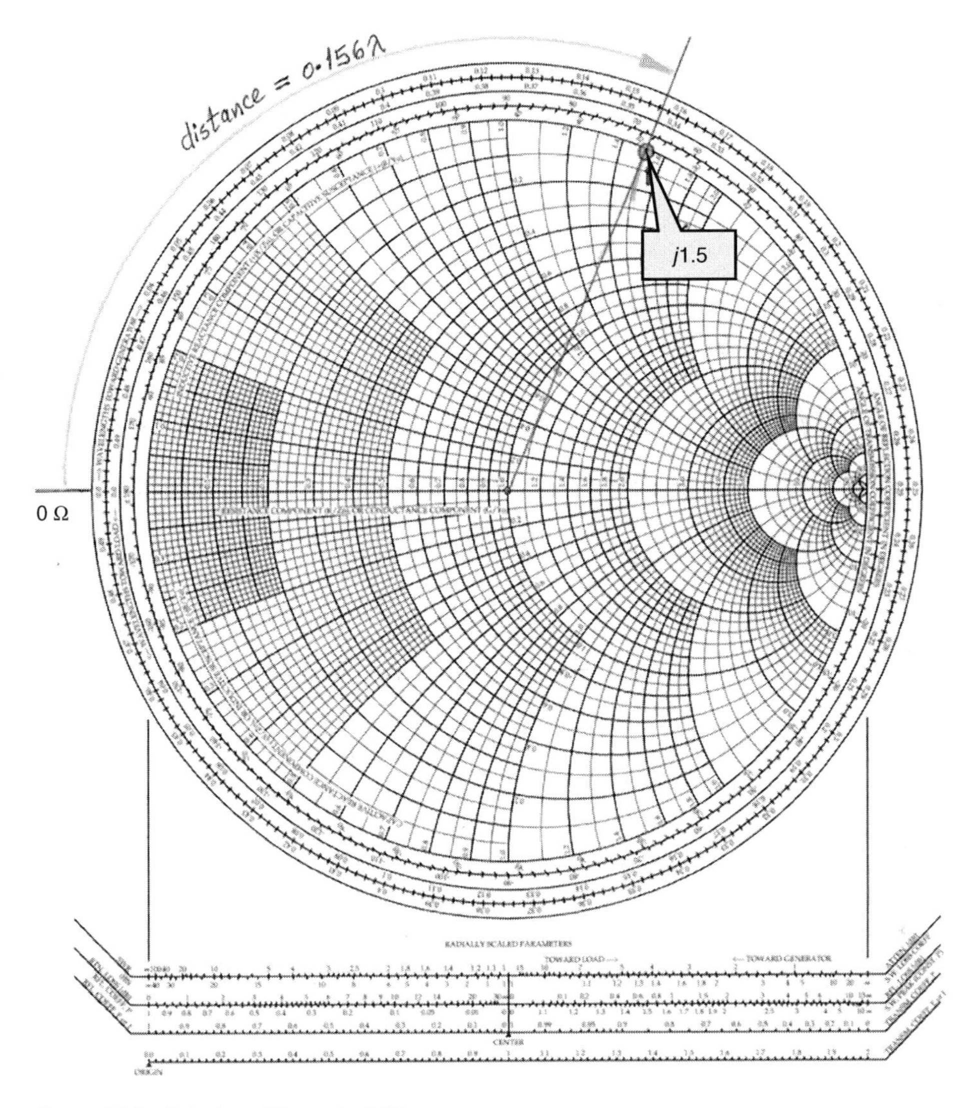

Figure 4.21 Solution of Example 4.23.

The length of the transmission line in wave lengths is

$$l = \frac{15}{75} = 0.2\ \lambda$$

The location of the open load on Smith chart is at the right side of the resistance line on the outer scale as shown in Figure 4.22. Thus, the distance from the open-circuit load to the input is ($0.25\ \lambda + 0.2\ \lambda = 0.45\ \lambda$).

From a Smith chart graph, the normalized input impedance at $0.2\ \lambda$ from the load is $Z_i/Z_o = -j\ 0.325$. Hence, the input impedance is

$$Z_i = -j0.325 \times Z_o = -j0.325 \times 50$$
$$= -j16.25\ \Omega$$

Figure 4.22 illustrates the solution of this Example.

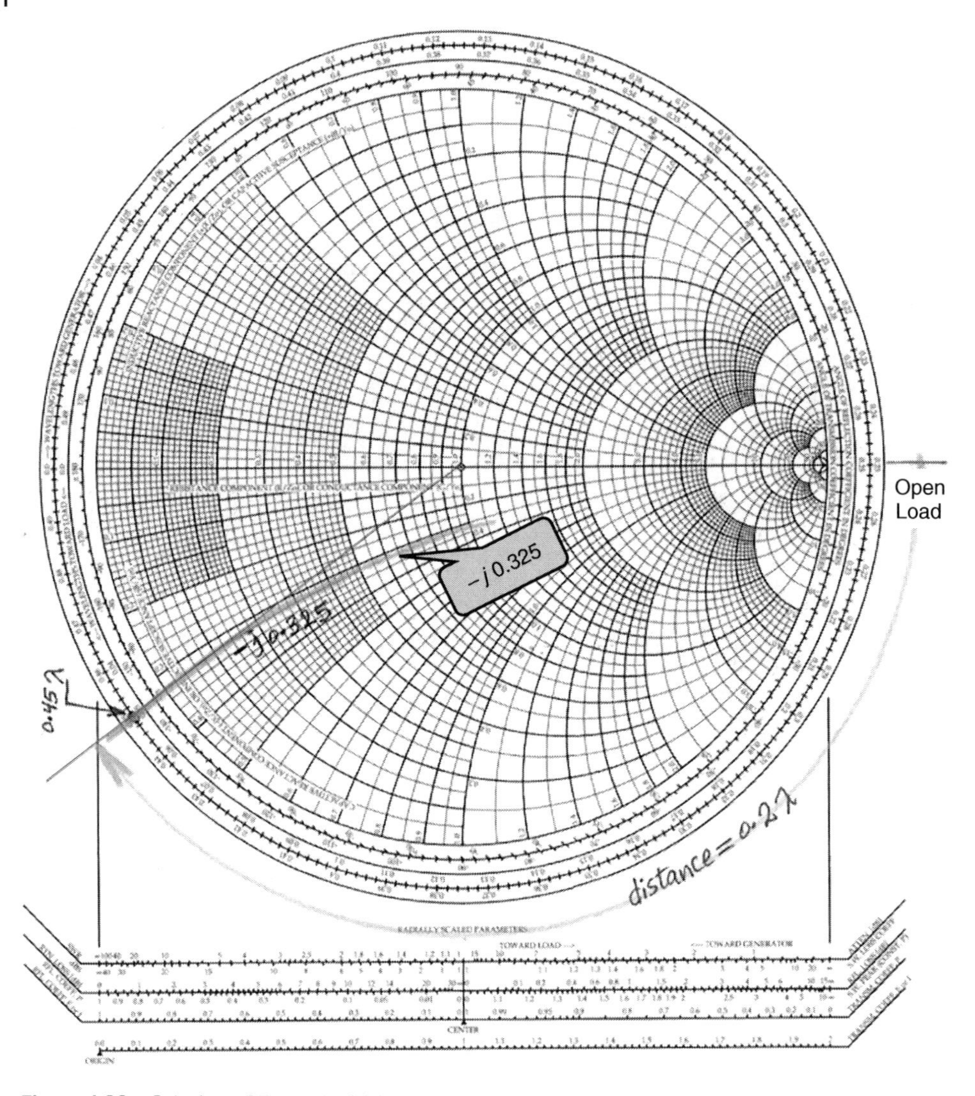

Figure 4.22 Solution of Example 4.24.

Example 4.25

Using Smith chart, determine the input impedance of the network shown in Figure 4.23.

Solution

Normalize the load impedance by dividing it by z_o,

$$\therefore\ z_n = \frac{100 - j25}{50} = 2 - j0.5$$

Locate z_n on the Smith chart as shown in Figure 4.24, then add $-j0.5$ (i.e., the normalized reactance of the series capacitor) to get the z_1, which is

$$z_1 = 2 - j1$$

From the center of the chart, draw a VSWR circle passing through z_1, and draw a line from the center of the chart through z_1 to the edge of the chart, which is at 0.287 λ.

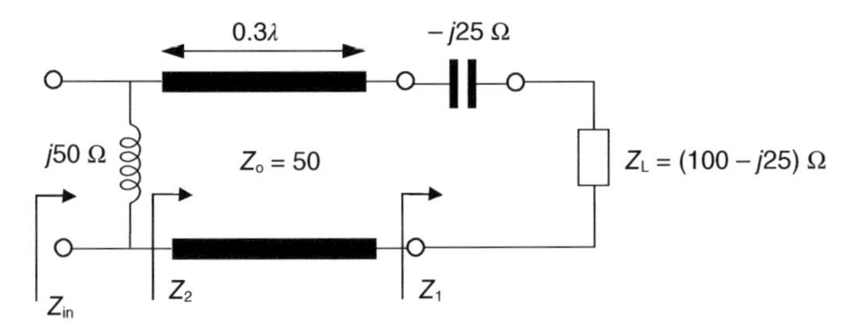

Figure 4.23 Example 4.25.

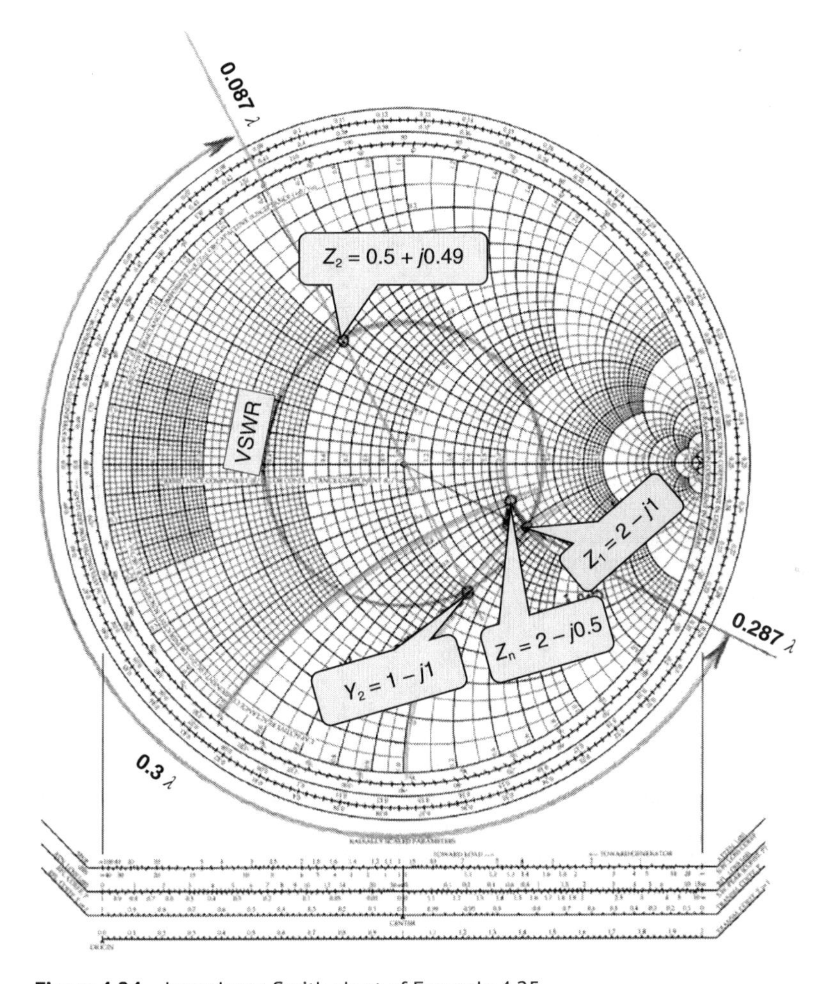

Figure 4.24 Impedance Smith chart of Example 4.25.

From the location of (0.287 λ), move toward generator 0.3λ (this means 0.287 λ + 0.3 λ which is 0.587 λ). Since one revolution of the chart is 0.5 λ, the location of 0.587 λ is at 0.087 λ on the chart.

Draw a line from the location of 0.087 λ to the center of the chart, the intersection of the line with the VSWR circle is the value of z_2. From the chart, $z_2 = 0.5 + j0.49$ and $y_2 = 1 - j1$.

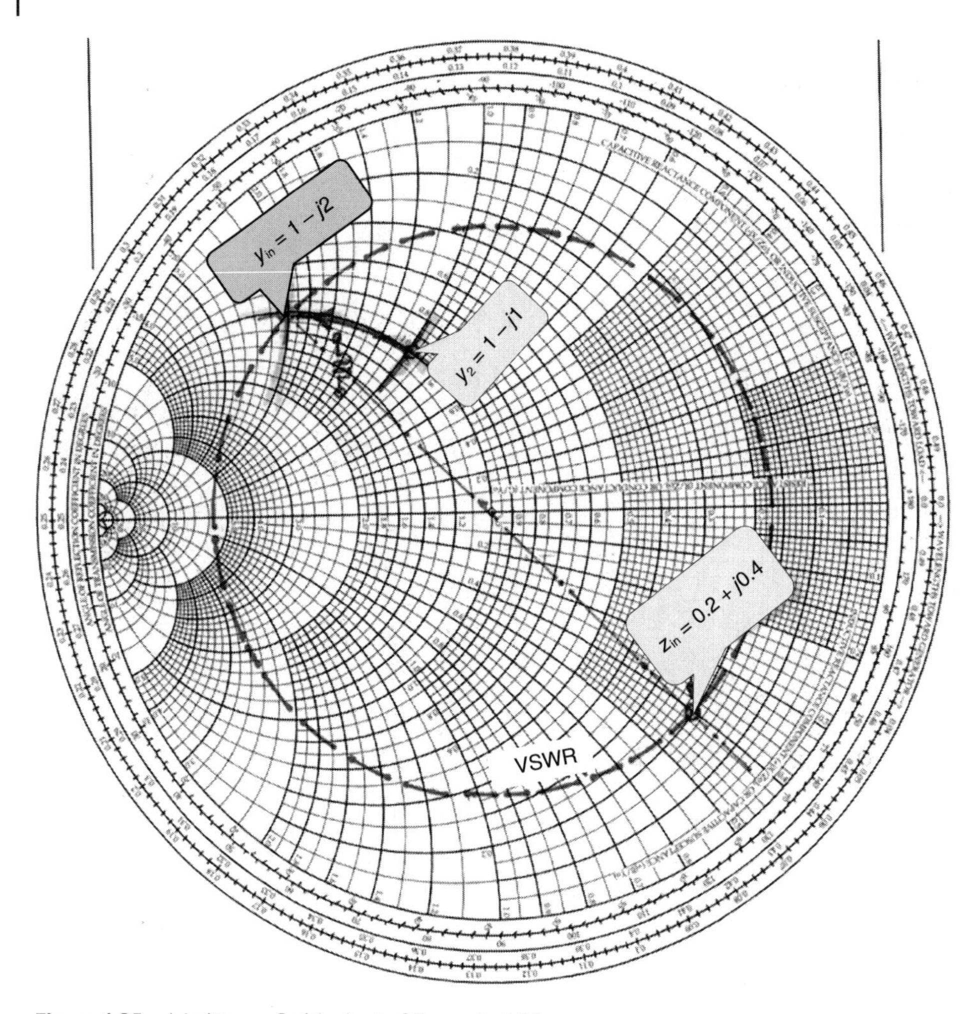

Figure 4.25 Admittance Smith chart of Example 4.25.

Locate y_2 on an admittance chart as shown in Figure 4.25, then add the normalized admittance of the shunt inductor ($-j1$). Thus, the normalized input admittance $y_{\text{in}} = 1 - j2$, and the normalized input impedance is $z_{\text{in}} = 0.2 + j0.4$

Deformalize z_{in} by multiplying it by z_{o} to get the input impedance Z_{in} which is

$$Z_{\text{in}} = 10 + j20\,\Omega$$

4.10 Impedance Matching Using Smith Chart

At microwave and mmWave frequencies, it is more efficient to implement reactive elements using transmission line stubs. A stub is a length of a transmission line terminated in either a short-circuit or an open-circuit; therefore, a stub should have purely reactive impedance. This section introduces single-stub and double-stub matching methods.

4.10.1 Single-Stub Matching

In the single-stub matching technique, the stub is placed in parallel with the load admittance at the distance, *d*, from the load as shown in Figure 4.26.

The concept of single-stub matching is based on moving the load admittance, y_L, on the SWR circle of the Smith chart to a point on the unit circle (i.e., the conductance $g = 1$), therefore, the admittance at a distance, *d*, from the load before adding the stub is given by

$$y_d = 1 \pm jb$$

where the imaginary part, jb (susceptance) of y_d (admittance at the stub's location before adding the stub) can be cancelled by adding either an open-circuit or short-circuit stub of length, l_s, to provide a perfect match given by

$$y_{11} = 1 + j0$$

However, the drawback of the single-stub matching is that it depends on the load impedance, when the load changes the location and the length of the stub change. This issue is resolved by using double-stub technique, which will be discussed in the next subsection.

Example 4.26

For the network shown in Figure 4.27, use the Smith chart to determine the length of an open-circuit stub, l_s, and its location at distance, *d*, from the load.

Solution

Normalize the load impedance by dividing it by z_o (i.e., $z_l = 0.5 - j1$) and locate it on the Smith chart, then, draw a SWR circle passing through the normalized z_L as shown in Figure 4.28.

Determine the normalized load admittance, y_L, by rotating 180° on the SWR circle from the normalized load impedance location. Once the admittance is located on the chart, all the values read on the chart are normalized admittances. From Figure 4.28, $y_L = 0.4 - j0.8$.

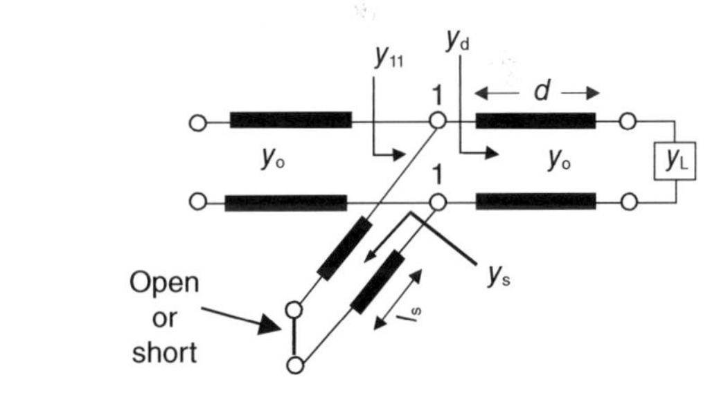

Figure 4.26 Illustration of single-stub matching.

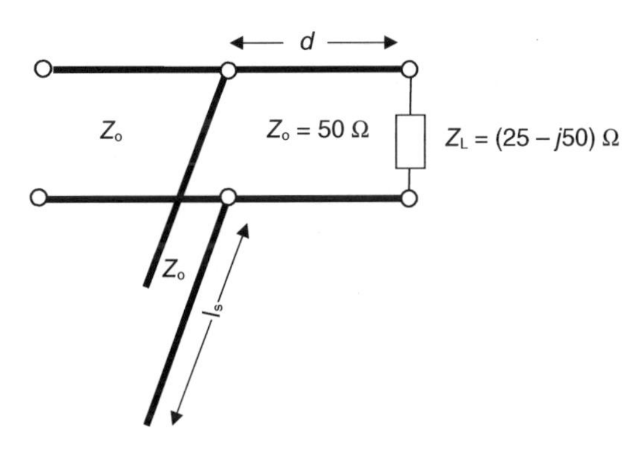

Figure 4.27 Example 4.26.

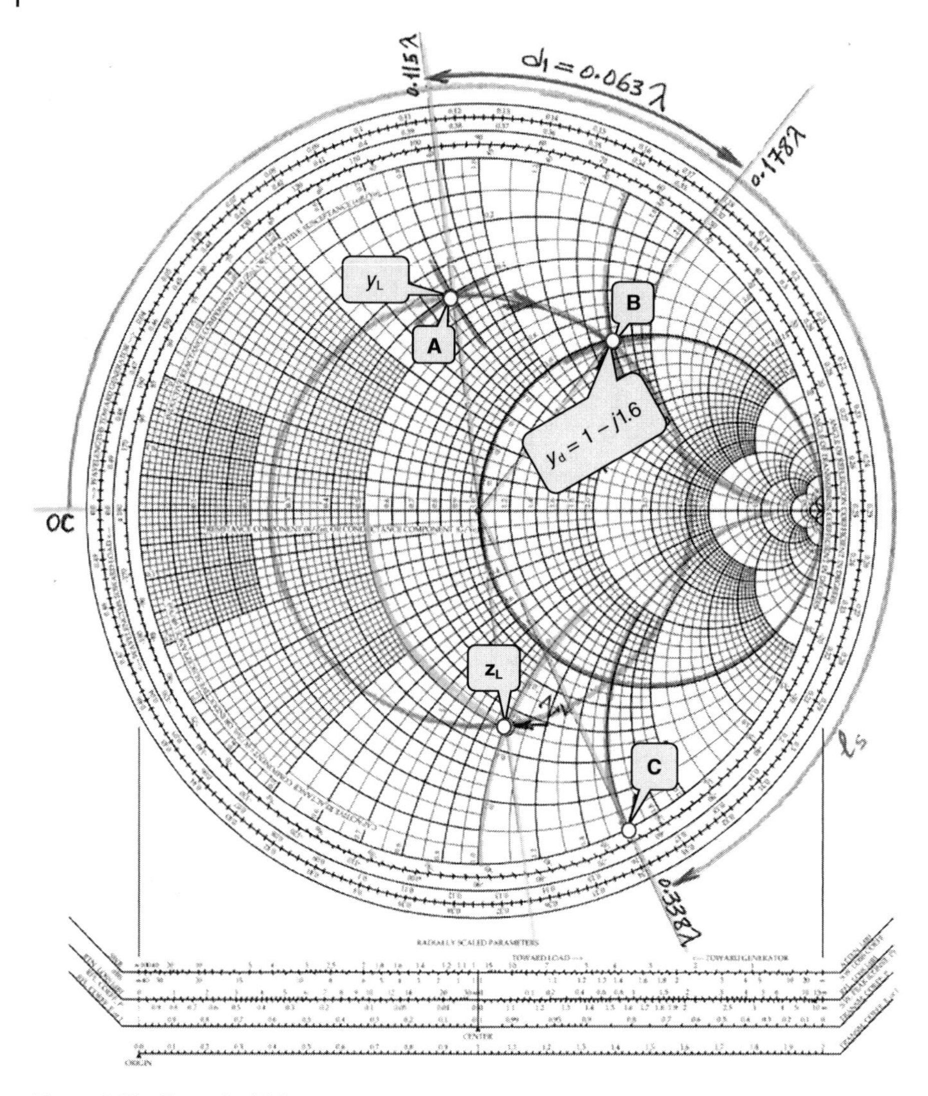

Figure 4.28 Example 4.26.

On the Y-chart, the reactive values on the upper half of the chart are negative, whereas the values on the lower half are positive. Also, the open circuit is at the chart's far edge on the left side of the chart center, and the short circuit is at the far edge on the right side of the chart.

Draw a line from the center of the chart through y_L location (**point *A***) to the edge of the chart and mark the location on the scale of the wavelength toward generator (the location is **0.115** λ).

Move on the SWR circle until it intersects with the unit circle at point ***B*** (i.e., $g = 1$), then read the admittance ($y_d = 1 + j1.55$) as shown in Figure 4.28. Notice that the SWR circle intersect with the unit circle at another point that can be considered as possible solution.

Draw a line from the center of the chart through **point *B*** to the edge of the chart and mark the location on the wavelength toward generator (from the chart it is **0.178** λ), The difference between the locations of point ***A*** and point ***B*** is the required location of the stub ($d1 = 0.178\ \lambda - 0.115\ \lambda =$ **0.063** λ).

To achieve a perfect match (i.e., $y_d = 1 + j0$), locate the opposite of the imaginary part of y_d at point $\boldsymbol{C}$ (i.e., $-j1.55$), then draw a line from the center of the chart through point $\boldsymbol{C}$ to the edge of the chart and mark the location in wavelength (**0.338 λ**). The distance from open-circuit location on the Y-chart to point $\boldsymbol{C}$ (toward generator) is the length of the open-stub 0.338λ. Therefore, the required stub's length and location are given by

$$\text{Stub length} = \underline{\mathbf{0.338}\ \lambda},$$

$$\text{Stub location} = \underline{\mathbf{0.063}\ \lambda}$$

Example 4.27

For the network shown in Figure 4.29, use Smith chart to determine the length of the short-circuit stub, l_s, and its location at a distance d from the load to achieve a perfect match.

Solution

Normalize the load impedance and locate it on the chart, then determine the load admittance as follows:

$$z_L = (75 - j100)/2 = 1.5 - j2$$

$$y_L = 1/z_L = 0.24 + j\,0.32$$

From the Smith chart in Figure 4.30, the location of the stub is the difference between the locations of pints $\boldsymbol{A}$ and $\boldsymbol{B}$, which is **0.129 λ** (i.e., 0.182 λ – 0.053 λ).

The length of the short-circuit stub is **0.084λ** (i.e., 0.33 λ – 0.25 λ), which is the distance between point $\boldsymbol{C}$ and the short-circuit location on the Smith chart as shown in Figure 4.30.

4.10.2 Double-Stub Matching

In the double-stub-matching technique, two stubs are used. The matching is realized by changing the lengths of the stubs, but their locations remain fixed. Figure 4.31 shows a double-stub-matching network.

By changing the lengths of the two stubs, the matching between the load impedance and the characteristic impedance of the transmission line can be achieved. In this technique, the stubs are inserted at predetermined locations. Thus, if the load impedance changes, the stubs need to be replaced with another set of stubs that have different lengths. The length of the first stub, l_{s1}, is chosen so that the admittance at the location where the second stub should be placed (i.e., before

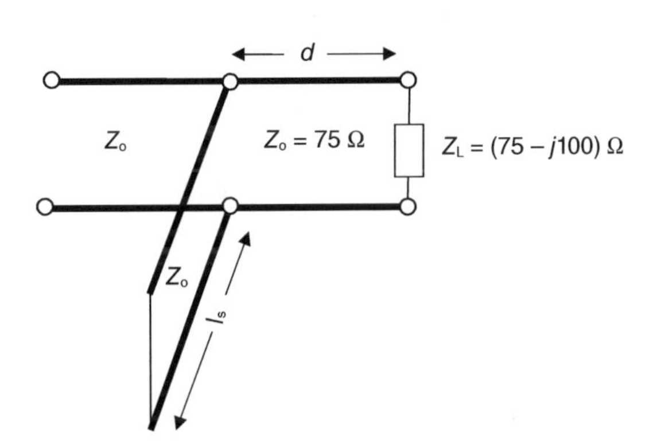

Figure 4.29 Example 4.27.

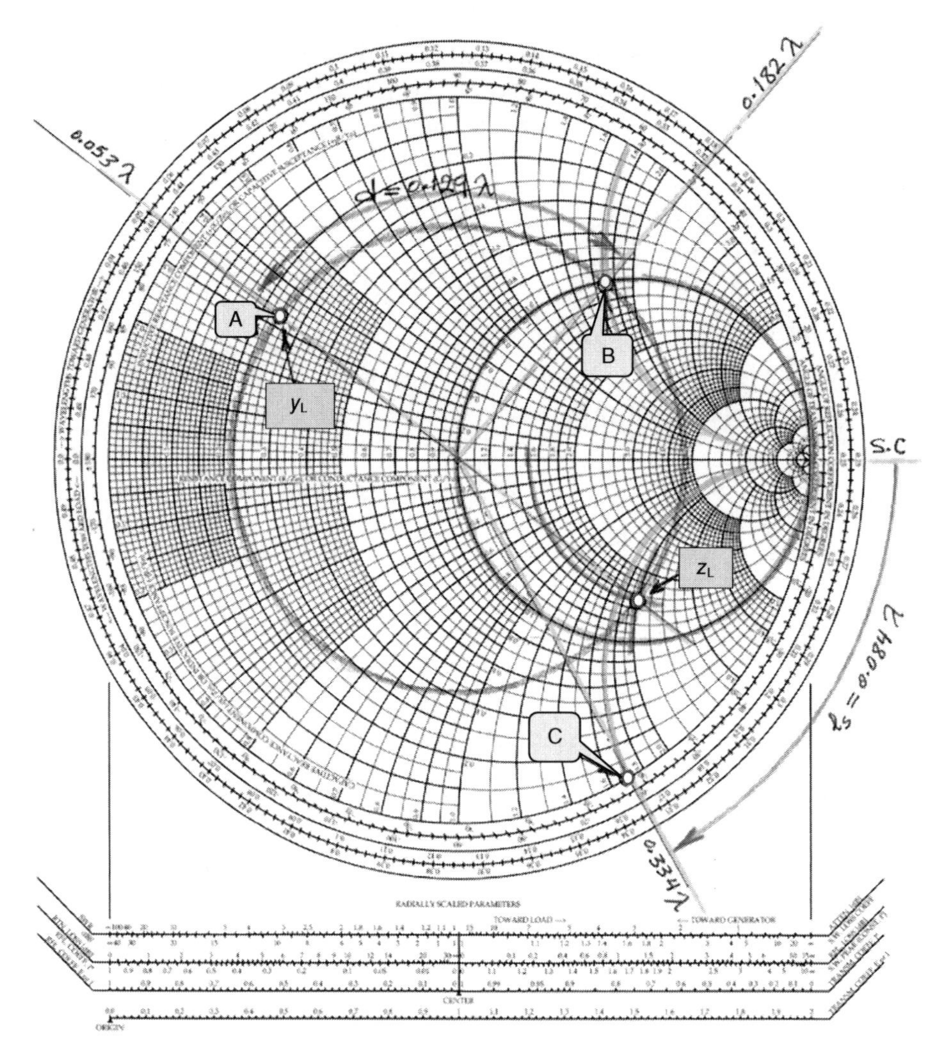

Figure 4.30 Solution of Example 4.27.

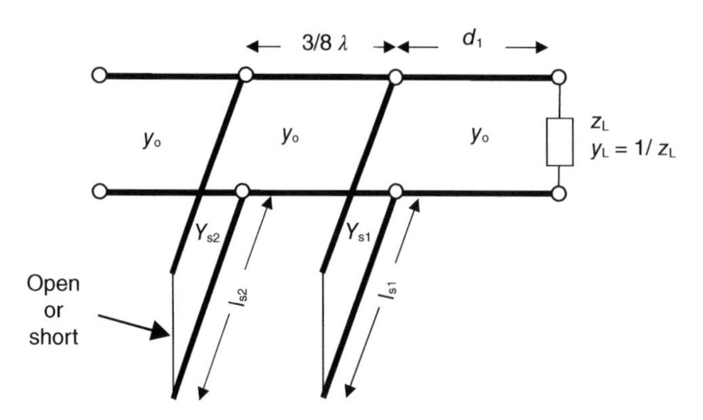

Figure 4.31 Illustration of double-stub matching.

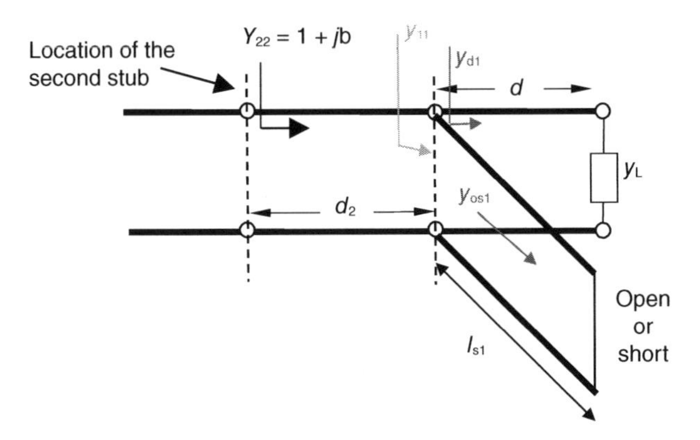

Figure 4.32 Admittance at the location of the second stub before it is inserted.

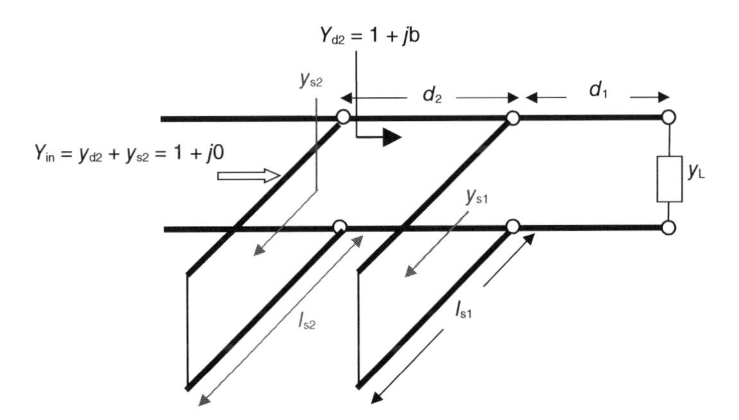

Figure 4.33 Illustration of the effect of the second stub on the input admittance.

adding the second stub) has a real part equal to the characteristic admittance of the line as illustrated in Figure 4.32. The length of the second stub is selected to eliminate the imaginary part of the admittance at the location of the stub insertion as illustrated in Figure 4.33.

To determine the distance between the stubs using the Smith chart, an auxiliary unit circle must be drawn on the chart. Figure 4.34 shows an auxiliary unit circle for a spacing of 3/8 λ between the stubs.

In Figure 4.34, the auxiliary unit circle is 270° (i.e., $\theta_{aux} = 270°$ corresponds to 3/8 λ spacing between the two stubs). The following are the steps needed to determine the lengths of the stubs in a double-stub matching.

Locate the normalized load admittance ($y_L = 1/z_L$) on the Smith chart (thus, the chart will be used as an admittance chart).

From the center of the chart, draw a SWR circle passing through the location of the normalized admittance. Note that the SWR circle intersects with the unit circle at two points, and either point can be considered as a possible solution.

Transfer the admittance at the location of the first stub, d_{s1}, to the location of the second stub, d_{s2}; this is done by moving on a VSWR (i.e., the VSWR of the transferred admittance) circle to intersect with the original unit circle.

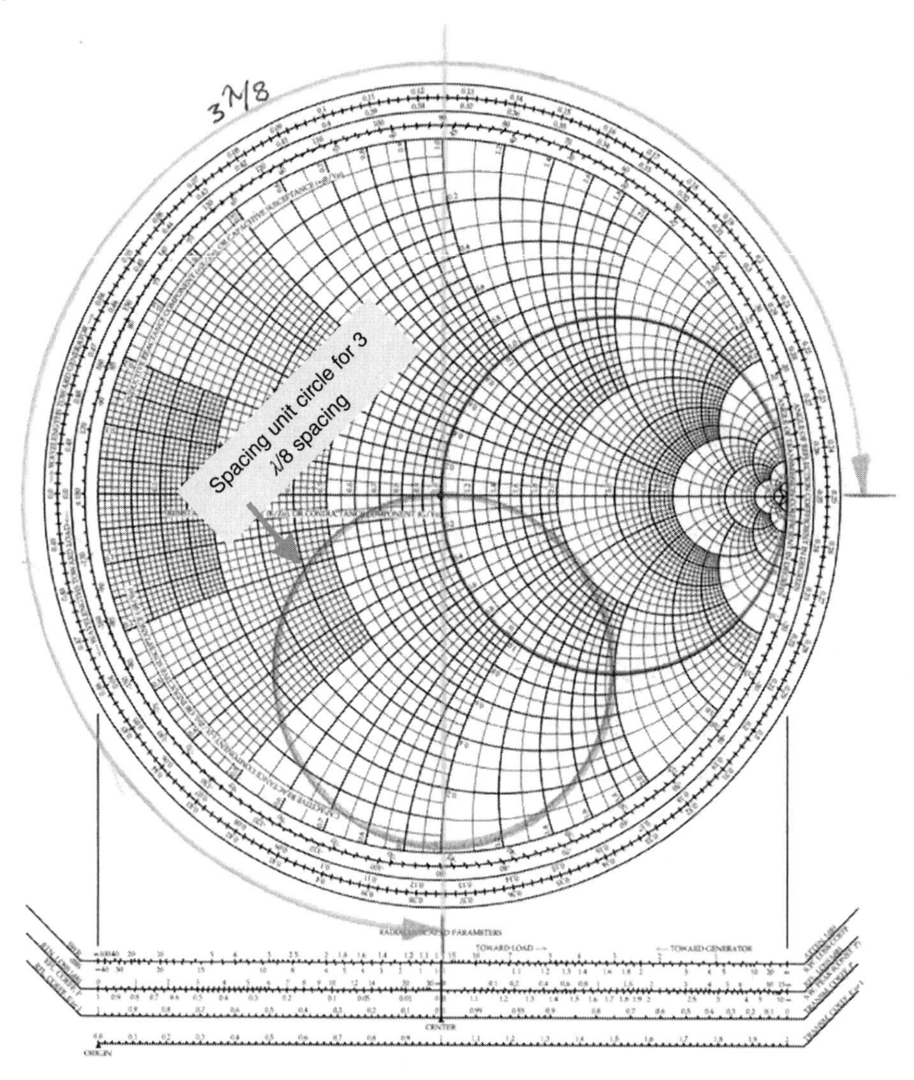

Figure 4.34 Illustration of the angle of rotation θ_{aux} (270°) corresponds to the distance between the stubs (3/8 λ).

Add the opposite of the second stub's admittance so that the parallel admittance at the location of the second stub becomes equal to the characteristic impedance of the transmission line (i.e., perfect match).

The disadvantage of this technique is that certain load impedances cannot be matched. However, this problem can be mitigated by using three stubs separated by $3/8\lambda$. Example 4.28 presented below provides the steps to determine the lengths of the stubs in the double-stub-matching technique.

Example 4.28

For the double-stub network shown in Figure 4.35, d_1 is the location of the first stub from the load, and d_2 is the distance between the first and second stub. Use the Smith chart to determine the lengths of the two short-circuit stubs, to achieve a perfect match.

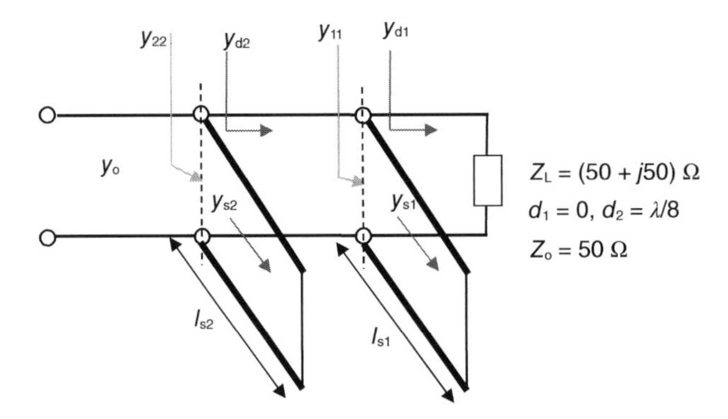

Figure 4.35 Example 4.28.

Solution
Locate the normalized load admittance on the chart.

$$z_L = (50 + j50)/50 = 1 + j1, \quad \text{and} \quad y_L = 1/z_L = 0.5 - j0.5$$

Draw an auxiliary unit circle with an auxiliary angle $\theta_{aux} = 90°$ as shown in Figure 4.36.

Locate the normalized load impedance, z_L, on Smith chart and then draw a VSWR circle and locate y_L (i.e., the inverse of z_L) at point ***A***.

From point ***A***, move on a conductance circle (clockwise) to intersect with the auxiliary unit circle at point ***B*** ($y_{11} = 1 + j0.15$), and draw a VSWR circle passing through point ***B***.

From point ***B***, move on the SWR circle to intersect with the original unit circle at pint ***C***. ($y_{d2} = 1 + j0.71$).

Add a susceptance value of $-j\,0.71$ (point **D**) to cancel out the imaginary part of y_{d2}. Hence,

$$y_{22} = (1 + j0.71) - j0.71 = 1, \text{which is a perfect match.}$$

The distance from short-circuited (***SC***) location of the Smith chart (i.e., the outermost point on the right side of the chart) to point ***D*** is the length of the second stub $l_{s2} = 0.1541$ (i.e., $0.404\ \lambda - 0.25\ \lambda$).

Determine the difference between the imaginary parts of y_{1d} ($j0.15$) and y_L ($-j0.5$) at points ***A*** and ***B***, then locate the value of the difference ($j0.65$) on the outermost scale of the chart at point ***E***.

The distance from the ***SC*** location to point ***E*** (clockwise) is the length of the first stub $l_{s1} = 0.342l$ (i.e., $25\ \lambda + 0.092\ \lambda$).

Example 4.29
In the double-stub network shown in Figure 4.37, use the Smith chart to determine the lengths of the two short-circuit stubs to achieve a perfect match.

Solution
Normalized the load impedance as follows:

$$z_L = (28 + j0)/60 = 0.467, \quad \text{and} \quad y_L = 1/z_L = 2.14$$

Locate y_L on the chart (point A) and draw a VSWR circle as in Figure 4.38.

Move on the VSWR (clockwise) $\lambda/18$ (i.e., $0.055\ \lambda$, point *B* on the chart) and draw a line from the center of the chart to the edge of the chart through point *B*.

Move on a constant conductance circle until it intersects with the auxiliary unit circle at point ***C***, then draw a VSWR circle through point C.

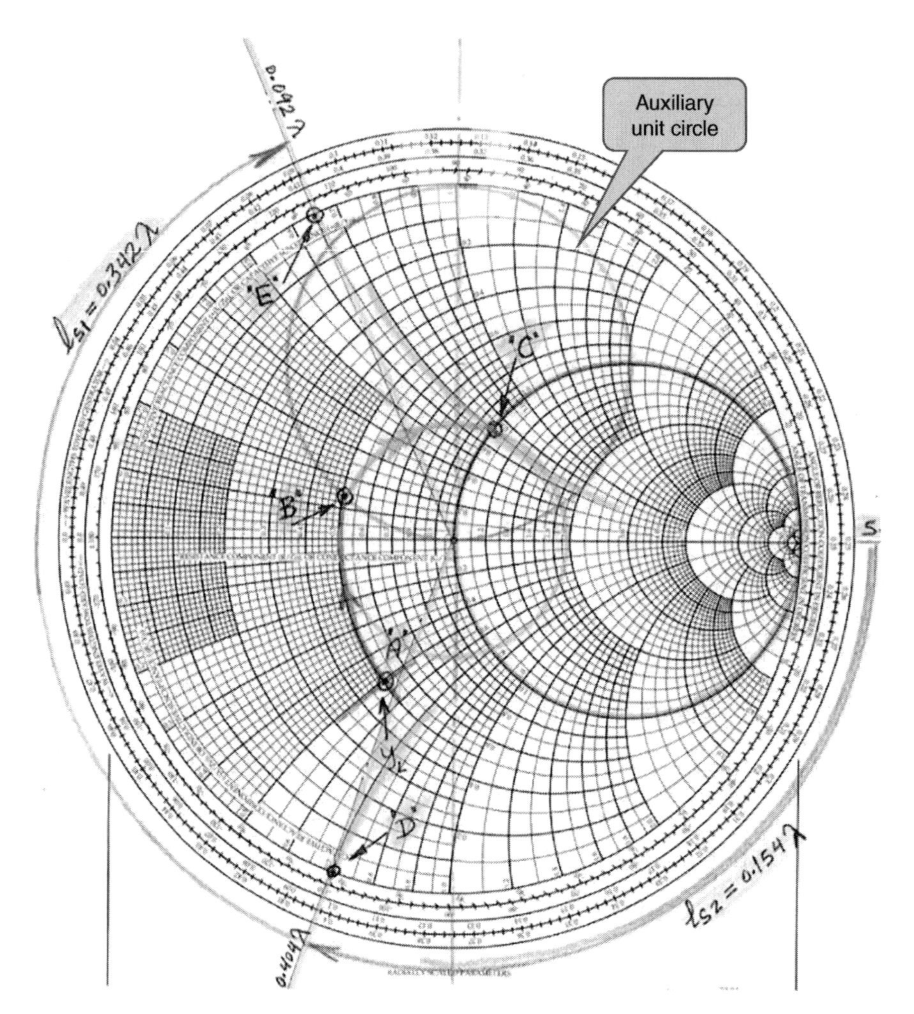

Figure 4.36 Solution of Example 4.28.

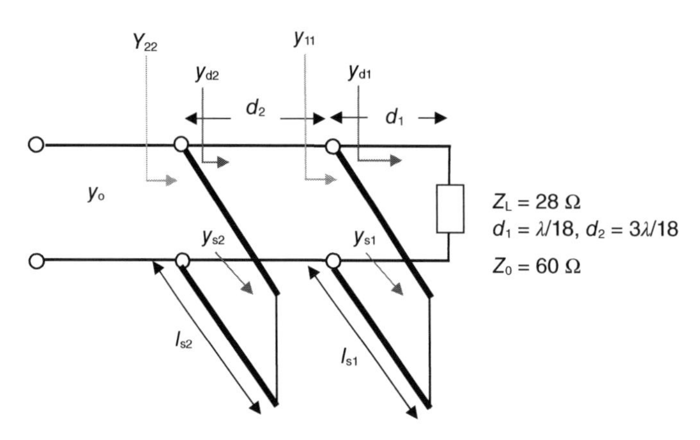

Figure 4.37 Example 4.29.

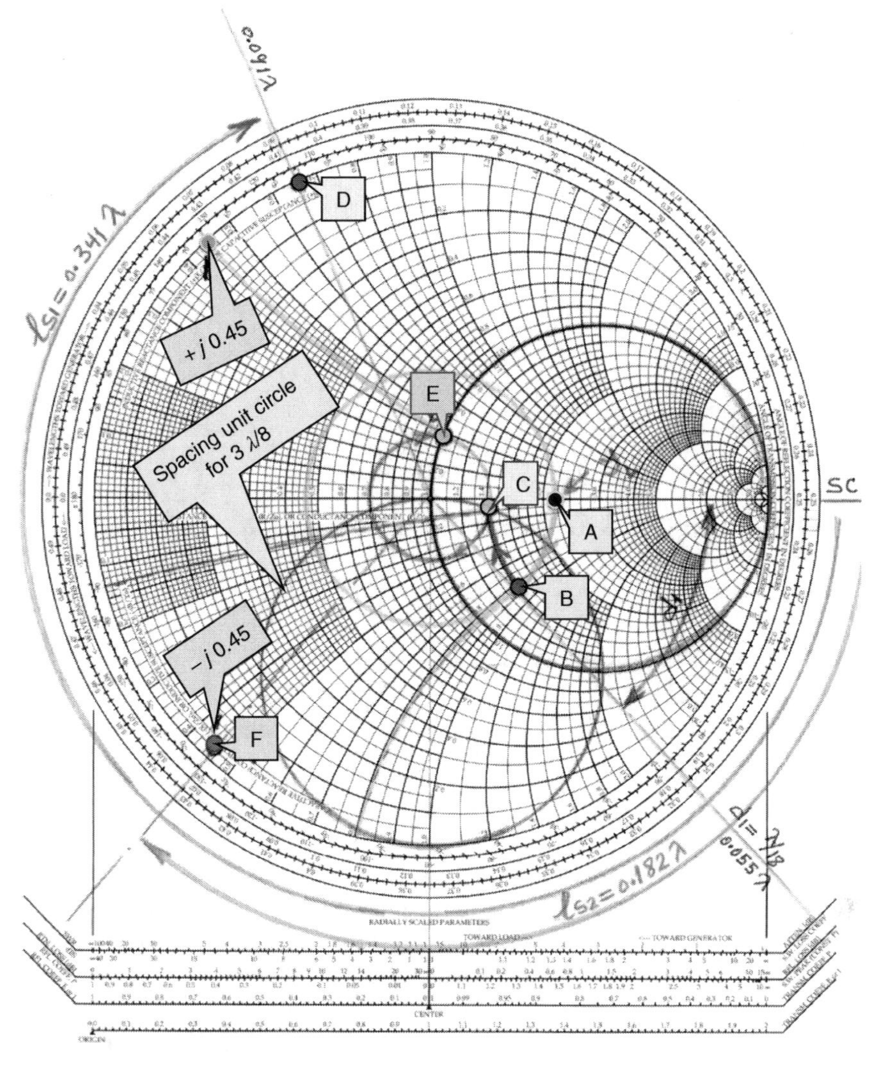

Figure 4.38 Solution of Example 4.29.

Determine the difference between the susceptance of points B and C, which is

$-j0.67\ [-j0.8 - (-j0.13) = -j0.67]$.

Locate the opposite value of the difference (i.e., $+j0.67$) on the chart (point ***D***). The distance from SC to point D is the length of the first stub which is 0.341 l.

Move on the VSWR circle through point C until it intersects with the original auxiliary unit circle at point ***E*** then mark the susceptance of this point ($j0.45$).

Locate the opposite value opposite value of $j0.45$ (i.e., $-j0.45$) on the chart (point F), the distance between SC and point F is the length of the second stub which is 0.182 λ.

Example 4.30

In the double-stub network shown in Figure 4.39, the distance between the first and the second stub is $\lambda/4$, and the first stub is located at the load. Use the Smith chart to determine the lengths of the stubs to achieve a perfect match.

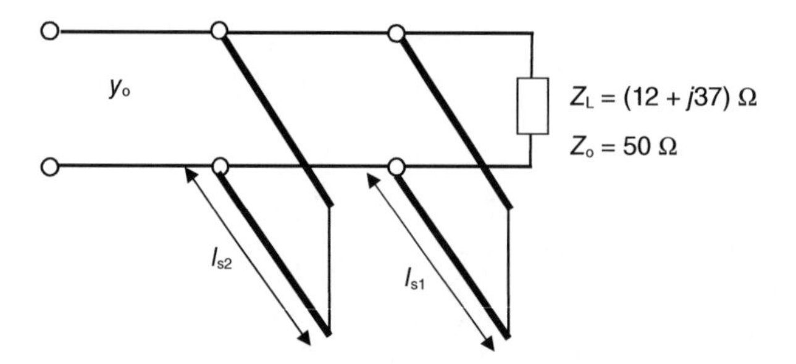

Figure 4.39 Example 4.30.

Solution
The normalized load impedance as follows:

$$z_L = (12 + j37)/50 = 0.24 + j0.74, \quad \text{and} \quad y_L = 1/z_L = 0.4 - j1.2$$

Since the spacing between the two stubs is $\lambda/4$, an auxiliary unit circle at 180° needs to be located on the Smith chart as shown in Figure 4.40.

The solution of this Example is shown in Figure 4.40. The same steps as those used in Example 4.29 can be used to solve this problem.

From Figure 4.40, the length of the second stub is $l_{s2} = (0.25 - 0.14)\ \lambda = 0.11\ \lambda$, and the length of the first stub is $l_{s1} = (0.25 + 0.097)\ \lambda = 0.347\ \lambda$.

4.11 ABCD Parameters

ABCD parameters [1] of a two-port network are also known as transmission, cascade, and chain parameters. They are a simple set of equations relating the input voltage and current to the output voltage and current. They are mainly used in 2-port network analysis when the networks are cascaded. Figure 4.41 represents a two-port network and a cascaded two-port network.

The *ABCD* matrix of cascaded connections of two or more two-port networks can be obtained by multiplying the *ABCD* matrices of the individual two ports. The ABCD matrix of a two-port network is defined in terms of voltages and currents as shown in Figure 4.41. The relationship between these parameters is given by the following two equations:

$$V_1 = AV_2 - BI_2 \tag{4.40}$$

$$I_1 = CV_2 - DI_2 \tag{4.41}$$

The ABCD parameters are determined as follows:

$$A = \left.\frac{V_1}{V_2}\right|_{I_2=0}, \quad B = \left.\frac{V_1}{I_2}\right|_{V_2=0}, \quad C = \left.\frac{I_1}{V_2}\right|_{I_2=0}, \quad D = \left.\frac{I_1}{I_2}\right|_{V_2=0} \tag{4.42}$$

where A is the open-circuit reverse voltage transfer ratio (unitless), B is the short-circuit reverse transfer impedance in ohms, C is open-circuit reverse transfer admittance in siemens, and D is the short-circuit reverse current transfer ratio (unitless). Equations (4.40) and (4.41) can be written as

$$\begin{bmatrix} V_1 \\ I_1 \end{bmatrix} = \begin{bmatrix} A & B \\ C & D \end{bmatrix} \begin{bmatrix} V_2 \\ I_2 \end{bmatrix} \tag{4.43}$$

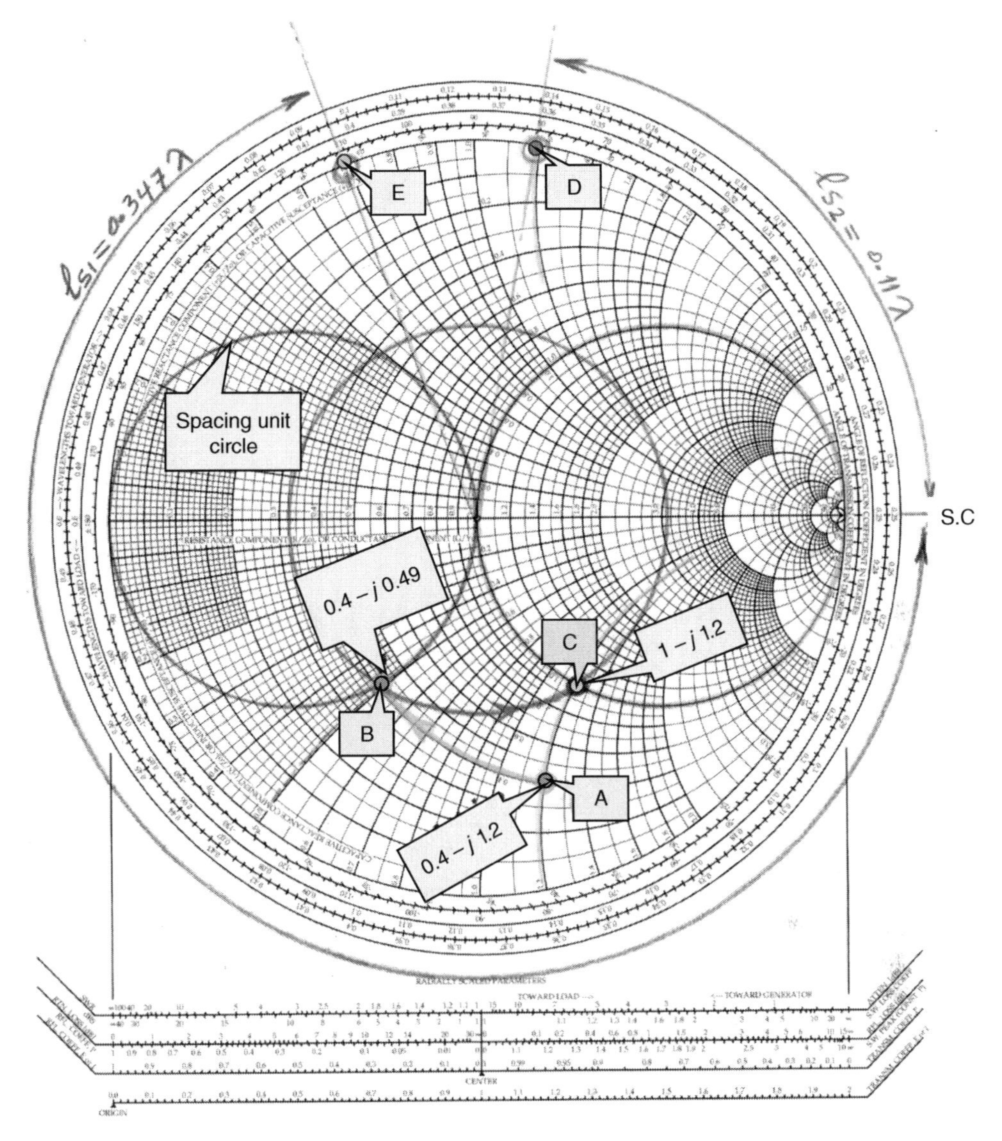

Figure 4.40 Solution of Example 4.30.

where V_1, V_2, I_1, and I_2 represent the voltages and currents at port 1 and port 2. In Figure 4.41b, the cascaded two-port networks can be represented by the following equations are

$$\begin{bmatrix} V_1 \\ I_1 \end{bmatrix} = \begin{bmatrix} A_1 & B_1 \\ C_1 & D_1 \end{bmatrix} \begin{bmatrix} V_2 \\ I_2 \end{bmatrix} \tag{4.44}$$

and

$$\begin{bmatrix} V_2 \\ I_2 \end{bmatrix} = \begin{bmatrix} A_2 & B_2 \\ C_2 & D_2 \end{bmatrix} \begin{bmatrix} V_3 \\ I_3 \end{bmatrix} \tag{4.45}$$

Substituting Eq. (4.45) into Eq. (4.44) gives the total cascaded matrix which is given by

$$\begin{bmatrix} V_1 \\ I_1 \end{bmatrix} = \begin{bmatrix} A_1 & B_1 \\ C_1 & D_1 \end{bmatrix} \begin{bmatrix} A_2 & B_2 \\ C_2 & D_2 \end{bmatrix} \begin{bmatrix} V_3 \\ I_3 \end{bmatrix} \tag{4.46}$$

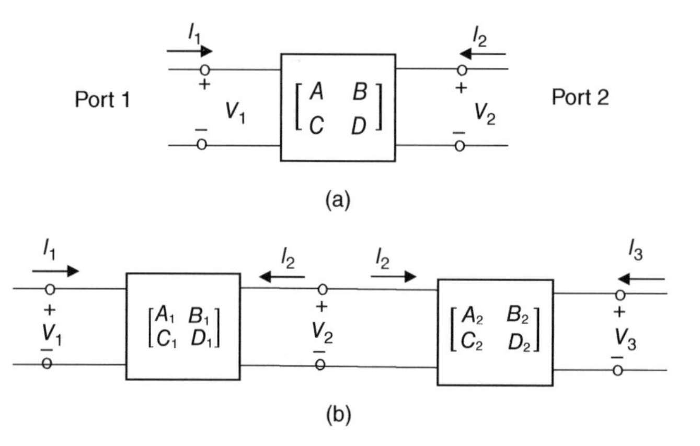

Figure 4.41 (a) Two-port network and (b) cascaded two-port networks.

The ABCD parameters can be derived in terms of the Z, Y, or S parameters of a two-port network.

Example 4.31

Determine the ABCD parameters of the two-port network shown in Figure 4.42.

Solution

The ABCD parameters' equations are given by

$$V_1 = AV_2 + BI_2, \qquad I_1 = CV_2 + DI_2$$

$$\therefore\ A = \left.\frac{V_1}{V_2}\right|_{I_2=0}$$

In Figure 4.43, when $I_2 = 0$, the voltage at node a is given by

$$V_a = \frac{V_1}{30 + (20/(10+10))} \times 10 = 0.25\ V_1$$

The voltage at node b is V_2 and is given by

$$V_2 = \frac{0.25V_1}{(10+10)} \times 10 = 0.125\ V_1$$

$$\therefore\ A = \frac{V_1}{V_2} = 8$$

The parameter C is given by

$$C = \left.\frac{I_1}{V_2}\right|_{I_2=0}, \quad I_2 = I_1\left(\frac{20}{20 + (10+10)}\right) = 0.5\, I_1$$

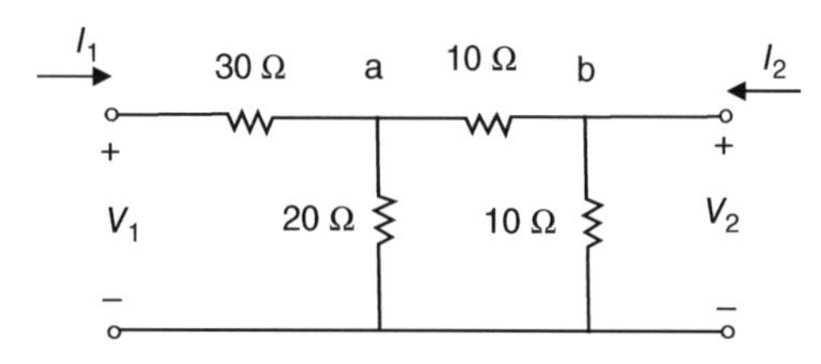

Figure 4.42 Example 4.31.

$$V_2 = V_b = I_2 \times 10 = 0.5\, I_1 \times 10 = 5\, I_1$$

$$\therefore\ C = 0.2$$

The parameter D is given by

$$D = \left.\frac{I_1}{I_2}\right|_{V_2=0}$$

at $V_2 = 0$,

$$I_2 = (-I_1)\left(\frac{20}{20+10}\right), \quad \frac{I_1}{I_2} = 1.5$$

$$\therefore\ D = 1.5$$

The parameter B is given by

$B = \left.\frac{V_1}{I_2}\right|_{V_2=0}$, at $V_2 = 0$, I_2 is given by

$$I_2 = I_1\left(\frac{20}{20+10}\right) = \frac{20}{30}I_1 = \frac{20}{30} \times \frac{V_1}{30 + (20||10)},$$

$$\therefore\ B = \frac{V_1}{I_2} = 54.99$$

The ***ABCD*** matrix of the network is written as

$$\therefore\quad \begin{bmatrix} A & B \\ C & D \end{bmatrix} = \begin{bmatrix} 8 & 55 \\ 0.2 & 1.5 \end{bmatrix}$$

Example 4.32

Determine the ABCD parameters of the two-port network shown in Figure 4.43.

Solution

The ABCD parameters' equations are given by

$$V_1 = AV_2 + BI_2, \qquad I_1 = CV_2 + DI_2$$

$$\therefore\ A = \left.\frac{V_1}{V_2}\right|_{I_2=0}$$

In Figure 4.43, when $I_2 = 0$ the voltage at node a can be calculated as follows:

$$V_a = V_2 = \frac{V_1}{1+5} \times 5 = \frac{5}{6}, \qquad \therefore\ A = \left.\frac{V_1}{V_2}\right|_{I_2=0} = 1.2$$

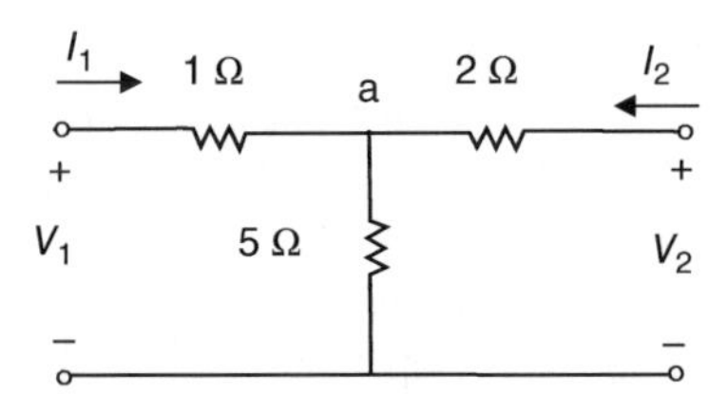

Figure 4.43 Example 4.32.

The parameter, C, is calculated as follows:

$$I_1 = \frac{V_1}{1+5}, \quad V_2 = \frac{V_1}{1+5} \times 5, \quad V_1 = \frac{6}{5} V_2$$

$$\therefore \; I_1 = \frac{1}{5} V_2$$

$$\therefore \; C = \left. \frac{I_1}{V_2} \right|_{I_2=0} = \frac{1}{5} \; \mho$$

The parameter, B, is calculated for $V_2 = 0$ as follows:

$$I_1 = \frac{V_1}{1+(5\|2)}, \quad I_2 = I_1 \times \frac{5}{7} = \frac{V_1}{2.43} \times \frac{5}{7}, \quad \frac{V_1}{I_2} = 3.4, \quad \therefore \; B = 3.4\,\Omega$$

The parameter, D, is calculated for $V_2 = 0$ as follows:

$$I_1 = \frac{V_1}{1+(5\|2)} = \frac{V_1}{2.43}, \quad I_2 = I_1 \times \frac{5}{5+2}, \quad \frac{I_1}{I_2} = \frac{7}{5}, \quad \therefore \; D = \left. \frac{I_1}{I_2} \right|_{V_2=0} = 1.4$$

Thus, the ABCD parameters of Figure 4.43 is given by

$$\therefore \quad \begin{bmatrix} A & B \\ C & D \end{bmatrix} = \begin{bmatrix} 1.2 & 3.4 \\ 0.2 & 1.4 \end{bmatrix}$$

4.11.1 ABCD Parameters of a Lossless Transmission Line

The *ABCD* matrix of a lossless transmission line (i.e., attenuation $\alpha = 0$) is given by

$$\begin{bmatrix} A & B \\ C & D \end{bmatrix} = \begin{bmatrix} \cos\beta l & jZ_0 \sin\beta l \\ jY_0 \sin\beta l & \cos\beta l \end{bmatrix} \tag{4.47}$$

where β is the phase constant, Y_0 is the characteristic admittance, Z_0 is the characteristic impedance, and l is the length of the line. From Eq. (4.47), the ABCD parameters of a transmission line depend on the length of line. For series impedance, Z, the ABCD matrix is given by

$$\begin{bmatrix} A & B \\ C & D \end{bmatrix} = \begin{bmatrix} 1 & Z \\ 0 & 1 \end{bmatrix} \tag{4.48}$$

and for shunt admittance, Y, the *ABCD* matrix is given by

$$\begin{bmatrix} A & B \\ C & D \end{bmatrix} = \begin{bmatrix} 1 & 0 \\ Y & 1 \end{bmatrix} \tag{4.49}$$

Understanding the ABCD parameters of transmission lines is useful for analyzing and characterizing transmission line structures.

4.12 *S*-parameters

S-parameters (*scattering parameters*) [8, 9] describe the input–output relationship between the ports of an electrical network. A network is a circuit with one or more ports; for Example, an antenna is a one-port network, an amplifier or filter is a two-port network; mixer, circulator, and directional coupler are three-port networks. The return loss of a device can be measured by injecting an RF signal into the input port and measuring the level of the signal at the output port.

S-parameters are labeled as "S_{ij}," where i represents the output port, and j represents the input port. If an RF signal is applied to port 1 of a device, the measured signal at port 2 of the device is named S_{21}, which represents the gain or the insertion loss of a device. Also, the measured reflected power from port 1 at port 1 represents the reflection coefficient S_{11}. For a one-port network, there is only one parameter which is S_{11}.

For a two-port network, S-parameters are represented by a 2×2 matrix and given by

$$[S] = \begin{bmatrix} S_{11} & S_{12} \\ S_{21} & S_{22} \end{bmatrix} \tag{4.50}$$

For a three-port network, S-parameters are given by

$$[S] = \begin{bmatrix} S_{11} & S_{12} & S_{13} \\ S_{21} & S_{22} & S_{23} \\ S_{31} & S_{32} & S_{33} \end{bmatrix} \tag{4.51}$$

Figure 4.44 Illustrates the S-parameters of a two-port network.

In Figure 4.44, a_1 is the incident normalized voltage wave $\frac{V_1^+}{\sqrt{Z_o}}$ at port 1, and b_1 is the reflected normalized voltage wave $\frac{V_1^-}{\sqrt{Z_o}}$ at port 1. The scattering parameters for a two-port network can be described by the following matrix:

$$\begin{bmatrix} b_1 \\ b_2 \end{bmatrix} = \begin{bmatrix} S_{11} & S_{12} \\ S_{21} & S_{22} \end{bmatrix} \begin{bmatrix} a_1 \\ a_2 \end{bmatrix} \tag{4.52}$$

where

S_{11} = the input reflection coefficient with the output terminated in a matched load
S_{21} = the forward transmission coefficient with the output terminated in a matched load
S_{22} = the output reflection coefficient with the input terminated in a matched load
S_{12} = the reverse transmission coefficient with the input terminated in a matched load

These parameters are calculated as follows:

$$S_{11} = \left.\frac{b_1}{a_1}\right|_{a_2=0}$$

$$S_{21} = \left.\frac{b_2}{a_1}\right|_{a_2=0}$$

$$S_{22} = \left.\frac{b_2}{a_2}\right|_{a_1=0}$$

$$S_{12} = \left.\frac{b_1}{a_2}\right|_{a_1=0}$$

S_{11} is measured with the output port terminated by a matched load (i.e., $a_2 = 0$), and S_{22} is measured with the input port terminated by a matched load ($a_1 = 0$). S-parameters are complex values with magnitude and phase. The advantage of S-parameters is that they are measured with matched termination; another advantage is that they can be cascaded to predict the overall system response. These parameters can be measured over their operating frequency using a vector network analyzer (VNA), Figure 4.45 shows a photo of VNA.

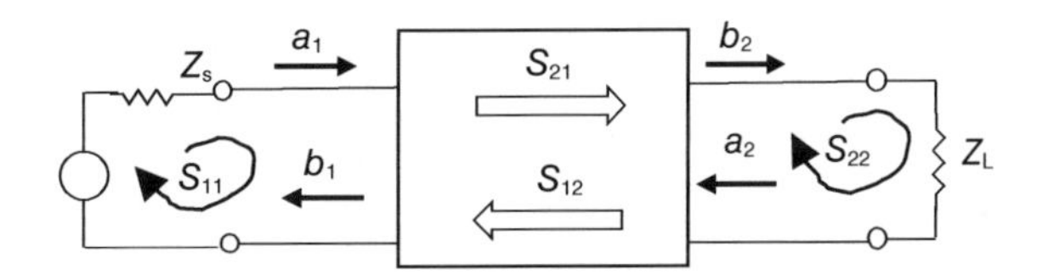

Figure 4.44 Illustration of S-parameters of a two-port network.

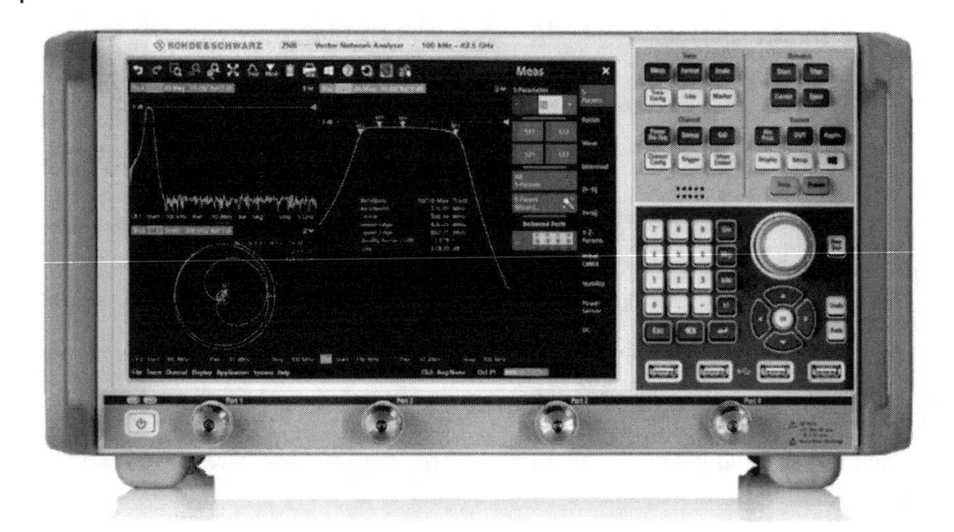

Figure 4.45 Display of measurements on a vector network analyzer (VNA). Source: Rohde & Schwarz Emirates L.L.C.

For an RF component such as a filter or amplifier, the measured insertion loss or gain is the measured S_{21} in dB, while the input and output return losses are the S_{11} and S_{22} in dB. The return loss can be expressed in dB as

$$RL = -20\log_{10}|S_{11}| \tag{4.53}$$

and the VSWR is expressed as

$$\text{VSWR} = \frac{1+|S_{11}|}{1-|S_{11}|}$$

Example 4.33

Determine the S-parameters of the network shown in Figure 4.46.

Solution

$$S_{11} = \Gamma_{\text{in}} = \frac{Z_{\text{in}} - Z_o}{Z_{\text{in}} + Z_o} \tag{4.54}$$

$$Z_{\text{in}} = Z_1 + [Z_2/(Z_3 + Z_o)] = 50\,\Omega$$

$$\therefore\ S_{11} = \frac{50-50}{50+50} = 0$$

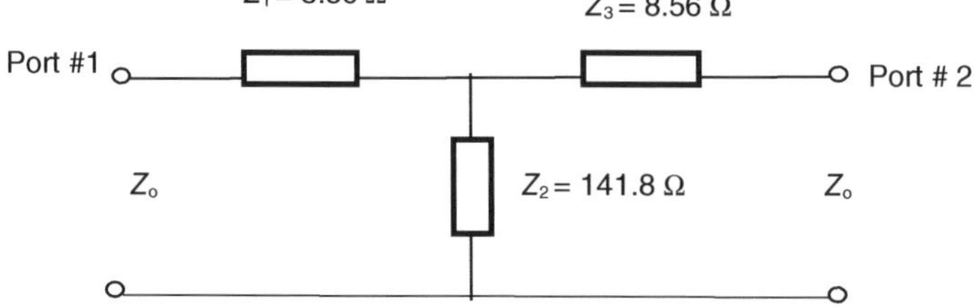

Figure 4.46 Example 4.33.

$S_{11} = 0$ means that the input is matched. Because of the circuit symmetry, $S_{22} = 0$.

To determine S_{21}, the voltage at port 2 should be calculated and the ratio of V_2 to V_1 gives S_{21}. From the Figure 4.46, V_2 is given by

$$V_2 = V_1 \left(\frac{z_2}{z_1 + (z_2 || (z_3 + z_o))} \right) \left(\frac{z_0}{z_2 + z_3 + z_o} \right) = 0.708\ V_1$$

$$\therefore\ S_{21} = \frac{V_2}{V_1} = 0.708$$

Because of the circuit symmetry, $S_{21} = S_{12}$. Thus, the scattering matrix of the T-section two-port network is

$$[S] = \begin{bmatrix} 0 & 0.708 \\ 0.708 & 0 \end{bmatrix}$$

Example 4.34

Determine the S-parameters of a series impedance connected to a transmission line with characteristic impedance z_o as shown in Figure 4.47a.

Solution

The input reflection coefficient $\mathbf{\Gamma}_{in}$ (i.e., $\boldsymbol{S}_{11}$) at port # 1 is given by

$$S_{11} = \Gamma_{in} = \frac{Z_{in} - Z_o}{Z_{in} + Z_o}$$

$$z_{in} = z + z_o$$

$$\therefore\ S_{11} = \frac{Z}{Z + 2Z_o}$$

From the symmetry of the network in Figure 4.47a, $S_{22} = S_{11}$. The ratio of V_2/V_1 gives S_{21}. The voltage at port #2, V_2 is given as

$$V_2 = V_1 \left(\frac{Z_2}{Z_1 + (Z_2 || Z_3 + Z_o)} \right) \left(\frac{Z_o}{Z_2 + Z_3 + Z_o} \right)$$

$$\therefore\ S_{21} = S_{12} = \left(\frac{Z_2}{Z_1 + (Z_2 || Z_3 + Z_o)} \right) \left(\frac{Z_o}{Z_2 + Z_3 + Z_o} \right)$$

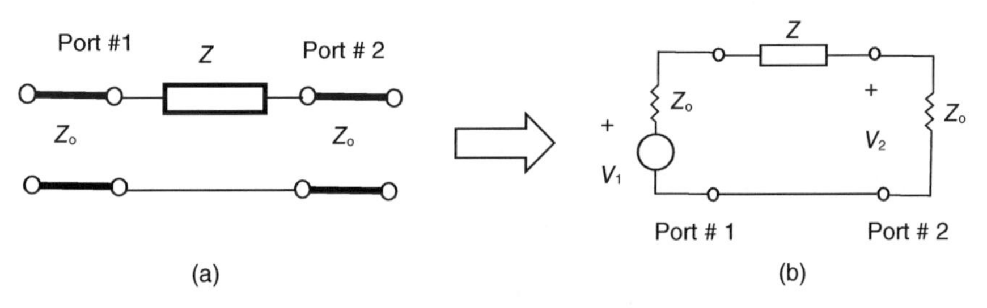

Figure 4.47 Example 4.34. (a) A two-port network of series impedance Z. (b) Two-port network excited by a source and terminated in Z_o.

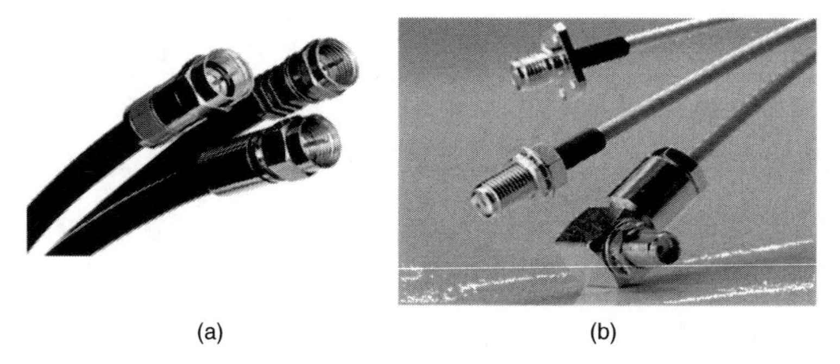

Figure 4.48 Transmission line connectors: (a) coax connectors and (b) SMA connectors. Source: (a) airborne77/Adobe Stock and (b) Frontlynk Inc.

4.13 Transmission Line Connectors

Transmission line connectors are used to facilitate the connection between different devices or the devices and test equipment; they influence the measured results. The choice of the right connector depends on the frequency of operation. Figure 4.48a shows coaxial cable connectors, and Figure 4.48b shows SMA connectors.

Review Questions

4.1 Briefly explain the difference between a coaxial cable and a microstrip line.

4.2 Compare and contrast microstrip lines, coplanar waveguides (CPWs), conductor-backed CPWs, and strip lines.

4.3 Explain the application of using a quarter-wavelength transmission line.

4.4 Explain the terms VSWR, SWR, and reflection coefficient.

4.5 Calculate the voltage reflection coefficient for a transmission line with a characteristic impedance of 50 Ω and terminated in a 100 Ω load.

4.6 For a transmission line with a characteristic impedance of 50 Ω, if the reflection coefficient is 0.175, determine the load impedance.

4.7 Calculate the skin depth of a copper conductor at 100 MHz.

4.8 For a transmission line that has the following parameters per unit length: $L = 0.25\ \mu\text{H/m}$, $C = 445\ \text{pF/m}$, $R = 5\ \Omega/\text{m}$, and $G = 0.01\ \mho/\text{m}$, calculate the complex propagation constant and the characteristic impedance of the line at operating frequency of 900 MHz.

4.9 A 0.3 λ lossless transmission line with a characteristic impedance $Z_0 = 75\ \Omega$ is terminated in a load impedance of $40 + j20\ \Omega$, determine the reflection coefficient, VSWR, and the input impedance.

4.10 Use Smith chart to verify the solution of problem.

4.11 For a transmission line of length 0.8 λ with characteristic impedance of 50 Ω and terminated in a load of $70 + j40\ \Omega$, determine the input impedance using Smith chart.

4.12 Calculate the input impedance of a 0.25 λ length transmission line with characteristic impedance of 50 Ω and terminated in a shorted load. Use Smith chart to verify your calculation.

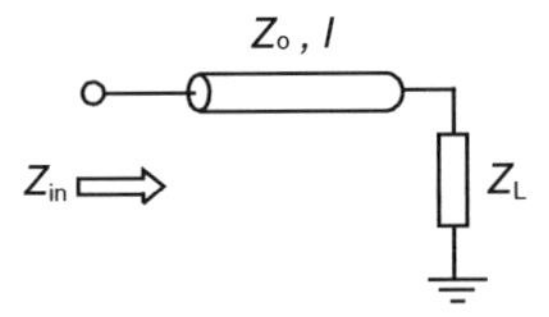

4.13 Determine the input impedance of a 0.3 λ transmission line that has a characteristic impedance of 50 Ω and terminated in a load impedance $Z_L = 10 + j20$

4.14 For a transmission line of length 2.3 cm and a characteristic impedance of 50 Ω, determine the following parameters for a loaded impedance $Z_L = (100 + j50)\ \Omega$ and operating at 3 GHz.

(a) VSWR

(b) Reflection coefficient

(c) Input impedances

4.15 Use Smit chart to find the length of a short-circuited 50 Ω transmission line stub to give the following input impedances

(a) $Z_{in} = 0$

(b) $Z_{in} = \infty$

(c) $Z_{in} = j50\ \Omega$

(d) $Z_{in} = -j50\ \Omega$

4.16 Explain the difference between single-stub and double-stub matching techniques.

4.17 For the transmission line shown later, use Smith chart to determine the VSWR on the line, the return loss, the load admittance, and the reflection coefficient at the load.

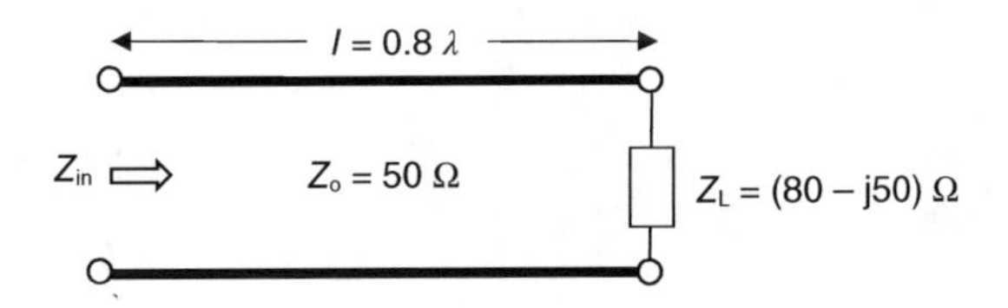

4.18 For the scattering matrix shown below, determine the following:

$$[S] = \begin{bmatrix} 0.13\angle 90^\circ & 0.42\angle 180^\circ & 0.12\angle 180^\circ \\ 0.40\angle 180^\circ & 0.21\angle 0^\circ & 0.55\angle 45^\circ \\ 0.40\angle 180^\circ & 0.55\angle 45^\circ & 0.21\angle 0^\circ \end{bmatrix}$$

(a) The return loss at each port when all other ports are terminated in a matched load.
(b) The insertion loss and phase between ports 2 and 1 when all other ports are terminated in a matched load.

4.19 In the double-stub network shown below, the distance between the first and the second stub is $\lambda/4$, and the first stub is located at the load. Use Smith chart to determine the lengths of the stubs to achieve a perfect match.

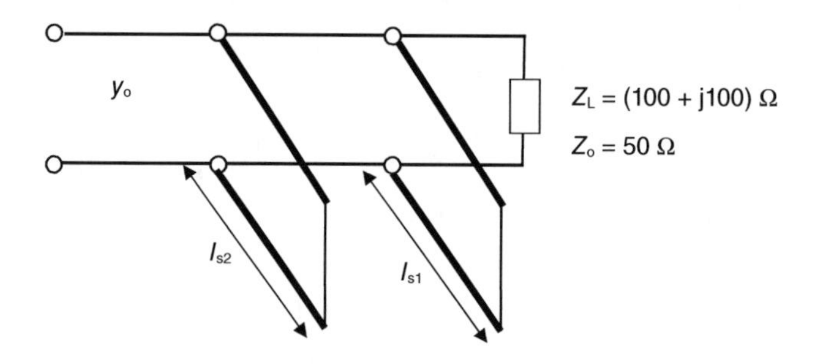

4.20 For the network shown use Smith chart to determine the length of the short-circuit stub, ls, and its location at a distance, d, to achieve a perfect match.

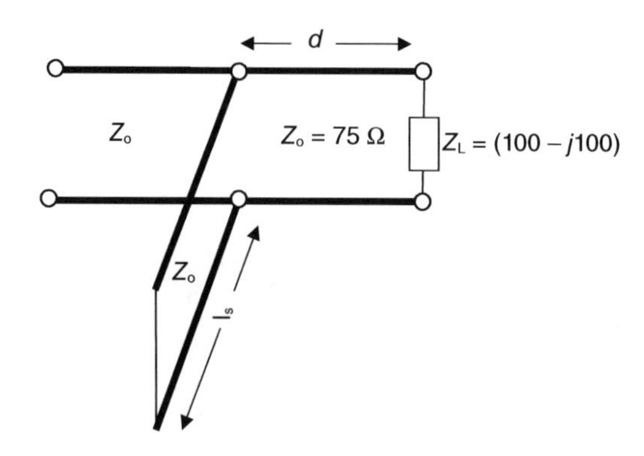

4.21 Determine the ABCD parameters of the two-port network shown below

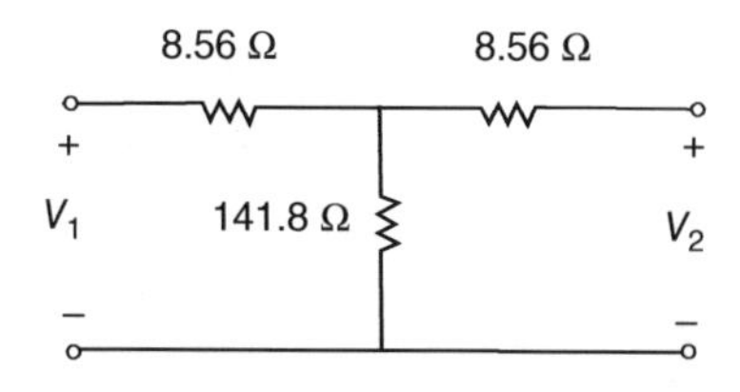

4.22 Determine the ABCD parameters of the two-port network shown below

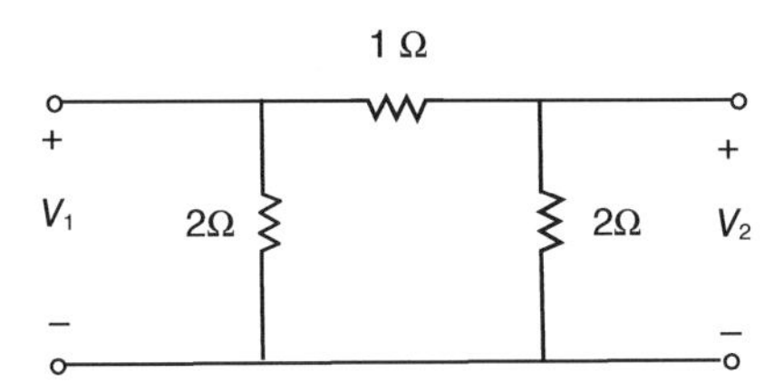

References

1 David M. Pozar, *Microwave Engineering*, 4th ed., Wiley, 2011.
2 Terry C. Edwards, Michael B. Steer, *Foundations for Microstrip Circuit Design*, Wiley, 2016.
3 Ramesh Garg, Bahi I. J., Maurizio Bozzi, *Microstrip Lines and Slotlines,* Artech House, 2013.
4 Bahl I. J. and Trivedi D. K., "A designer guide to microstrip line," *Microwaves*, 1977.
5 David K. Cheng, *Field and Wave Electromagnetics*, 2nd ed., Pearson India, 2014.
6 Advanced Design System (ADS), http://www.keysight.com.
7 Ansys HFSS, https://www.ansys.com.
8 "S-parameters design," *Hewlett-Packard Application Note 154*, 1972.
9 Guillermo Gonzalez, *Microwave Transistor Amplifiers: Analysis and Design*, 2nd ed., Pearson, 1996.

Suggested Readings

Chandan Yadav, Sebastien Fregonese et al., "S-parameter measurement and EM simulation of electronic devices towards THz frequency range," *IEEE 34th International Conference on Microelectronic Test Structures (ICMTS)*, 2022.

Francis Rodes, Xavier Hochart, "Measurement of extreme impedances with the S parameters," *IEEE Instrumentation & Measurement Magazine*, Volume: 25, Issue: 7, 2022.

Ibrahim Haroun, Jim Wight, Calvin Plett, Aly Fathy, "Experimental characterization of EC-CPW transmission lines and passive components for 60-GHz CMOS radios," *IEEE MTT-S International Microwave Symposium,* IEEE, 2010.

Ibrahim Haroun, Ta-Yeh Lin, Da-Chiang Chang, Calvin Plett, "A reduced-size, low-loss 57–86 GHz IPD-based power divider using loaded modified CPW transmission lines," *Asia Pacific Microwave Conference Proceedings*, IEEE, 2012.

Ibrahim Haroun, Yuan-Chia Hsu, Da-Chiang Chang, Calvin Plett, "A novel reduced-size 60-GHz 180° coupler using LG-CPW transmission lines," *Asia-Pacific Microwave Conference*, IEEE, 2011.

Iulia Andreea Mocanu, Norocel Codreanu, Mihaela Pantazică, "Design and analysis of hybrid couplers using lumped elements and microstrip topology," *IEEE 28th International Symposium for Design and Technology in Electronic Packaging (SIITME)*, 2022.

Kangtai Zheng, Jimhong He, "Miniaturization of Wilkinson power divider using zigzag combination of microstrip and CPW lines," *IEEE International Workshop on Electromagnetics: Applications and Student Innovation Competition (iWEM)*, 2022.

Mingming Ma, Fei You et al., "A generalized multiport conversion between S parameter and ABCD parameter," *International Conference on Microwave and Millimeter Wave Technology (ICMMT)*, 2022.

Neeraj Kumar Maurya, Max J. Ammann, Patrick McEvoy, "Dual-band dual-polarized microstrip array for mm-Wave and sub-6 GHz applications," *International Workshop on Antenna Technology (iWAT)*, 2022.

Qiaowei Yuan, "S-parameters for calculating the maximum efficiency of a MIMO-WPT system: applicable to near/far field coupling, capacitive/magnetic coupling," *IEEE Microwave Magazine*, Volume: 24, Issue: 4, 2023.

Karimov S. M., Karimov A. A., Ravshanov D. Ch., "Development of a microstrip coupler with reduced dimensions," *Systems of Signals Generating and Processing in the Field of on Board Communications*, 2022.

Satoshi Ono, Tsuyoshi Narita et al., "Design and fabrication of stripline type of a broadband quadrature hybrid couplers with transitions of coaxial-microstrip-stripline," *Asia-Pacific Microwave Conference (APMC)*, 2022.

Wei-Chen Lee, Rei-Si Hong, Jhan-Li Wu, "Development of a low-cost probe based on the SMA daptor for S-parameter measurement," *8th International Conference on Applied System Innovation (ICASI)*, 2022.

Xiao-Fang Li, Jian-Kang Xiao, "Dual-band bandpass filter based on suspended coplanar waveguide-microstrip hybrid," *IEEE Transactions on Circuits and Systems I: Regular Papers*, 2023.

Xiaoqing Wu, Lin-Ping Shen, "Compact ultra-wideband microstrip 3dB branch-line coupler using coupled-lines," *IEEE International Symposium on Antennas and Propagation and USNC-URSI Radio Science Meeting (AP-S/URSI)*, 2022.

Yi-Fan Tsao, Arpan Desai, Heng-Tung Hsu, "Dual-band and dual-polarization CPW fed MIMO antenna for fifth-generation mobile communications technology at 28 and 38 GHz," *IEEE Access*, Volume: 10, 2022.

Zhen-Ge Zhang, Maomi Feng et al., "A compact single-layer broadband coplanar waveguide to coplanar strip line transition," *IEEE MTT-S International Microwave Workshop Series on Advanced Materials and Processes for RF and THz Applications (IMWS-AMP)*, 2022.

Ziheng Zhou, Yuehe Ge et al. "Wideband MIMO antennas with enhanced isolation using coupled CPW transmission lines," *IEEE Transactions on Antennas and Propagation*, Volume: 71, Issue: 2, 2023.

Zulfi, Joko Suryana, Achmad Munir, "Phase reconfigurable hybrid coupler implemented using capacitor-loaded transmission lines," *14th International Conference on Computational Intelligence and Communication Networks (CICN)*, 2022.

5

RF Subsystem Blocks

This chapter provides the essential background for understanding the performance of the RF building blocks that are used in wireless communication systems. Such blocks include low-noise amplifiers, mixers, filters, frequency synthesizers, RF power amplifiers, duplexers, circulators/isolators, directional couplers, attenuators, RF switches, RF phase shifters, and power combiners/splitters. The chapter focuses on the blocks' theory of operation and the impact of the blocks' parameters on the system performance, rather than on the circuit design. The provided references at the end of the chapter cover the circuit design of RF components. Review questions are provided to help the reader understand the chapter's contents. A list of references and suggested readings that present the recent research activities that are related to the chapter's topics are provided.

5.1 Chapter Objectives

On reading this chapter, the reader will be able to:

- Draw a block diagram of an RF transceiver frontend and briefly explain the function of each block.
- List the performance parameters of low-noise amplifiers, mixers, filters, frequency synthesizers, RF power amplifiers, duplexers, circulators/isolators, directional couplers, attenuators, RF switches, RF phase shifters, and power combiners/splitters, and explain how to characterize these parameters.
- Compare and contrast the Y-factor and cold noise methods for measuring the noise figure of an RF amplifier.
- Explain the basic concept of balanced and double-balanced mixers.
- Describe the basic concept of cavity filters.
- Explain the basic concept of frequency synthesizers and their applications in wireless systems.
- Explain and illustrate the operation of a feedforward linearization system.
- Explain the basic concept of RF switches and phase shifters.

5.2 Introduction to RF Building Blocks

The continuous increase in the development and deployment of wireless communication systems has created significant interest and need for RF building blocks that fulfill the demand for

Essentials of RF Front-end Design and Testing: A Practical Guide for Wireless Systems, First Edition. Ibrahim A. Haroun.

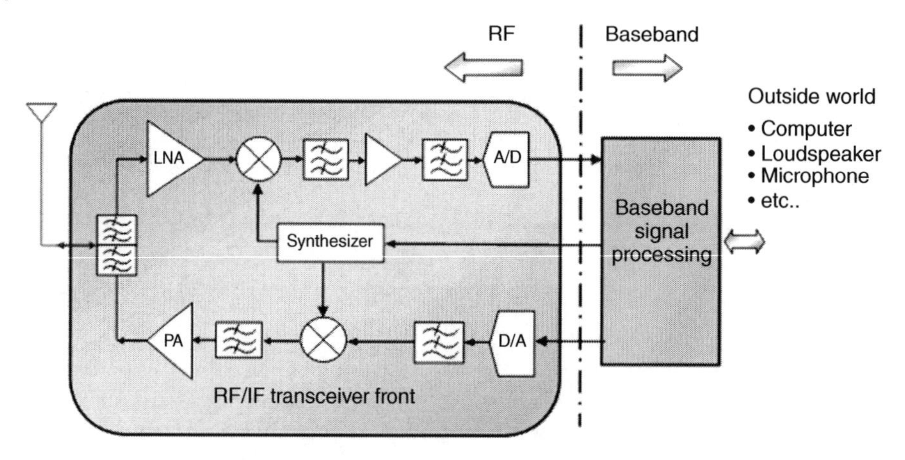

Figure 5.1 Block diagram of RF transceiver system.

low-cost, low-power consumption, and reliable communication systems. Such blocks are available as surface-mount packages and commercial off-the-shelf (COTS) connectorized modules. For COTS modules, the input and output of the module (i.e., block) are connoted with other blocks or test equipment using RF cables. Using such modules shortens the development cycle of proof-of-concept systems and helps adjust the system's performance using the most suited modules. In RF system development, a proof-of-concept is essential before high-volume production. Figure 5.1 shows the location of the RF building blocks in an RF transceiver, which is used in wireless communication systems.

5.3 Low-Noise Amplifiers

Low-noise amplifiers (LNA) [1–3] are the most critical building blocks in radio receivers. An LNA is the first gain block in the receiver's chain and influences the receiver's noise figure. The noise figure impacts the receiver's sensitivity and bit-error rate. Such amplifiers must provide sufficient gain to minimize the cascaded noise figure (NF) of the receiver.

5.3.1 Noise Figure (NF)

The noise figure of an LNA over its operating frequency band is defined as the amplifier's contribution to the thermal noise at its output. The NF is given by

$$\mathrm{NF} = 10 \log \left[\frac{(S/N)_i}{(S/N)_o} \right] \tag{5.1}$$

where $(S/N)_i$ and $(S/N)_o$ are the signal-to-noise ratios at the input and output of the receiver, respectively. In an RF receiver, more than one device (e.g., amplifiers, filters, etc.) is cascaded, and the cascaded noise factor F_{cascaded} is given by

$$F_{\mathrm{cascaded}} = F_i + \sum_{i=2}^{N} \frac{F_i - 1}{\prod_{j=1}^{i-1} G_j} \tag{5.2}$$

Figure 5.2 Commercial low-noise amplifier products. Source: Mouser Electronics, Inc.

where N is the number of stages in the RF chain, and G_j is the gain of the jth stage. The cascaded noise figure is expressed in dB and given by

$$NF_{\text{cascaded}} = 10 \log (F_{\text{cascaded}}) \tag{5.3}$$

The higher the gain of the first block in a receiver, the lower the cascaded noise figure and the better the receiver sensitivity, which enables the receiver to detect weak signals. Noise figures are measured using a noise-figure meter or spectrum analyzer. Low-noise amplifiers are available as commercial off-the-shelf (COTS) products and as surface-mount packages for use in system-on-board development. Figure 5.2 shows commercial LNA products.

An important consideration in low-noise amplifiers is amplifier stability. An unstable amplifier will oscillate and generate undesired frequencies that might interfere with other components in the system. Conversely, a low-noise amplifier will be unconditionally stable if it does not oscillate for any input or output impedances. For an amplifier to be unconditionally stable, it must satisfy the following condition:

$$K > 1, \quad \text{and} \qquad |S_{11}S_{22} - S_{21}S_{12}| < 1,$$

where k is the stability factor and is given by

$$k = \frac{1 - |S_{11}|^2 - |S_{22}|^2 + |S_{11}\, S_{22} - S_{21}\, S_{12}|^2}{2|S_{21}\, S_{12}|} \tag{5.4}$$

where S_{11}, S_{21}, S_{12}, and S_{22} are the input reflection coefficient, forward transmission coefficient (i.e., gain), reverse transmission coefficient (i.e., isolation), and output reflection coefficient, respectively. The scattering parameters (S-parameters) of a low-noise amplifier are measured using a vector network analyzer (VNA). Figure 5.3 shows the test setup for measuring the S-parameters of a low-noise amplifier using VNA.

Figure 5.3 Test setup for measuring LNA's S-parameters.

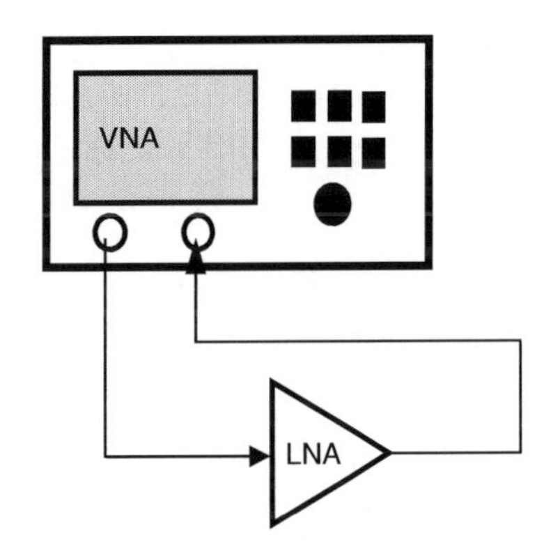

5.3.2 Noise Figure Measurement Methods

There are two primary methods for measuring noise figure, and these methods are called *Y-Factor* and *Cold Noise*.

5.3.2.1 *Y*-Factor Method for Measuring NF

Most noise figure meters and analyzers use the Y-Factor method for measuring noise figures. In this method, a noise source is applied to the device's input, and then the device is powered ON (i.e., hot) and OFF (cold). Each time the device is turned ON and OFF, a power measurement at the output of the device under test (DUT) is performed. The Y-Factor is defined as the ratio of "HOT" to "COLD" measured noise power (in Watts), Y is expressed as

$$Y = \frac{P_{\text{hot}}}{P_{\text{cold}}} \tag{5.5}$$

The term "HOT" refers to the state of the noise source being powered ON and adding noise to the device, and "COLD" refers to the noise source being powered off, but still connected to the input of the DUT. Almost all noise sources in their "OFF" or "cold" state provide a 50-ohm termination to the input of the DUT.

The noise source has an associated parameter called *excess noise ratio* (ENR), which is the power level difference between hot and cold states, compared to the thermal noise power at the standard reference temperature, T_0 (290 K). The *ENR* is given by

$$\text{ENR} = \frac{T_\text{h}}{T_\text{o}} - 1 \tag{5.6}$$

Once the ENR is determined, the noise figure can be estimated using the following equation:

$$\text{NF} = 10\log_{10}\left[\frac{\text{ENR}}{\text{Y}-1}\right] \tag{5.7}$$

where ENR and Y are unitless values. Equation (5.7) can be written as

$$\text{NF}_{\text{dB}} = \text{ENR}_{\text{dB}} - (P_{\text{hot, dBm}} - P_{\text{cold, dBm}}) \tag{5.8}$$

The equivalent noise temperature, T_e, of a device under test can be calculated using the Y-factor and expressed as

$$T_\text{e} = \frac{T_1 - YT_2}{Y-1} - 1 \tag{5.9}$$

and the noise factor can be written as

$$F = 1 + \frac{T_\text{e}}{T_\text{o}} \tag{5.10}$$

where T_o is room temperature 290 K. The noise figure in dB is expressed as

$$\text{NF}_{\text{dB}} = 10\ \log(F)$$

Example 5.1

A Y-factor method is used to measure the equivalent noise temperature of a device with a hot load of $T_1 = 320\,\text{K}$ and a cold load of $T_2 = 77\,\text{K}$. If the measured Y-factor is 0.608 dB, determine the device's noise figure.

Solution

The NF can be calculated using the following equations

$$F = 1 + \frac{T_e}{T_o}, \quad \text{and} \quad T_e = \frac{T_1 - YT_2}{Y - 1} - 1$$

Converting the Y-factor to a linear scale gives $Y = 1.15$, thus,

$$\therefore T_e = \frac{320 - 1.15 \times 77}{1.15 - 1} - 1 = 1.540$$

and

$$F = 1 + \frac{1.54}{290} = 6.31, \quad NF = F_{dB} = 8\text{ dB}$$

5.3.2.2 Cold Noise Method for Measuring NF

The *Cold Noise* method is another technique for measuring noise figures. It relies on measuring just the cold noise power of the DUT when a 50-ohm termination is applied to its input. This method requires measuring the gain of the device under test. Thus, only one measurement (i.e., noise power) needs to be made. Having the gain and noise power, the noise figure is calculated as

$$\mathrm{NF_{dB}} = 10\log_{10}\left[\frac{\mathrm{P_{cold}}}{\mathrm{KTBG}}\right] \tag{5.11}$$

$$\mathrm{NF_{dB}} = P_{\mathrm{cold,dBm}} - (-174\,\mathrm{dBm/Hz}) - 10\log_{10}B - G_{\mathrm{dB}} \tag{5.12}$$

where B is the bandwidth over which the cold noise power is measured. The value of −174 dBm/Hz is the thermal noise power associated with the temperature of 290 K; it is the product of k (1.38×10^{-23} J/K) and T (290 K) converted to logarithmic format in dBm.

5.3.3 Intermodulation Distortion

When two or more signals that are separated in frequency are applied to the input of a low-noise amplifier operating in its nonlinear region due to high-level input signals, the output of the amplifier would contain additional frequency components called *intermodulation products.* To quantify the linearity of an RF amplifier, it is important to determine its one dB compression point. The *one dB compression point* ($\mathrm{P_{1dB}}$) *is* the point at which the output power is one dB below the output produced by a perfectly linear amplifier. Figure 5.4 illustrates the 1 dB compression point of an amplifier.

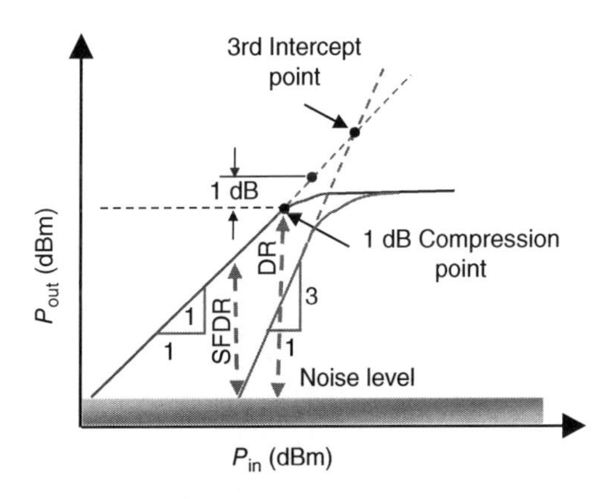

Figure 5.4 Illustration of $\mathrm{P_{1dB}}$ and third-order intercept of an amplifier.

If the input signal $V_{\text{in}}(t)$ to an amplifier is given by

$$v_{\text{in}}(t) = A[\cos(2\pi f_1 t) + \cos(2\pi f_2 t)] \tag{5.13}$$

the output signal $v_o(t)$ of the amplifier can be expressed by a power series and is given by

$$v_o(t) = a_o + a_1 v_{\text{in}}(t) + a_2 v_{\text{in}}^2(t) + a_3 v_{\text{in}}^3(t) + \ldots \tag{5.14}$$

The output signal would have frequency components that include f_1, f_2, $2f_1$, $2f_2$, $f_1 \pm f_2$, $2f_1 \pm f_2$, $2f_2 \pm f_1$, $3f_1$, and $3f_2$, where $2f_1$, $2f_2$ are the 2nd harmonics, $3f_1$, $3f_2$ are 3rd harmonics, $f_1 \pm f_2$ are the second-order intermod, and $2f_1 \pm f_2$ and $2f_2 \pm f_1$ are the third-order intermod. All harmonics beyond the fundamental components represent *harmonic distortion (HD)* that is given by

$$\text{HD} = \frac{\sqrt{\sum_{i=2}^{N} v_i^2}}{v_i} \tag{5.15}$$

where v_i is the voltage of the ith frequency component. Figure 5.5 shows the output spectrum of the second- and third-order two-tone intermodulation products.

From Figure 5.5, the output third-order intercept point is estimated as

$$\text{OIP}_3 = A + \frac{\Delta}{2} \tag{5.16}$$

where "A" is the output signal amplitude in dBm, Δ is the difference between the desired signal and the undesired third-order intermodulation product in dB. The input third-order intercept point is given by

$$\text{IIP}_3 = \text{OIP}_3 - G \tag{5.17}$$

where "G" is the amplifier's gain. Figure 5.6 shows the test setup for measuring the intermodulation products of an amplifier.

In a low-noise amplifier, all the second-order products are undesired and can be easily filtered from the output because they are far from the fundamental frequency components f_1 and f_2. The third-order products cannot be easily filtered from the passband of the amplifier because they are close to the fundamental components. Such intermodulation products distort the output signal; this effect is called *third-order intermodulation distortion*. From Figure 5.4, both the linear and third-order responses exhibit compression at high input powers; because the linear and third-order responses have different slopes, they intersect at a hypothetical point called third-order intercept point (IP3). A practical approximation rule assumes that the IP3 is 12 to 15 dB greater than the one dB compression point. The one dB compression point is a critical parameter because it is one of two parameters which determine the amplifier's dynamic range DR, which is given by

$$\text{DR} = P_{1\text{dB}} - P_{\text{MDS}} \tag{5.18}$$

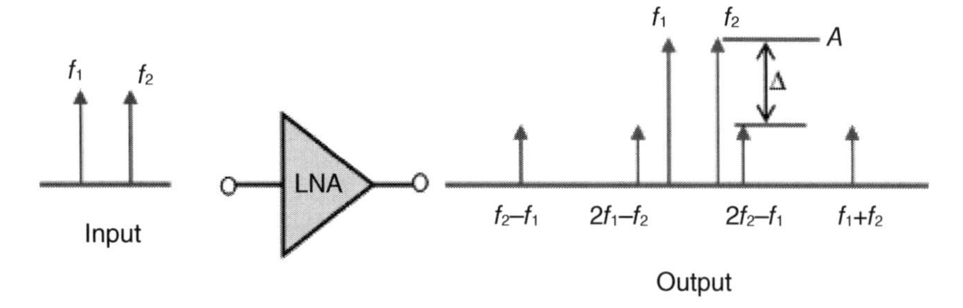

Figure 5.5 Output spectrum of the second- and third-order two-tone Intermodulation products.

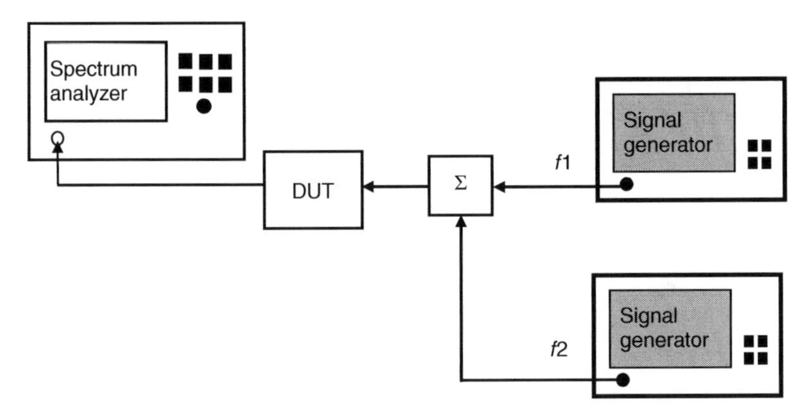

Figure 5.6 Test setup for measuring LNAs third-order intercept point.

where P_{MDS} is the minimum detectable signal; thus, higher P_{1dB} would increase the amplifier's dynamic range. For low-noise amplifiers, the operation is limited by the noise level at the low end and the maximum power level at which the intermodulation distortion becomes unacceptable. This operating range is called *spurious-free dynamic range* and is given by

$$\text{SFDR} = \frac{2}{3}\,(\text{IP3} - P_{\text{MDS}}) \tag{5.19}$$

5.3.4 LNA Performance Parameters

The performance parameters of a low-noise amplifier that should be characterized to ensure they meet the system specifications are:

- Frequency range of operation
- Noise figure
- Gain (S_{21})
- Gain flatness
- Input return loss (S_{11})
- Output return loss (S_{22})
- One dB compression point P_{1dB}
- Third-order intercept point (IP3)
- DC voltage

5.4 RF Mixers

Mixers [1] are RF building blocks needed for up and down frequency conversions in radio receivers and transmitters. A mixer is a three-port device that uses a nonlinear elements (i.e., diodes or transistors) to achieve frequency conversion. An ideal mixer produces an output that consists of the sum and difference frequencies of the two input signals and products of the input frequencies. The operation of RF mixers is based on the nonlinearity of the diodes or the transistors that are used in the mixer's circuit. A nonlinear component generates various harmonics and other products of the input frequencies. Thus, filtering must be used at the mixer's output to filter out the undesired frequency components.

5.4.1 Frequency Conversion

In radio transmitters, the intermediate frequency (IF) signal is connected to the mixer's input port, and the LO (i.e., local oscillator) signal is connected to mixer's LO port as shown in Figure 5.7a. The IF and LO signals are expressed as

$$v_{\mathrm{IF}}(t) = \mathrm{V}_{\mathrm{IF}} \cos \omega_{\mathrm{IF}}(t) \tag{5.20}$$

$$v_{\mathrm{LO}}(t) = V_{\mathrm{LO}} \cos \omega_{\mathrm{LO}}(t) \tag{5.21}$$

The output of an upconverter mixer is the product of the LO and IF signals and is given by

$$v_{\mathrm{RF}}(t) = v_{\mathrm{LO}}(t) \times v_{\mathrm{IF}}(t) \tag{5.22}$$

$$v_{\mathrm{RF}}(t) = \frac{1}{2} V_{\mathrm{LO}} V_{\mathrm{RF}} \left[\cos(\omega_{\mathrm{LO}} - \omega_{\mathrm{IF}})t + \cos(\omega_{\mathrm{LO}} + \omega_{\mathrm{IF}})t\right] \tag{5.23}$$

From Eq. (5.23), the output of the mixer is the sum and difference of the input signal's angular frequencies; thus, the f_{RF} is given by

$$f_{\mathrm{RF}} = f_{\mathrm{LO}} \pm f_{\mathrm{IF}} \tag{5.24}$$

The sum and differences of the IF and LO frequencies are called the sidebands, the $f_{\mathrm{LO}} + f_{\mathrm{IF}}$ is referred to as the upper sideband (USB) and the $f_{\mathrm{LO}} - f_{\mathrm{IF}}$ is the lower sideband (LSB). In a transmitter, a high-pass filter must be used at the mixer's output to pass only the USB signal to be transmitted. In an RF receiver, the mixer is used as down converter to down-convert the RF signal to an IF signal, as illustrated in Figure 5.7b.

5.4.2 Image Frequency

In radio receivers, the RF signal is delivered to the receiver's input via an antenna; thus, other interfering signals may reach the receiver's input and get down-converted. If an interferer signal

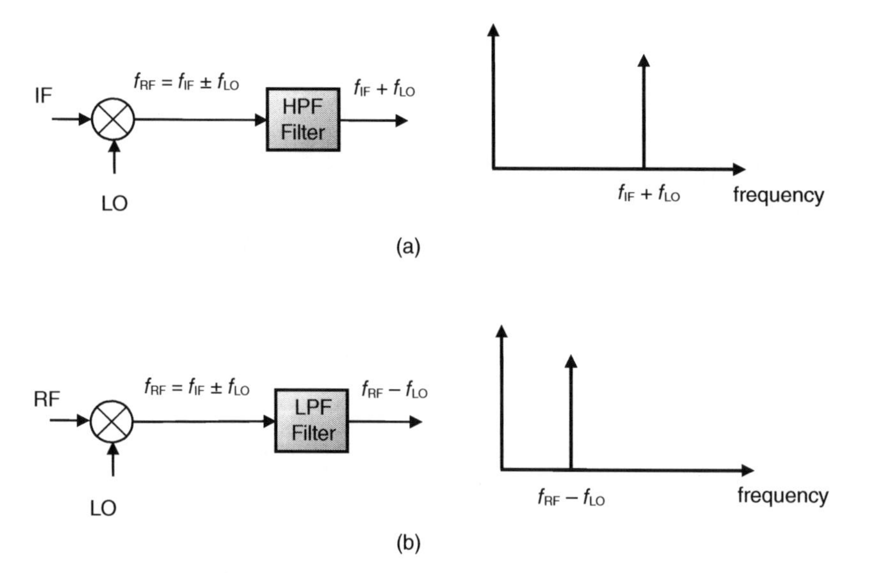

Figure 5.7 Illustration of mixer's frequency conversion. (a) Frequency up-conversion. (b) Frequency down-conversion.

mixes with the LO signal, resulting in a frequency equal to the desired IF, the frequency of the interferer will be called the image frequency f_{im} and is given by

$$f_{im} = f_{LO} - f_{IF} \tag{5.25}$$

The image response is critical because the down-converted image frequency is indistinguishable from the desired down-converted signal. Image rejection mixers enable overcoming the image frequency problem.

5.4.3 Image Rejection Mixers

Image rejection mixers utilize phasing techniques to cancel out the unwanted mix products by using two balanced mixers and quadrature (90°) hybrids, as shown in Figure 5.8.

5.4.4 Double-Balanced Mixers

A double-balanced mixer has both its inputs applied to differential circuits, so neither of the input signals appears at the mixer's output (i.e., the input RF and local oscillator signals are balanced out), and only the product signal appears at the output. Figure 5.9 shows a schematic diagram of a double-balanced mixer that uses four diodes.

Double-balanced mixers require higher LO drive levels than single-balanced mixers. These mixers have better suppression of spurious products (all even order products of the LO and RF inputs are suppressed); also, they have better isolation between all ports in comparison with a single-balanced mixer (i.e., a single device such as a diode or transistor). Figure 5.10 shows a commercial mixer module and surface-mount mixer package.

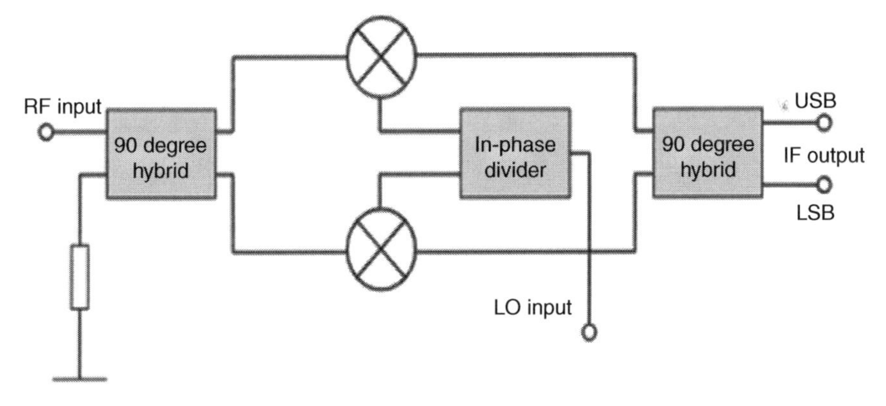

Figure 5.8 Block diagram of an image rejection mixer.

Figure 5.9 Schematic diagram of a double-balanced mixer.

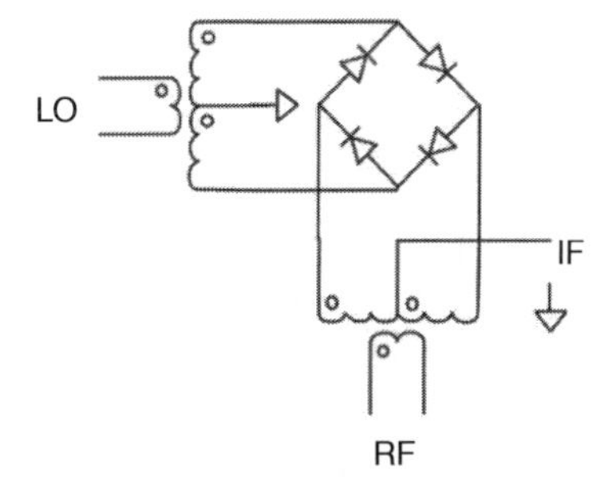

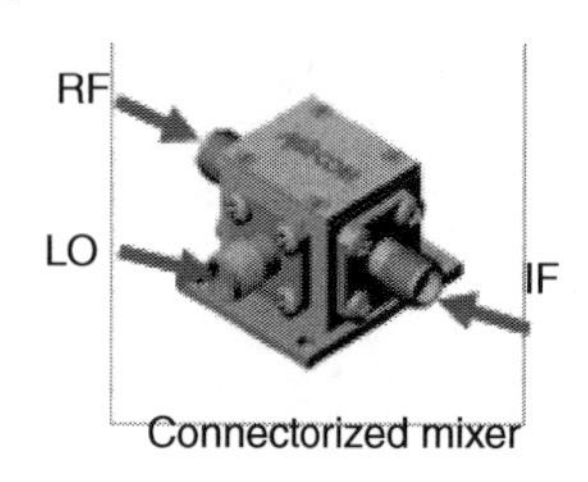

Figure 5.10 Commercial mixer products. Source: Mouser Electronics, Inc.

Mixer modules are used to develop proof-of-concept prototypes, whereas surface-mount components are used to develop system-on-board. Also, mixers are developed on chips as part of a system-on-chip (SoC).

5.4.5 Mixer's Conversion LOS

A figure of merit for an RF mixer is the conversion loss because it impacts the gain of the transmit or receive chains in a radio transceiver. In a radio receiver, the conversion loss (L_c) is given by

$$L_c = 10\log_{10}\left(\frac{\text{RF input power}}{\text{IF output power}}\right) \tag{5.26}$$

The conversion loss should be small because it impacts the receiver noise figure and sensitivity. Transistor mixers have conversion gain and better noise figure performance than diode mixers. The conversion loss is impacted by the local oscillator power.

5.4.6 Mixer's Noise Figure

The noise at the output of a mixer is due to the conversion of the input noise at the RF frequency and the image frequency. If N_i is noise at the mixer's input and N_A is noise added by the mixer, the total noise at the mixer's output (N_o) can be expressed as

$$N_o = GN_i + N_A \tag{5.27}$$

The noise factor of a mixer is defined as the ratio of the total noise at the mixer's output to the noise at the mixer's output due to the input signal. Thus,

$$\text{Noise Factor} = \frac{GN_i + N_A}{GN_i} \tag{5.28}$$

where G is the conversion gain of the mixer; the noise factor in dB (i.e., noise figure) is given by,

$$\text{NF} = N_{o(\text{dB})} - GN_{i(\text{dB})} \tag{5.29}$$

The noise figure of SSB (single-sideband) mixer is 3 dB higher than that of the DSS (double-sideband) mixer.

The LO power of a mixer determines the dynamic range of the mixer. Spurious performance of a mixer is influenced by the LO power level; a high LO drive improves the IP3 but at the expense of degrading the noise performance, which impacts the performance of radio receivers.

5.4.7 Mixer's Performance Parameters

The mixer's parameters that impact the system performance are:

- Operating frequency range
- IF bandwidth

- Conversion loss/gain
- Noise figure
- Isolation between ports LO-RF, LO-IF, and RF-IF
- LO VSWR, IF VSWR, and RF VSWR
- Input third-order intercept point IP3
- Input one dB compression
- Temperature range

Conversion Loss is a measure of the mixer's frequency conversion efficiency expressed in dB. It is the ratio of the IF output power to the input RF signal power and depends on the impedance match at the RF and IF ports and the mixer type.

Isolation is a measure of the feed through between the mixer ports including LO-to-RF, LO-to-IF, and RF-to-IF. High isolation provides low leakage.

Conversion Compression is a measure of the maximum RF input level over which the mixer will have a constant conversion gain. The compression point of the mixer depends on the LO drive level.

Dynamic Range is a measure of the power range over which the mixer provides a useful operation. The upper limit of the dynamic range is limited by the conversion compression point, while the lower limit is limited by the mixer's noise figure.

Third-Order Intercept Point IP3 is a measure of the mixer's ability to convert the RF signal with minimum distortion. Higher IP3 is an indication of better linearity and reduces intermodulation spurious; this can be obtained by using a high level of LO power. The IP3 of a mixer is typically 4–5 dB above the LO power.

Noise Figure, for a passive mixer, the noise figure NF can be assumed equal to the conversion loss. However, for active mixers, the noise figure and the conversion loss are independently related.

5.5 Filters

Filters [2] are used in communication systems to control the frequency response by selecting the desired signals and attenuating undesired signals. In radio receivers, one fundamental filter function is noise rejection at the receiver input. Filters set the receiver sensitivity by limiting the total noise power seen by the LNA and subsequent RF blocks. Also, they are used to attenuate the undesired mixer's frequency products and set the IF bandwidth of the receiver. In transmitters, filters are used to set the desired sidebands, limit the transmitted signal's bandwidth, and attenuate the mixers' spurious response. Filters are described by their frequency characteristics, such as low pass, high pass, band pass, or band stop (band reject). Figure 5.11 shows ideal and practical filter responses of a bandpass filter.

5.5.1 Low-Pass Filters

A low-pass filter (LPF) passes low frequencies up to a certain cutoff frequency and attenuates any frequencies beyond the cutoff. The cutoff frequency f_c is defined as the frequency at which the filter's response drops by 3 dB from the insertion loss value. An LPF can be achieved using a lossless *L–C* network or constant impedance structure, as shown in Figure 5.12. Figure 5.12 shows some commercial products of low-pass filters and the filter's frequency response. From Figure 5.12, at low frequencies, the impedance of the series inductor ($X_L = j\omega L$) is low, while the impedance of

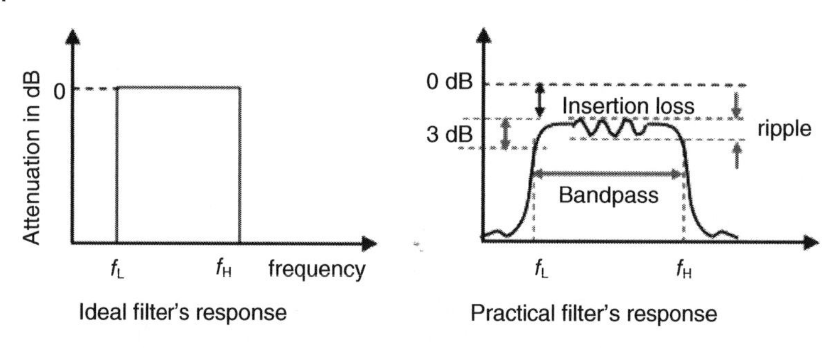

Figure 5.11 Filter response characteristics.

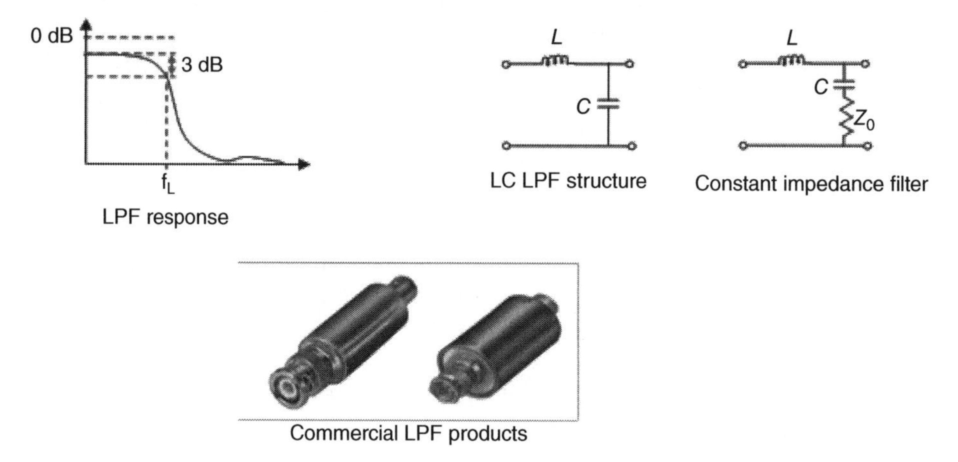

Figure 5.12 LPF structure and commercial LPF products.

the parallel capacitor ($X_c = 1/j\omega C$) is high. At high frequencies, the impedance of the series inductor becomes high; therefore, it attenuates the signal, and the impedance of the capacitor becomes low and short-circuits what is left of the signal.

5.5.2 High-Pass Filters

A high-pass filter (HPF) passes frequencies above the cutoff frequency and attenuates frequencies below the f_c. Figure 5.13 shows the structure of an HPF and its frequency response. In this structure, the series capacitor has high impedance at low frequency, while the inductors have low

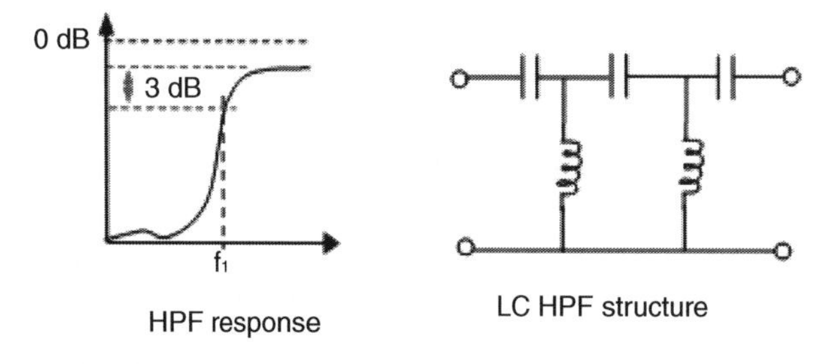

Figure 5.13 HPF structure and its frequency response.

impedance. Therefore, the signal becomes attenuated by the first capacitor and short-circuited by the first inductor; the situation is reversed at high frequency.

5.5.3 Bandpass Filters

A bandpass filter (BPF) only passes frequencies within its passband and rejects all the frequencies outside it. Figure 5.14 shows the structure of a BPF and its frequency response. In Figure 5.14, frequencies below the passband are rejected by the series capacitor and short-circuited by the parallel inductor, which has low impedance. Frequencies above the passband are rejected by the series inductor (high impedance) and short-circuited by the parallel capacitor (low impedance).

5.5.4 Bandstop Filters

A bandstop filter passes frequencies outside its passband and rejects frequencies in its passband. It is constructed with shunt-connected series resonant circuits and series-connected parallel resonant circuits. These filters are called band rejects or notch filters. Figure 5.15 shows the structure of a bandstop filter and its frequency response.

5.5.5 Cavity Resonator Filters

A cavity filter is an RF filter that operates on the principle of resonance. It is a resonator with a "tuning screw" (to fine-tune the frequency) inside a "conducting box." An RF or microwave resonator is a closed metallic structure (i.e., waveguides with both ends terminated in a short circuit). The resonator oscillates with higher amplitude at a specific set of frequencies referred to as resonant frequencies. When an RF signal passes through the cavity filter, a resonator acts as a bandpass filter and passes RF signals at the resonant frequencies while blocking other frequencies. Figure 5.16 shows a photo of a cavity resonator filter.

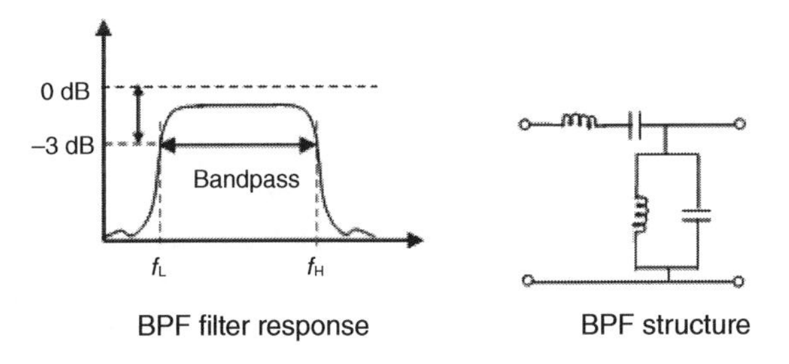

Figure 5.14 BPF structure and frequency response.

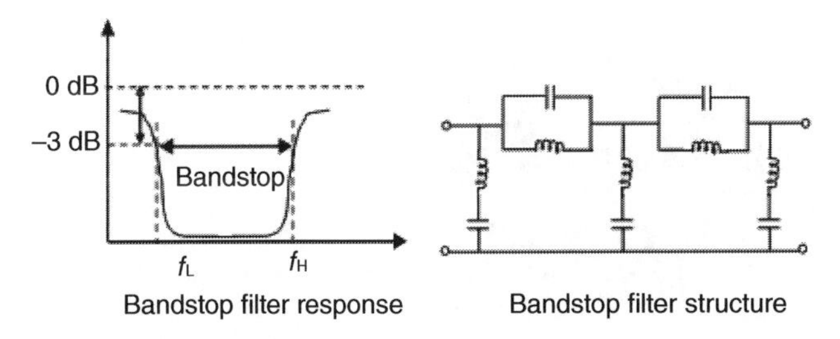

Figure 5.15 Bandstop filter structure and its frequency response.

Figure 5.16 Cavity resonator filter. Source: everything RF.

The resonant frequency of the cavity resonator depends on its dimension (length, width, height), mode number, dielectric constant (ε_r), and magnetic permeability (μ_r) of the filter's material. The resonator is fitted with a screw to tune the frequency by modifying the physical length (inner space length) and its capacitance to the ground.

Cavity resonator filters are commonly used in microwave and millimeter-wave systems that need filters with high-Q factor, lower insertion loss, and temperature stability. They are used in 5G NR base station radios. More than one cavity filter is grouped in series with each other to increase filter effectiveness.

5.5.6 Duplexers

Duplexers are two combined bandpass filters (transmit filter and receive filter) in a single package. These filters are used in wireless communication systems to allow the use of a single antenna for both the transmit and receive operation. The isolation between the transmit and receive ports of a duplexer is a crucial performance parameter to protect the receiver from the carrier leak and to prevent receiver desensitization that increases the system bit-error rate. Figure 5.17 shows a simplified block diagram of a microwave transceiver system and a picture of the COTS duplexer.

The duplexer's performance parameters include operating frequency, return loss at each port (i.e., S_{11}, S_{22}, and S_{33}), isolation between the transmit and receive ports (i.e., S_{23}), insertion loss between the antenna and transmit ports (i.e., S_{13}) which impacts the transmit power, insertion loss between the antenna and receive ports (S_{12}) that influences the receiver noise figure and the receiver sensitivity, transmit filter's out-of-band rejection to meet the requirements for spurious emissions, and the power handling capability. The duplexer can be characterized using a VNA. When two ports of the duplexer are connected to the VNA, the third duplexer's port must be terminated.

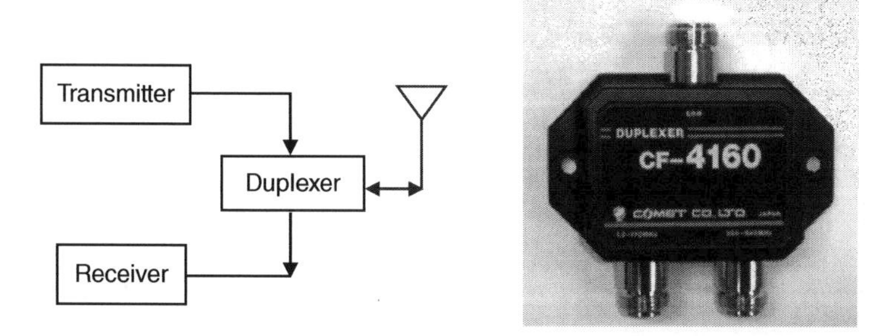

Figure 5.17 A block diagram of a transceiver and a picture of a COTS duplexer. Source: ML&S.

5.5.7 RF Filter's Specifications

The following are the filter's parameters that impact the system performance:

- Frequency range
- Insertion loss
- Attenuation in the out-of-band
- Return loss
- In-band ripple
- Group delay
- Input power handling
- Input and output Impedances
- Temperature range

Insertion loss: is the amount of the signal attenuation in the filter's passband, measured in dB.

Shape factor: It is the ratio of the bandwidth measured at an attenuation of 60 dB greater than the insertion loss to the bandwidth measured at 6 dB greater than the insertion loss. The steeper the attenuation curve of the filter, the closer the shape factor will be to 1.

Ripple: It is the maximum deviation of the attenuation from the insertion loss in the filter's passband.

Out-of-band rejection: It is the measure of the attenuation outside the passband of the filter.

Group delay: It is the change in phase with respect to the frequency. The group delay impacts the bit-error rate of radio systems and becomes more severe at the edges of the filter, and it is measured in nanoseconds.

5.6 Frequency Synthesizers

Frequency synthesizers are the heart of any wireless communication system. They generate the oscillator signals that are needed for up and down frequency conversions in the radio front end of RF transceivers. They have the advantage of being controlled by microprocessors. A frequency synthesizer is basically a *phase-locked loop* (PLL) and frequency divider. Another type of synthesizer is called *Direct Digital Synthesizer* (DSS). A DSS synthesizer has limited frequency ranges, but it could be used as part of a hybrid DDS/PLL scheme to increase the frequency range.

5.6.1 Phase-Locked Loop (PLL) Frequency Synthesizer

A PLL is a closed-loop frequency control system which uses feedback to allow a voltage-controlled oscillator [1] to precisely track the phase and frequency of a stable reference oscillator. When PLL is combined with a feedback frequency divider, it forms a frequency synthesizer. Phase-locked loops have excellent frequency accuracy and phase noise performance, which are required in high-level modulation such as 16-QAM and 64-QAM. Figure 5.18 shows a basic block diagram of a phase-locked loop frequency synthesizer.

In Figure 5.18, the phase detector compares the signal from the external reference oscillator with the feedback signal from the voltage-controlled oscillator and provides an output (i.e., error) signal, which is proportional to the phase difference between its two input signals. The loop filter is a low-pass filter that passes the dc signal only to make corrections in the frequency of the voltage-controlled oscillator (VCO) and to align the phase of the VCO with that of the reference

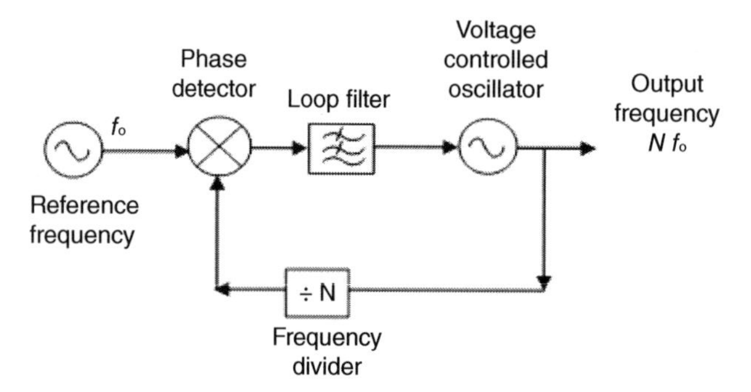

Figure 5.18 Block diagram of a phase-locked loop frequency synthesizer.

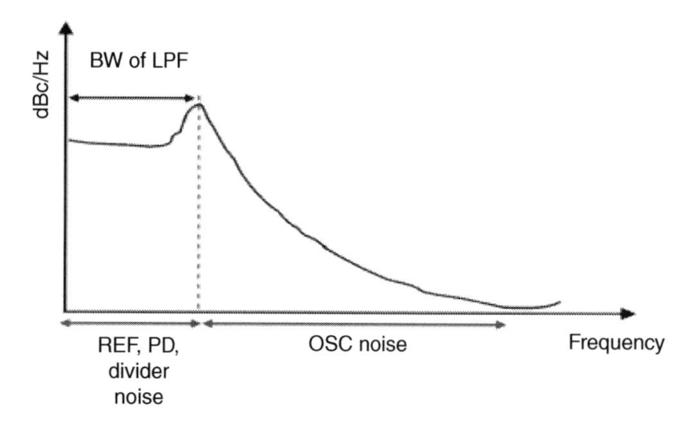

Figure 5.19 Illustration of noise performance of a PLL frequency synthesizer.

oscillator. The VCO's output is divided by N to match the frequency of the reference oscillator and to allow the PLL to track the input signal in frequency (reference frequency) and to be locked on it. The output of a PLL has phase noise characteristics similar to the reference oscillator but at a higher frequency. Figure 5.19 shows the noise performance of a PLL frequency synthesizer.

The key characteristics of a PLL include the *capture range*, which is the range of input frequency over which the loop can acquire locking; the *lock range,* which is the input frequency range over which the loop remains locked; the *settling time,* which is the time required for the loop to lock on to a new frequency. The performance parameters of a PLL are:

- Frequency range
- Frequency step size
- Settling time
- Output power
- Output phase noise
- Harmonic suppression
- Reference frequency
- Power supply
- Output impedance

The phase noise that results from the PLLs or oscillators could severely degrade the system's bit-error rate performance. Phase noise describes rapid, short-term, and random fluctuations in the frequency (or phase) of an oscillator signal. Such fluctuation introduces uncertainty in the detection process of digitally modulated signals. The factors that contribute to the phase noise include leakage of the reference frequency to the VCO, and the harmonics of the VCO. The single sideband (SSB) phase noise $L_{\phi}(f_{\mathrm{m}})$ is defined as the ratio of the power in a 1 Hz bandwidth at a frequency f_m far from the carrier to the power in the carrier itself, and it is expressed in dBc/Hz.

5.7 RF Oscillators

RF oscillators [1] are used to generate RF signals that are required for mixers or phase-locked loops. An LC resonant circuit can be used to generate a sinusoidal signal, which decays exponentially due to the internal losses of the resonator. Figure 5.20 shows an LCR resonator and its oscillation signal.

In Figure 5.20, when the capacitor C is charged to a voltage V_{c}, if the switch is connected to the inductor L, the capacitor C will discharge through the inductor, and consequently, current will flow in the inductor, causing a magnetic field to be generated. After the capacitor is discharged and no more current flows, the magnetic field will start to collapse and induce a current in the inductor. The induced current will cause the capacitor to be charged with the opposite polarity of its original charge. After the capacitor is charged again and no more current flows, the capacitor will start to discharge through the coil, causing the generation of magnetic field again. The repetitive process produces a damped oscillation, as shown in Figure 5.20. The resonance frequency f_{r} is given by

$$f_{\mathrm{r}} = \frac{1}{2\pi\sqrt{LC}} \tag{5.30}$$

where L and C are the inductance and capacitance of the resonance circuit, respectively. For the resonant circuit to function as an oscillator, an amplifier is needed to restore the lost energy and to provide constant amplitude output. In practice, oscillators do not need a switch because the positive feedback supports continuous oscillation. Figure 5.21 shows a generalized functional block diagram of an oscillator, which includes an amplifier with gain $A(f)$, and a feedback network with gain $\beta(f)$.

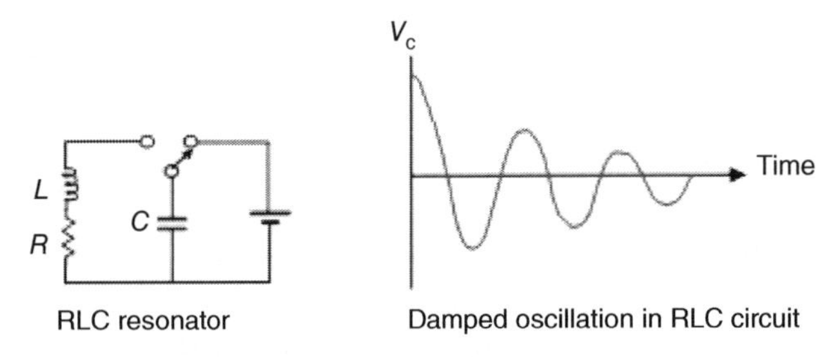

Figure 5.20 Illustration of RLC resonator.

Figure 5.21 Generic block diagram of an oscillator.

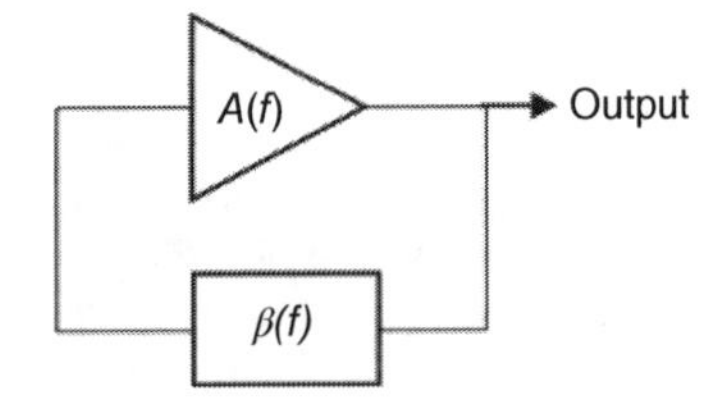

Any gain block can be made to oscillate if it has positive feedback that meets the Barkhausen criteria. This criterion includes the following:

- Loop gain $A(f) \times \beta(f) = 1$
- Loop phase shift $= n \times 360$, $n = 1, 2, 3, \ldots$

Any initial noise or surge of dc power in the oscillator circuit results in growing positive feedback, which creates a sinusoidal voltage in the resonant circuit at its resonant frequency. This voltage gets amplified, feedback to the input, and amplified repeatedly until the amplifier reaches operation in the saturation region, resulting in constant output. The transfer function of the oscillator system shown in Figure 5.21 is given by

$$\frac{v_o(t)}{v_i(t)} = \frac{A(f)}{1 - A(f) \times \beta(f)} \tag{5.31}$$

If the denominator of a transfer function of a system becomes zero (i.e., $A(f) \times \beta(f) = 1$), the system will oscillate. The choice of an oscillator depends on the following:

- Frequency of operation
- Frequency stability
- Output power

5.7.1 Oscillator's Frequency Stability

Frequency stability is an essential parameter in communication systems. In radio transmitters, if the frequency of a transmitter is not stable, that could result in harmful interference to other systems. Frequency stability is classified as short-term (due to fluctuation in dc supply voltage), and long-term (due to component aging and change in the temperature and humidity). The frequency or phase stability of an oscillator is usually considered in the long-term stability case. One should never underestimate the effects of these variations on the frequency of operation. The higher the circuit Q (quality factor) of the oscillator, the better the ability to filter out undesired harmonics and noise. The phase noise can be reduced by maximizing the Q of the tank circuit and choosing an active device with a low-noise figure. The frequency stability is quantified in terms of *parts per million* (ppm); for example, if the stability is $\pm 0.001\%$ which is $\pm 10/10^6$, this can be expressed as ± 10 parts per million. Frequency stability is a key system performance parameter in wireless communication systems.

5.7.2 Voltage-Controlled Oscillator

A VCO is an oscillator that can change its output frequency by adjusting the control voltage; it is used in phase-locked loop circuits. In a voltage-controlled oscillator, the frequency-determining element is a voltage-sensitive device such as a varactor diode whose capacitance changes as a function of the control voltage. Hence, the change in the control voltage results in a change in the frequency of the oscillator. The variation of the frequency as a function of the control voltage is given by

$$k = \frac{\Delta f}{\Delta v} \tag{5.32}$$

where k is the VCO sensitivity (Hz/volt), Δf is the change in frequency, and Δv is the change in the control voltage that causes change in frequency Δf. Figure 5.22 shows a commercial voltage-controlled oscillator module.

Figure 5.22 Voltage-controlled oscillator module.

The performance of a VCO is limited by the following factors:

- ***Frequency pushing***: This defines the change in the output frequency of the VCO due to the change in the supply voltage.
- ***Load pulling***: This describes the change in the output frequency of the VCO due to the change in the load of the VCO.
- ***Frequency drift***: This is the change in the output frequency of the VCO due to the change in the control voltage.

5.7.3 VCO Performance Parameters

The performance parameters of a VCO are:

- Frequency range
- Tuning voltage
- Tuning sensitivity
- Output phase noise
- Harmonic suppression
- Output power
- Operating temperature
- Output impedance
- Frequency pushing
- Frequency pulling

5.8 RF Power Amplifiers

Power amplifiers (PA) [4–6] are used as the final stage in radio transmitters to increase the signal output power before it is transmitted by the antenna. They impact the system's communication range; Figure 5.23 shows a commercial power amplifier module.

Figure 5.23 RF power amplifier module. Source: eBay Inc.

Table 5.1 Common power amplifier classifications.

Class	Max efficiency (%)
A	50
B	78.5
AB	50–78
C	>78

The linearity of a PA is a key characteristic, particularly when the signal to be transmitted contains both amplitude and phase modulation. Nonlinearities distort the amplified signals, resulting in considerable out-of-band radiation and interference to adjacent channels, which degrade the system performance. The efficiency η of a power amplifier is given by

$$\eta = \frac{P_{\text{out}}}{P_{\text{dc}}} \times 100\% \tag{5.33}$$

where P_{out} is the RF output power, and P_{dc} is the dc powers. A better measurement of the power amplifier efficiency is that takes into consideration the effect of the input power, which is referred to as power added efficiency (PAE); PAE is given by

$$\text{PAE} = \frac{(P_{\text{out}} - P_{\text{in}})}{P_{\text{dc}}} \times 100\% = \eta\left(1 - \frac{1}{G}\right) \tag{5.34}$$

where P_{out} and P_{in} are the output and input RF power, respectively, and G is the power gain of the amplifier.

It is important to quantify the adjacent channel power ratio (ACPR), which is a measure of the spectral regrowth of the power amplifier; ACPR is given by

$$\text{ACPR} = \frac{\text{Power spectral density in the main channel}}{\text{Power spectral density in the offset channel}} \tag{5.35}$$

Power amplifiers are classified into different types or classes such as A, B, AB, C, D, E, and F. In any amplifier class, there is a trade-off between the efficiency and linearity, higher efficiency means degraded linearity, and vice versa. High linearity in power amplifiers can be achieved by implementing linearization techniques. Table 5.1 summarizes the efficiencies of Class A, B, AB, and C amplifiers.

5.9 Power Amplifier Linearization Techniques

Power amplifiers are indispensable building blocks in any wireless communication system, and they are inherently nonlinear. The amplifier's nonlinearity produces spectral regrowth, which leads to adjacent channel interference and violations of the out-of-band emission requirements that are mandated by the regulatory. Also, nonlinearity causes in-band distortion, which degrades the system's bit-error-rate (BER) performance. To minimize the nonlinearity, the power amplifier can be backed off to operate within its linear range, but this impacts its efficiency. To improve

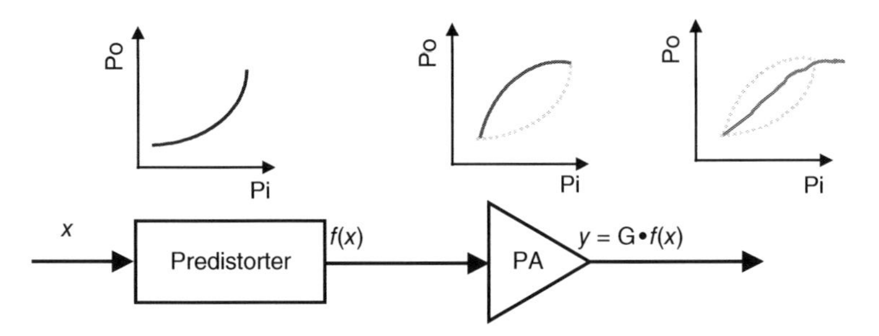

Figure 5.24 Illustration of linearization process.

the power amplifier efficiency without compromising its linearity, power amplifier linearization should be used.

Different linearization techniques [4–7] are used to improve the power amplifier's linearity and efficiency in mobile communication systems. The common linearization methods that are used in wireless communication systems are:

- Predistortion (analog and digital)
- Feedforward
- Cartesian feedback

Linearization is a corrective action to provide an opposite reaction to the nonlinearity to compensate for the nonlinearity effect. The correction can be made before or after the power amplifier, and is referred to as predistortion or postdistortion. Figure 5.24 illustrates the basic concept of predistortion linearization process.

In Figure 5.24, the predistorter block has a characteristic that is opposite to the characteristic of the power amplifier; thus, when the output of the predistorter is applied to the input of a power amplifier the output should be linear. The predistortion technique requires knowing the power amplifier characteristics.

5.9.1 Feedforward Linearization Method

The *feedforward linearization* is based on detecting and amplifying the nonlinearity and then subtracting this nonlinearity from the output of the power amplifier. Figure 5.25 shows a simplified block diagram of the feedforward linearization technique. The principle is simple, but it is hard to implement because the second amplifier must be perfectly linear.

Figure 5.25 Simplified block diagram of a feedforward linearization.

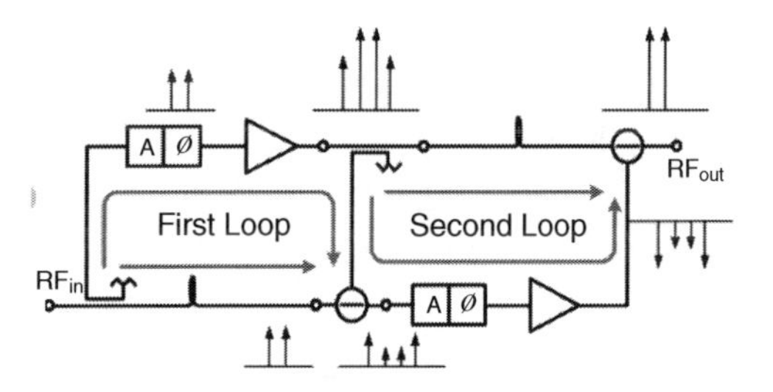

The linearization system shown in Figure 5.25 involves two cancellation loops; the purpose of the first loop (*error loop*) is to sample the distortion introduced by the first amplifier, and the second loop (*distortion cancellation loop*) uses the distortion sample obtained from the error loop to subtract the distortion component from the amplified input signal. In this ideal case, the output signal will have the input signal amplified by the power amplifier but without any distortion.

5.9.2 Power Amplifier Performance Parameters

The power amplifier performance parameters include the following:

- Frequency range
- Power gain
- Output power at 1 dB compression
- Input VSWR
- Output third-order intercept point
- Operating temperature

5.10 Circulators/Isolators

Circulators are three-port devices where the power flows in only one direction, from port #1 to port #2, port #2 to port #3, and port #3 to port #1. They are used in radio transceivers to allow the use of a single antenna for both transmit and receive operation. Figure 5.26 shows a photo of a commercial circulator and a block diagram of a transceiver frontend using a circulator.

Circulators use the property of ferrite material, which rotates the polarization of the signal's electric field when it paths through it. If the signal entering a circulator is rotated to become perpendicular to a resistive element inside the circulator, the signal will path through with minimal insertion loss. At the same time, the reflected signal in the opposite direction will be rotated to become parallel to the resistive element, resulting in significant attenuation. Typically, the reflected signal could be attenuated by 25 dB.

If one port of the circulator is connected to the ground via a load equal to the port's characteristic impedance, the circulator becomes an isolator. Isolators are 2-port devices that only pass the RF power in one direction while blocking the signal in the opposite direction. They are used to prevent any reflection from a load impedance from reaching the source. Figure 5.27 shows a commercial RF isolator module.

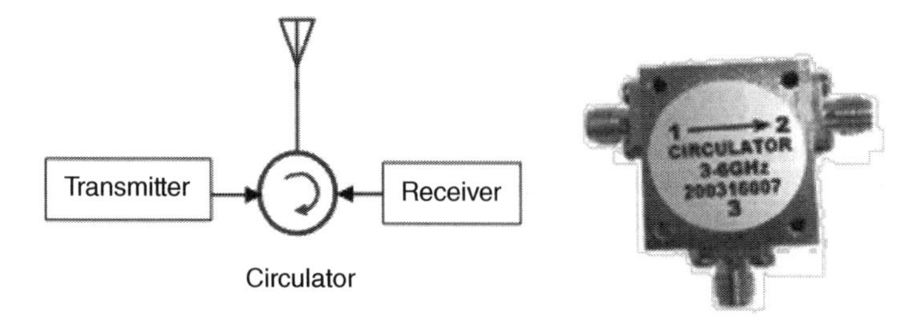

Figure 5.26 Photo of a commercial circulator and a diagram of a transceiver frontend using a circulator. Source: Infinite Electronics, Inc.

Figure 5.27 RF isolator module. Source: NWS.

The performance parameters of the circulators and isolators include, power handling, isolation, frequency of operation, and insertion loss.

5.11 Directional Couplers

Directional couplers are 4-port passive devices that are used to sample a small amount of input signal power for measurement purposes. Also, they are used to sample the output of power amplifiers in transceivers to control the output power level. They have low insertion loss and high directivity (isolation). Figure 5.28 shows a microstrip directional coupler and a waveguide directional coupler. In directional couplers, Port 1 is referred to as the input port, port 2 is the output port, port 3 is the coupled port, and port 4 is the isolated/terminated port.

A microstrip coupler consists of two sections called the main and coupler and is separated by a distance S in wavelength; the spacing distance between the main and coupler sections determines the amount of coupling between the ports. The two sections form the 4-port device. The length of the couple line L must be an odd multiple of $\lambda/4$ at the center frequency of operation.

In a directional coupler, the signal flows in the main section from port 1 to port 2 with low insertion loss. The signals at ports 2, 3, and 4 are measured relative to the signal of port 1. The directivity of a directional coupler is an indication of its ability to allow the signal to flow in only one direction. The performance parameters of the coupler are the power handling capabilities, frequency of operation, coupling factor, and directivity.

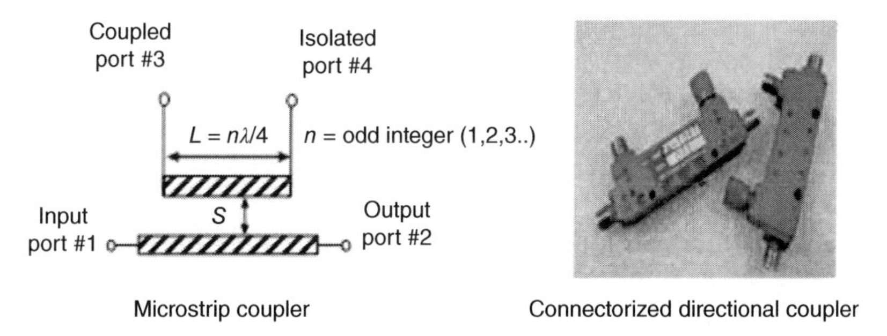

Figure 5.28 Directional couplers. Source: eBay Inc.

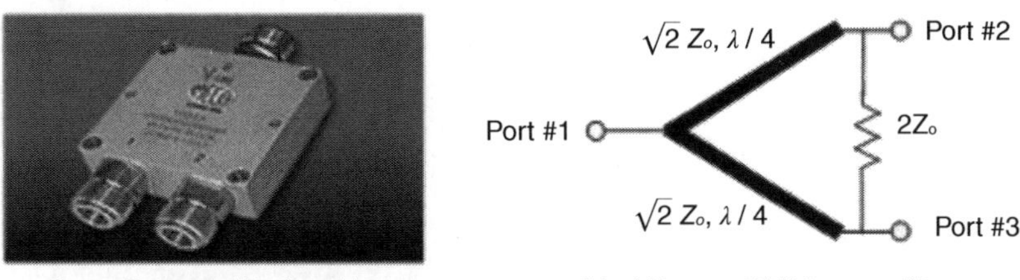

Figure 5.29 Photograph of a commercial power splitter/combiner and schematic of a Wilkinson splitter. Source: MECA Electronics.

5.12 Power Splitter/Combiner

Power splitters/combiners are indispensable building blocks in radio systems because of their use in many circuits including frequency distribution circuits, power amplifiers, and antenna feed networks. A power splitter is a device used to equally split the signal applied to the input port between the n output ports. Also, it could combine the signals from the n-ports and provide the combined signal at the summing port. The isolation between the ports and the phase balance between the ports are essential characteristics of combiner/splitter devices. Figure 5.29 shows a commercial power splitter/combiner, and an ideal Wilkinson splitter circuit.

The performance parameters of a power splitter/combiner include the frequency range of operation, return loss at each port, insertion loss from the input port to output ports, amplitude and phase imbalance between the output ports, and power handling. The device can be characterized using a vector network analyzer; the device's port that is not connected to the analyzer must be terminated.

5.13 Attenuators

Attenuators have many applications in RF/microwave systems. In radio transmitters, they are used to adjust the cascaded gain of the transmit chain and to attenuate any reflected signal from one stage to another in the chain. The power handling capabilities, frequency of operation, and attenuation are the main characteristics of the attenuator. Figure 5.30 shows commercial attenuator products.

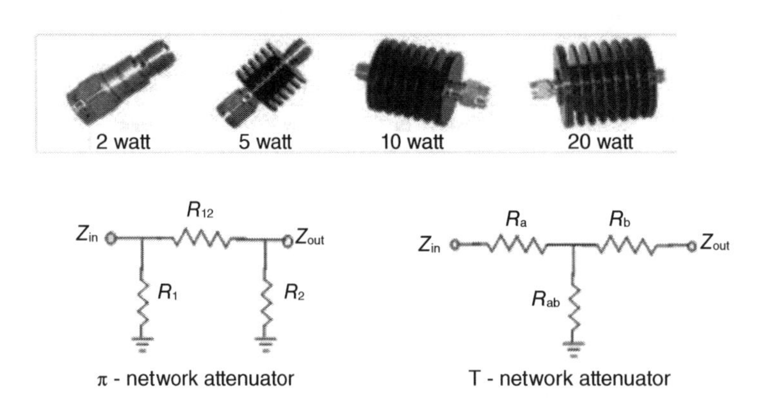

Figure 5.30 Commercial attenuator products.

In radio frontends, attenuators are implemented using surface-mount resistive components with π-networks or T-networks, as shown in Figure 5.30. The values of a π-network components can be calculated using the following equations:

$$R_1 = R_2 = \frac{1}{\left[\frac{10^{L/10}+1}{Z_i(10^{L/10}-1)} - \frac{1}{R_{12}}\right]} \tag{5.36}$$

$$R_{12} = \frac{1}{2}(10^{L/10} - 1)\sqrt{\frac{Z_i Z_o}{10^{L/10}}} \tag{5.37}$$

where

Z_i = input impedance of the network
Z_o = input impedance of the network
R_{12} = is the series resistance of the pi-network
$R_1 = R_2$ = is the shunt resistances of the pi-network

5.14 RF Phase Shifters

Phase Shifters [8] are critical components in RF, Microwave, and mmWave systems. They are widely used in beamforming, phased array antenna, and multiple-input multiple-output (MIMO) systems. Tunable phase shifters are used with analog/digital attenuators in phased array beamforming networks, where the insertion loss and phase deviation must be as small as possible to steer the lobes and nulls of antennas. A phase shifter is a two-port device, which changes the transmission phase angle. There are two types of phase shifters: analog phase shifters and digital phase shifters.

Analog phase shifters can be implemented using varactor diodes, whereas digital phase shifters are implemented using PIN diodes or MISFET devices and they are controlled by two-state switches. Digital phase shifters are more immune to noise, Figure 5.31 shows a phase shift module.

For a 6-bit digital phase shifter as represented in Figure 5.32, if the first bit "A" is logic "1" and the rest of the bits are logic "0," the input signal will experience 5.625 degrees.

If bits "E" and "F" are set to logic 1, and the rest of the bits are set to zeros, the input signal will experience 270 degrees phase shift. Figure 5.33 shows a diagram for the application of phase shifters in phased array antennas.

Figure 5.31 RF phase shifter module. Source: Analog Devices, Inc.

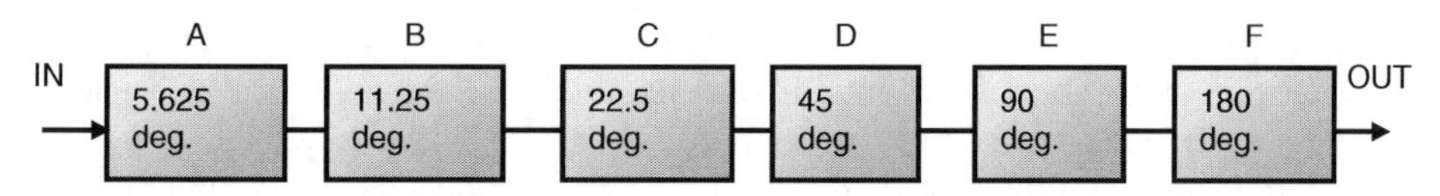

Figure 5.32 6-Bit phase shifter.

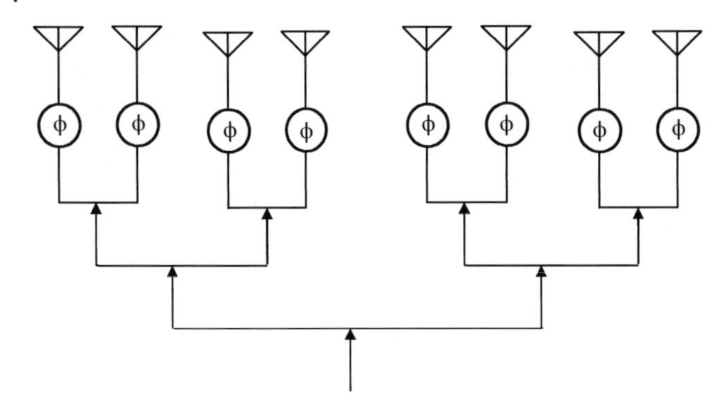

Figure 5.33 Application of phase shifters in phased array antenna.

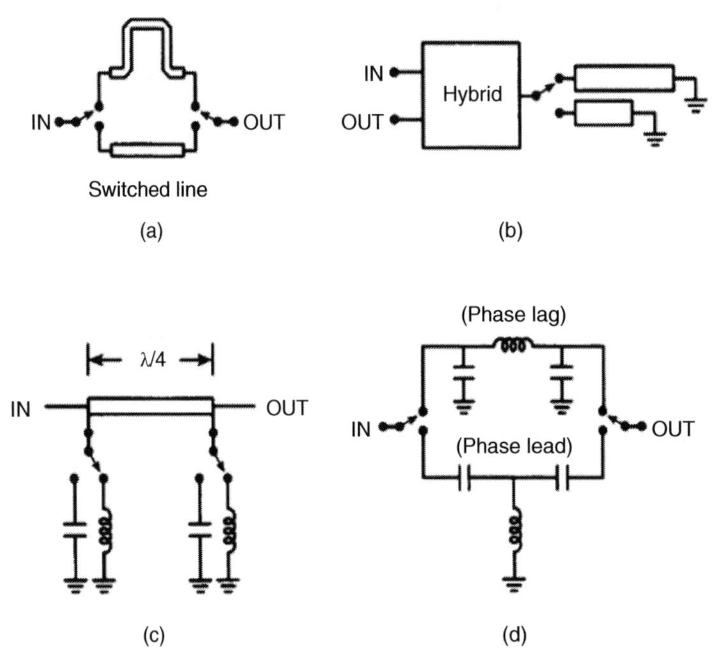

Figure 5.34 Basic types of phase shifters. (a) Switched line phase shifter. (b) Hybrid coupler phase shifter. (c) Loaded transmission line phase shifter. (d) High-pass/low-pass filter phase shifter.

The performance specifications of a phase shifter include the phase range (degrees) and phase accuracy, insertion loss, input IP3, input/output return losses, operating frequency, and amplitude variation with frequency. Figure 5.34 shows four basic types of phase-shifters, these are switched line, reflection (hybrid coupler), loaded-line, and low-pass/high-pass configurations. Figure 5.34a is a switched line phase shifter, which depends only on the lengths of the lines used. The lower transmission line is labeled as a "reference" line, and the top line as a "delay" line. The phase shift is the difference in electrical lengths of the two lines. The advantage of this type is that the phase shift will be approximately a linear function of frequency. The phase shift created is dependent only on the length of the transmission lines, which makes the Phase Shifter very stable over time and temperature. In Figure 5.34b, the hybrid coupler phase shifter provides a 90 degree phase shift. Figure 5.34c is a loaded transmission line phase shifter. By changing the load on the transmission line, its electrical length changes which represents change in the phase shift. Figure 5.34d is a high-pass/low-pass filter phase shifter. In the low-pass filter, the output of the filter lags the input, whereas in the high-pass filter the output leads the input.

5.14.1 Digital Phase Shifters

Digital phase-shifters are usually used in conjunction with phased array antennas.

The switched-line and low-pass-/high-pass phase shifters are wideband, whereas the reflection and loaded-line phase shifters are narrowband.

5.15 RF Switches

RF switches are used in many applications such as transmit/receive switching in time division duplexing (TDD) radio transceivers, tunable filters, and automation testing of simultaneous multiple RF devices. Also, they are used extensively in microwave test systems for signal routing between instruments and devices under test, and to remotely route high-frequency signals through transmission paths. There are different types of RF switches for different applications; a single-pole/double-throw (SPDT or 1:2) switch route signals from one input to two output paths; a single-pole/multiple-throw (SPnT) switch allows a single input to multiple (i.e., three or more) output paths.

RF switches can be either absorptive or reflective type. An absorptive switch (also known as a terminated switch) is a switch whose output ports are terminated with a 50 Ω load to match the system's impedance. Thus, when the switch is OFF, the signal is absorbed by the termination/load at the port and not reflected into the switch. This results in low VSWR. When the switch is turned ON, the signal propagates through the switch, and the terminated port is isolated out. Therefore, the switch has a low VSWR in both the on state and off state.

A *reflective switch* has no termination at the output ports when the switch is not active. This results in the input signal being reflected to the source, resulting in a poor return loss. These switches are easier to design, are lower in cost, and can handle a higher amount of power compared to absorptive switches. Figure 5.35 shows the RF switch for transmit and receive switching in radio transceivers.

5.15.1 PIN Diode RF Switch

PIN diodes (i.e., positive-intrinsic-negative semiconductors) are used in different configurations to function as RF switches. Figure 5.36 shows a transmit/receive RF switch using PIN diodes.

In Figure 4.36, when the diodes D1 and D2 are turned on by the control voltage (V_{CNTRL}), the RF signal passes from the transmitter to the antenna port 1 but will be blocked by the quarter-wavelength transmission line section that is shortened through D2; the resonant

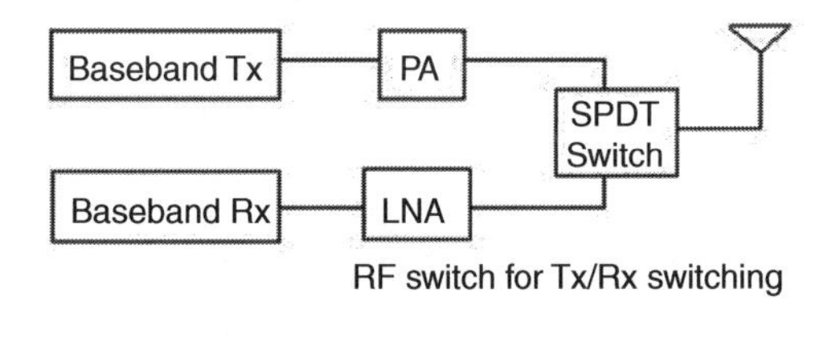

RF switch module

Figure 5.35 RF switch for Tx/Rx switching. Source: Analog Devices, Inc.

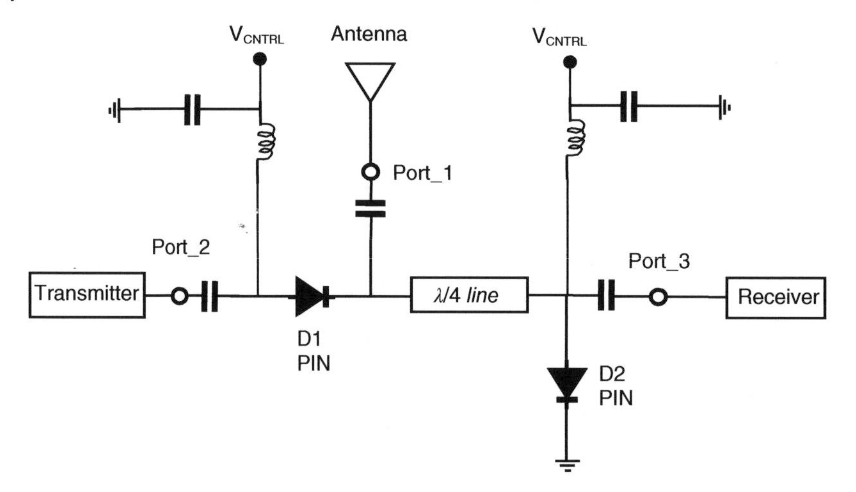

Figure 5.36 Transmit/Receive switch using PIN diodes.

quarter-wave shorted-stub acts as an open circuit to the RF signal, when the PIN diodes D1 and D2 are turned off, both D1 and D2 become open-circuits, allowing the RF signal to pass from the antenna port 1 to the receiver port 3, whereas the RF signal is blocked from entering the transmitter port 2. Therefore, the switch acts as an SPDT switch.

5.15.2 RF Switches Performance Parameters

The performance parameters of RF switches are:

- Frequency range
- Insertion loss
- Return loss
- Isolation
- Switching time (i.e., time required to change from "on" state to "off" state and vice versa)
- Settling time
- Power handling

The insertion loss of an RF switch is very important in receiver applications because it impacts the sensitivity and dynamic range of the receiver.

Review Questions

5.1 Draw a block diagram of an RF transceiver frontend and briefly explain the function of each block.

5.2 List the performance parameters of low-noise amplifiers.

5.3 What is the impact of the noise figure on the receiver's performance?

5.4 Draw a test setup for measuring the S-parameters of a low-noise amplifier.

5.5 What is the condition for having an unconditionally stable amplifier?

5.6 Explain the Y-Factor method of measuring noise figure.

5.7 Explain the noise method of measuring the noise figure.

5.8 Explain and illustrate the terms intermodulation, one dB compression point, and third-order intercept point in an amplifier.

5.9 Explain and draw the test setup for measuring the third-order intercept point of an amplifier.

5.10 What is meant by the spurious-free dynamic range (SFDR) of an amplifier?

5.11 List the performance parameters of an RF mixer, and explain how to measure the conversion loss of a mixer.

5.12 Compare and contrast double-balanced mixers and single-balanced mixers.

5.13 Explain what is meant by the noise factor of a mixer.

5.14 Describe the difference between low-pass filter, high-pass filter, band rejection filter, and bandpass filter.

5.15 List the performance parameters of RF filters and explain how to measure the frequency response of a filter.

5.16 Explain the basic concept of a cavity resonator.

5.17 Explain the application of duplexers and list their performance parameters.

5.18 Draw a block diagram of a phase-locked loop and describe its operation.

5.19 Explain the basic concept of a voltage-controlled oscillator and list its performance parameters.

5.20 What is the impact of the oscillators' phase noise in a radio transceiver on the receiver's performance?

5.21 Explain how the frequency stability of an oscillator is quantified.

5.22 Explain the term power added efficiency of a power amplifier.

5.23 How does the power amplifier's nonlinearity impact the performance of a radio transmitter?

5.24 Illustrate and explain the basic concept of the feedforward linearization technique that is used in power amplifiers.

5.25 List the performance parameters of a circulator and explain how it can be used as an isolator.

5.26 List the performance parameters of a directional coupler and describe how to measure its directivity.

5.27 Explain how to characterize a power combiner/splitter.

5.28 Compare and contrast digital phase shifters and analog phase shifters.

5.29 Describe the use of RF switches and list their performance parameters.

References

1 George D. Vendelin, Anthony M. Pavio, Ulrich L. Rohde, Matthias Rudolph, *Microwave Circuit Design Using Linear and Nonlinear Techniques*, 3rd ed., Wiley, 2021.

2 David M. Pozar, *Microwave Engineering*, 4th ed., Wiley, 2011.

3 Guillermo Gonzalez, *Microwave Transistor Amplifiers: Analysis and Design*, 2nd ed., Pearson, 1996.

4 Ashish Kumar, Meenakshi Rawat, "Comparative study based simple analog predistortion design with improved linearization," *IEEE Microwaves, Antennas, and Propagation Conference (MAPCON)*, IEEE, 2022.

5 Tian-Wei Huang, Ho-Ching Yen et al., "A 19.7–38.9-GHz ultrabroadband PA with phase linearization for 5G in 28-nm CMOS process," *IEEE Microwave and Wireless Components Letters*, Volume: 32, Issue: 4, 2022.

6 Minjae Jung, Seonjong Min, Byung-Wook Min, "A quasi-balanced power amplifier with feedforward linearization," *IEEE Microwave and Wireless Components Letters*, Volume: 32, Issue: 4, 2020.

7 Dinna Davis, R. Dipin Krishnan, "Power amplifier linearization using hybrid optimization techniques," *International Conference on Communication and Signal Processing (ICCSP)*, IEEE, 2020.

8 Yechen Tian, Junjie Gu, Hao Xu, Weitian Liu, Zongming Duan, Hao Gao, Na Yan, "A 26-32GHz 6-bit bidirectional passive phase shifter with 14dBm IP1dB and 2.6° RMS phase error for phased array system in 40nm CMOS," *IEEE/MTT-S International Microwave Symposium*, 2023.

Suggested Readings

Abdelhafid ES-SAQY, Maryam ABATA et al., "Very low phase noise voltage controlled oscillator for 5G mm-wave communication systems," *1st International Conference on Innovative Research in Applied Science, Engineering and Technology (IRASET)*, 2020.

Aniello Franzese, Nebojsa Maletic et al., "55% Fractional-bandwidth doherty power amplifier in 130-nm SiGe for 5G mm-wave applications," *16th European Microwave Integrated Circuits Conference (EuMIC)*, 2021.

Arjuna Marzuki, Chien Hsien Lee et al., "Design of 0.13-µm CMOS voltage-controlled oscillator (VCO) for 2.45 GHz IoT application," *3rd International Conference for Emerging Technology (INCET)*, IEEE, 2022.

Beomsoo Bae, Juhui Jeong et al., "24–40 GHz Dual-band highly linear CMOS up-conversion mixer for mmWave 5G NR FR2 cellular applications," *IEEE Microwave and Wireless Components Letters*, Volume: 32, Issue: 8, 2022.

Beomsoo Bae, Junghwan Han, "24–40 GHz gain-boosted wideband CMOS down-conversion mixer employing body-effect control for 5G NR applications," *IEEE Transactions on Circuits and Systems*, Volume: 69, Issue: 3, 2022.

Chaokui Hannachi, Ke. Wu, "Dual-mode RF mixer for low-power direct-conversion receiver," *IEEE Microwave and Wireless Components Letters*, Volume: 32, Issue: 6, 2022.

Chenguang Li, Ruitao Wang et al., "A compact broadband power amplifier covering 23-39 GHz for 5G mobile communication," *International Symposium on Radio-Frequency Integration Technology* (RFIT), IEEE, 2022.

Chengzhu Dong, Guoqing Dong, Sanming Hu, "A 5G millimeter low power phase-locked loop in low-cost 0.18-μm CMOS," *IEEE MTT-S International Wireless Symposium (IWS)*, 2020.

Dezhi Zhang, Zhe Du, Ming Cheng et al., "Demonstration and trial of a new CWDM and circulator integrated semi-active system for 5G fronthual," *Optical Fiber Communications Conference and Exhibition (OFC)*, 2022.

Ding He, Minyuan Yu et al., "An 18–50-GHz double-balanced GaAs mixer using novel ultrawideband balun," *IEEE Microwave and Wireless Technology Letters*, Volume: 33, Issue: 6, 2023.

Dushyant K. Sharma, Ravi T. Bura, "A novel and compact wideband doherty power amplifier architecture for 5G cellular infrastructure," *4th 5G World Forum (5GWF)*, IEEE, 2021.

Huiting Yu, Yongle Wu et al. "IPD millimeter-wave bandpass filter chip based on stepped-impedance coupled-line dual-mode resonator for 5G pplication," *IEEE Transactions on Circuits and Systems*, Volume: 69, Issue: 69, 2022.

Hyohyun Nam, Chae Jun Lee et al., "A D-band high-linearity down-conversion mixer for 6G wireless communications," *IEEE Microwave and Wireless Technology Letters*, Volume: 33, Issue: 5, 2023.

Hyun Bae Ahn, Hong-Gu Ji et al., "25–31 GHz GaN-based LNA MMIC employing hybrid-matching topology for 5G base station applications," *IEEE Microwave and Wireless Technology Letters Year: 2023*, Volume: 33, Issue: 1, 2023.

Iman Ghotbi, Baktash Behmanesh, Markus Törmänen, "A reconfigurable RF filter with 1%–40% fractional bandwidth for 5G FR1 receivers," *IEEE Solid-State Circuits Letters*, Volume: 6, 2023.

Ioannis Dimitrios Psycharis, Grigorios Kalivas, "A 40 GHz low phase noise VCO in 40 nm CMOS," *Panhellenic Conference on Electronics & Telecommunications (PACET)*, IEEE, 2022.

Jiajin Li, Lin Peng et al., "A mm-wave parallel-combined power amplifier supporting balanced/unbalanced mode for 5G NR FR2 applications," *Microwave and Wireless Technology Letters*, Volume: 33, Issue: 5, 2023.

Jiaxuan Li, Yang Yuan et al., "A broadband LNA with multiple bandwidth enhancement techniques," *IEEE Microwave and Wireless Technology Letters*, Volume: 33, Issue: 5, 2023.

Yechen Tian, Junjie Gu, Hao Xu, Weitian Liu, Zongming Duan, Hao Gao, Na Yan, "A 26-32GHz 6-bit bidirectional passive phase shifter with 14dBm IP1dB and 2.6° RMS phase error for phased array system in 40nm CMOS," *IEEE/MTT-S International Microwave Symposium*, 2023.

L. Letailleur, M. Villegas et al., "Characterization of GaN power amplifier using 5G mm-wave modulated signals," *Topical Conference on RF/Microwave Power Amplifiers for Radio and Wireless Applications*, IEEE, 2023.

Li Gao, Gabriel M. Rebeiz, "Wideband bandpass filter for 5G millimeter- wave application in 45-nm CMOS silicon-on-insulator," *IEEE Electron Device Letters*, Volume: 42, Issue: 8, 2021.

Liang Chen, Chao Li et al., "A PLL synthesizer for 5G mmW transceiver," *MTT-S International Wireless Symposium (IWS)*, 2020.

Mohamed A. ElBadry, Mohamed Mobarak, Mohamed A. Y. Abdalla, "A 24-41.5 GHz LNA with enhanced IP1dB in 65-nm bULK CMOS for 5G applications," *24th International Microwave and Radar Conference (MIKON)*, IEEE, 2022.

Mohammad Chahardori, Md, Aminul Hoque et al., "A compact Mm-wave multi-band VCO based on triple-mode resonator for 5G and beyond," *International Symposium on Radio-Frequency Integration Technology (RFIT)*, IEEE, 2022.

Nagarajan Mahalingam, Yisheng Wang et al., "A 24.6-32.5 GHz millimeter-wave frequency synthesizer for 5G wireless and 60 GHz applications," *IEEE MTT-S International Microwave Symposium (IMS)*, 2021.

Peta Guruprakashkumar, Darshak Bhatt, "A wideband noise cancelling mixer in 130-nm BiCMOS process," *IEEE Microwaves, Antennas, and Propagation Conference (MAPCON)*, IEEE, 2022.

Ren Imanishi, Hideyuki Nosaka, "Digital-controlled high-linearity phase shifter using vernier ladder network for beyond 5G phased array antenna," *52nd European Microwave Conference (EuMC)*, 2022.

Shahid Jamil, Muhammad Usman et al., "28-32 GHz wideband LNA for 5G applications," *1st International Conference on Microwave, Antennas & Circuits (ICMAC)*, IEEE, 2021.

Shuichi Sakata, Marie Taguchi et al., "A 3.4-4.1GHz broadband GaN doherty power amplifier module for 5G massive-MIMO base-stations," *Asia-Pacific Microwave Conference (APMC)*, 2022.

Sokha Khim, Sovuthy Cheab, "Design of miniaturized S-band high isolation strip-line circulator for 5G application," *International Conference on Future Trends in Smart Communities (ICFTSC)*, 2022.

Xi Zhu, Zeyu Ge et al., "Millimeter-wave CMOS passive filters for 5G applications," *MTT-S International Microwave Filter Workshop (IMFW)*, IEEE, 2021.

Xiaoyue Xia, Zuojun Wang, "A 26/38-GHz dual-band filtering balanced power amplifier MMIC for 5G mobile communications," *IEEE Microwave and Wireless Technology Letters*, Volume: 33, Issue: 4, 2023.

Yunbo Rao, Huizhen Jenny Qian et al., "Miniaturized 28-GHz packaged bandpass filter with high selectivity and wide stopband using multilayer PCB technology," *IEEE Microwave and Wireless Components Letters*, Volume: 32, Issue: 6, 2023.

Ziad Elkhatib, Sherif Moussa et al., "RF LNA with simultaneous noise-cancellation and distortion-cancellation for wireless RF systems," *International Conference on Electrical and Computing Technologies and Applications (ICECTA)*, IEEE, 2022.

Zongming Duan, Qiang Ma, Ying Liu, Yuefei Dai, "A 6-bit CMOS phase shifter with active balun and three-stage poly-phase filter for phased arrays," *International Conference on Microwave and Millimeter Wave Technology (ICMMT)*, 2020.

Zuojun Wang, Debin Hou et al., "A linearity-enhanced 18.7–36.5-GHz LNA with 1.5–2.1-dB NF for radar applications," *IEEE Microwave and Wireless Components Letters*, Volume: 32, Issue: 8, 2022.

6

Basics of RF Transceivers

This chapter discusses different receiver and transmitter architectures with focus on the system performance rather than on the circuit design. It covers superheterodyne, direct conversion (i.e., zero-IF), low-IF, software-defined radio (SDR), and direct-conversion transmitter. The receiver parameters that are discussed include receiver sensitivity and selectivity, receiver intermodulation, and dynamic range. The transmitter system parameters that are covered include output transmitter power, spurious emission, frequency error, error vector magnitude (EVM), occupied bandwidth (OBW), adjacent channel leakage power ratio (ACLR), operating band unwanted emission (OBUE), spurious emission, and transmitter intermodulation. The impact of the transmitter and receiver parameters on the system performance is also addressed. RF block-level budget analysis and examples of the receiver RF block-level budget analysis are presented to help the readers apply the covered topic to develop an RF block-level budget. Sufficient solved problems are presented to show how to calculate the transceiver performance parameters. Review questions and problems, and suggested readings that present the recent research and development in radio transceivers are provided at the end of the chapter.

6.1 Chapter Objectives

On reading this chapter, the reader will be able to:

- Draw a block diagram of a radio transceiver and explain the function of each block.
- Explain and illustrate the problem of image frequency in a full-duplex transceiver and how to mitigate such a problem.
- Discuss and explain the receiver's sensitivity, selectivity, intermodulation, and spurious-free dynamic range.
- Discuss and explain the receiver's sensitivity, receiver selectivity, receiver intermodulation, and receiver spurious-free dynamic range.
- Draw a block diagram of a dual-conversion receiver and explain its advantages.
- Draw a block diagram of the direct-conversion receiver and explain its advantages.
- Develop a block-level budget for a receiver to determine the receiver sensitivity and spurious-free dynamic range.
- Calculate the receiver sensitivity and dynamic range for a given bit-error rate.
- Explain the key transmitter performance parameters and their impact on the system performance including transmitter output power, EVM, spurious emissions, and transmitter intermodulation.

Essentials of RF Front-end Design and Testing: A Practical Guide for Wireless Systems, First Edition. Ibrahim A. Haroun.

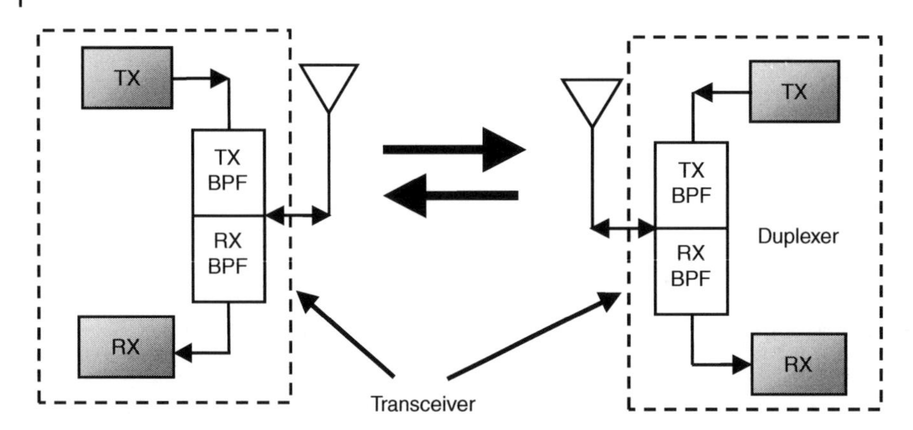

Figure 6.1 Illustration of a communication link with two transceivers.

6.2 RF Transceivers

RF transceivers [1–6] are the interface between the antennas and the baseband processing in wireless communication systems. They perform the transmitter and receiver functions using a single antenna and share some RF building blocks such as a duplexer, RF switches, and frequency synthesizers.

In RF transceivers, the transmitter and receiver are combined in the same enclosure. Figure 6.1 illustrates a communication link with two transceivers. In Figure 6.1, the TX block is the transmitter, and the RX block is the receiver.

6.3 Superheterodyne Receiver Architecture

Superheterodyne receivers are radio technology that has dominated the market for over 100 years. These receivers are widely used in domestic broadcast and professional communication equipment and are also called superhet receivers. Such receivers use a mixer and local oscillator to down-convert the received signal to a fixed intermediate frequency (IF) signal that can be more easily processed. The use of IF provides better selectivity, which enables the receiver to select the intended channel and reject adjacent channels. In addition, the IF improves the receiver's signal-to-noise ratio (SNR), which reduces the bit-error rate in the receiver. The receiver's noise figure (i.e., noise added by the receiver's components) is a key performance parameter that impacts the receiver's sensitivity (i.e., the receiver's ability to detect weak signals). Figure 6.2 shows a block datagram of a superheterodyne receiver.

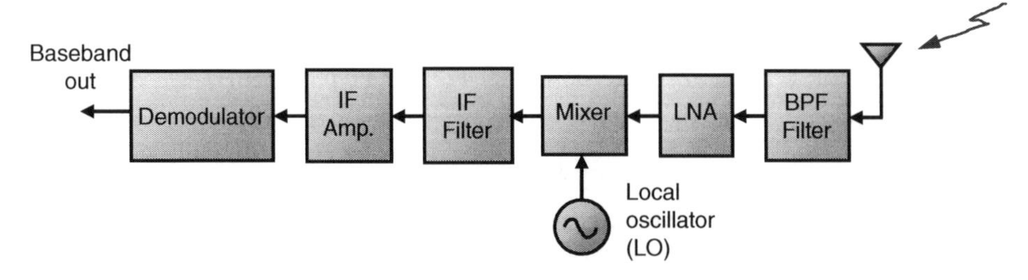

Figure 6.2 Block diagram of a superheterodyne receiver.

In Figure 6.2, the receiver's antenna converts the electromagnetic waves to RF signals. The band pass filter (BPF) provides some selectivity by filtering out the undesired frequencies and passes only the desired frequency range. The BPF is followed by a low-noise amplifier (LNA) that amplifies the RF signal and compensates for the filter's insertion loss. The output of the LNA is then applied to a mixer. The mixer output is a composite signal of $(f_{RF} - f_{LO})$ and $(f_{RF} + f_{LO})$ frequencies and other frequencies that include $f_{RF}, f_{LO}, 2f_{RF}, 2f_{LO}, 2f_{RF} - f_{LO}, 2f_{LO} - f_{RF}$, etc. The IF filter following the mixer passes only the difference frequency between the RF and LO (local oscillator) frequencies $(f_{LO} - f_{RF})$. The IF amplifier increases the signal level to a level that enables the demodulator to recover the baseband signal. The frequency translation (i.e., conversion) processes may be performed more than once to ensure the frequency image is not very close to the RF signal.

A superheterodyne receiver consists of three sections which are the RF, IF, and baseband. Figure 6.3 shows a block diagram of a superheterodyne transceiver.

In Figure 6.3, the receiver's RF chain includes part of the duplexer, which is the receive BPF filter, LNA, RF bandpass filter (BPF), and an RF-to-IF down-converter (mixer). After the down-converter, the I/Q (In-phase/Quadrature) demodulator is the second frequency converter, which down-converts the signal frequency from IF to baseband. The demodulator contains two mixers, converting the IF signal into I and Q signals. The low-pass filters (LPF) after the I and Q channel mixers filter out the unwanted mixing products. The filtered I and Q signals get amplified and then applied to the analog-to-digital converters (ADCs), which convert the analog signals to digital signals for further processing in the digital-signal-processing section of the receiver chain to optimize the receiver performance.

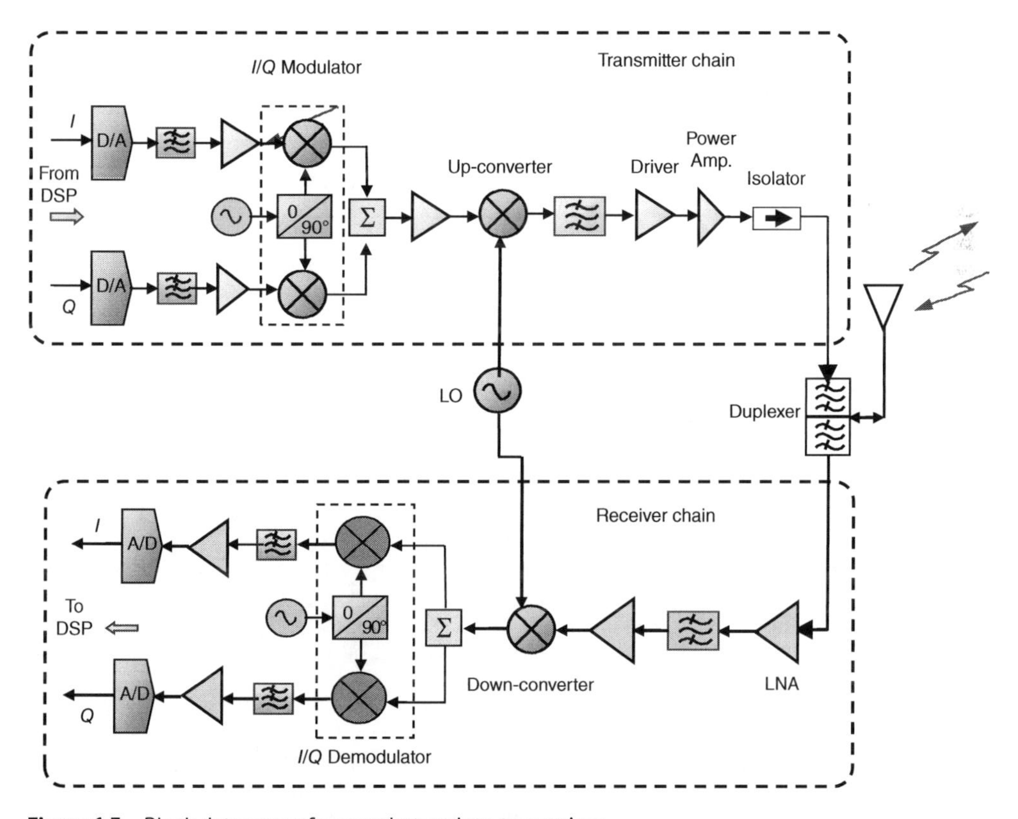

Figure 6.3 Block datagram of a superheterodyne transceiver.

6.4 Receiver System Parameters

The key performance parameters of a radio receiver are sensitivity, intermodulation, adjacent channel and alternate channel selectivity, and dynamic range. These parameters are discussed in the following subsections and their test setups are covered in Chapter 9.

6.4.1 Receiver Sensitivity

Receiver sensitivity is a fundamental property that affects the system's bit-error rate. It is defined as the minimum input signal level that enables the receiver to produce a specific signal-to-noise ratio at its output to support the required bit error rate. The sensitivity, $S_{\min}$, is given by

$$S_{\min} = -174\,\text{dBm} + \text{NF}_{\text{RX}} + 10 \cdot \log(B) + \left(\frac{S}{N}\right)_o \tag{6.1}$$

where

NF_{RX} = the receiver's noise figure in dB
B = receiver's bandwidth in Hz
S/N = Signal-to-noise at the demodulator's output in dB

The sensitivity also is defined in terms of the bit rate and given by

$$S_{\min} = -174\,\text{dBm} + \text{NF}\,[\text{dB}] + 10 \cdot \log(R_b) + \frac{E_b}{N_o}\,[\text{dB}] \tag{6.2}$$

where

R_b = bit rate [bps]
E_b/N_o = energy per bit to noise power spectral density in dB.

From Eqs. (6.1) and (6.2), the signal-to-noise ratio can be written as

$$\left(\frac{S}{N}\right)_{\min} = \frac{B}{R_b} \cdot \frac{E_b}{N_o} \tag{6.3}$$

Example 6.1
A transceiver system uses QPSK with a bit rate of 46.6 kbps, and the receiver has a bandwidth of 30 kHz, and noise figure of 8 dB. If the bit error rate is 10^{-5}, determine the receiver sensitivity.

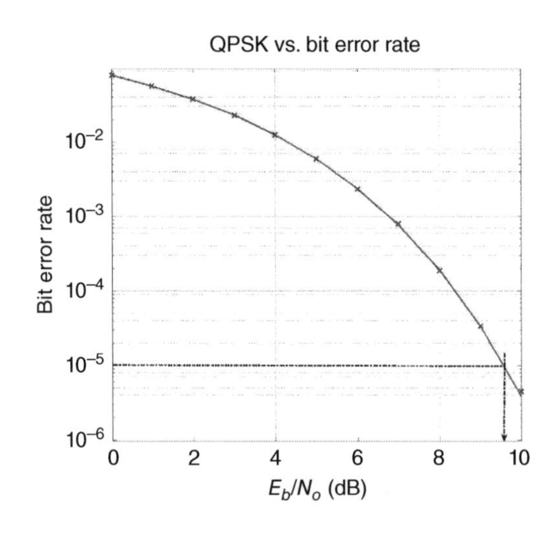

Solution

From the graph of QPSK vs. bit error rate, at bit error rate of 10^{-5}, the E_b/N_o is about 9.8 dB. Thus, The receiver sensitivity can be calculated using the following equation:

$$S_{\text{min}} = -174\,\text{dBm} + NF_{\text{dB}} + 10 * \log(R_b) + \frac{E_b}{N_o}$$

$$S_{\text{min}} = -174\,\text{dBm} + 8 + 10 * \log(30 \times 10^3) + 9.7$$

$$S_{\text{min}} = -109.61\,\text{dBm}$$

6.4.2 Noise Figure

The noise figure represents the contribution of a circuit component to the thermal noise. It is a measure of the degradation of the SNR between the input and out of the component. When noise and a signal are applied to the input of a noiseless component, both get amplified or attenuated by the same factor. However, if the component adds noise, the output noise power will be increased at the output, resulting in a reduced SNR. The noise factor, F, is a measure of the degradation in SNR and is given by

$$F = \frac{(S/N)_i}{(S/N)_o} = \frac{S_i}{N_i} \cdot \frac{N_o}{S_o} \geq 1, \tag{6.4}$$

where S_i and N_i are the input signal and noise powers, and S_o and N_o are the output signal and noise powers. The noise power from a matched load at $T_o = 290$ K is expressed as

$$N_i = k\,T_o B \tag{6.5}$$

$$\therefore F = \frac{N_o}{Gk\,T_o B} \tag{6.6}$$

where G is the gain, k is Boltzmann's constant (1.38×10^{-23} J/K), and B is the bandwidth in Hz. The noise factor is expressed in dB and called noise figure (NF), which is given by

$$\text{NF} = 10\,\log(F) \tag{6.7}$$

The output noise power is the sum of the amplified input noise power and the internally generated noise. Thus,

$$N_o = kGB\,(T_o + T_e) \tag{6.8}$$

where T_e is the called the equivalent noise temperature. Thus, Eq. (6.4) can be written as

$$\therefore F = \frac{S_i}{k\,T_o B} \cdot \frac{kGB\,(T_o + T_e)}{GS_i} = 1 + \frac{T_e}{T_o} \geq 1 \tag{6.9}$$

$$T_e = (F - 1)\,T_o \tag{6.10}$$

For a lossy transmission line or passive component, the NF is the same as the line's loss (L). Lossless components do not generate thermal noise; at room temperature $F = L$.

For a cascaded 2-por networks, T_e is given by

$$T_e = T_1 + \frac{T_2}{G_1} + \frac{T_3}{G_1 G_2} + \cdots + \frac{T_n}{G_1 G_2 \ldots\ldots G_{n-1}} \tag{6.11}$$

The cascaded noise factor F_{cascaded} is given by

$$F_{\text{cascaded}} = F_i + \sum_{i=2}^{N} \frac{F_i - 1}{\prod_{j=1}^{i-1} G_j} \tag{6.12}$$

$$F_{\text{cascaded}} = F_1 + \frac{F_2 - 1}{G_1} + \frac{F_3 - 1}{G_1 G_2} + \frac{F_4 - 1}{G_1 G_2 G_3} + \cdots + \frac{F_N - 1}{G_1 G_2 \cdots G_{N-1}} \tag{6.13}$$

where N is the number of stages in the system, and G_j is the gain of the jth stage. The cascaded NF is F_{cascaded} in dB. Equation (6.13) indicates that the higher the gain of the first stage in a receiver, the lower the cascaded NF, and the better the receiver sensitivity, which results in a reduced bit error rate.

Example 6.2
Determine the noise temperature for a receiver that has 10 dB NF at room temperature.

Solution

$$\text{NF} = 10 \log\left(1 + \frac{T_e}{T_o}\right), \text{thus}$$

$$1 + \frac{T_e}{T_o} = (10)^{\text{NF}/10}, \text{and } T_e = T_o\,[(10)^{10/10} - 1]$$

$$\therefore T_e = 2610\,^{\circ}\text{K}$$

Example 6.3
For a QPSK system with a bit rate of 46.6 kbps, if the receiver bandwidth is 30 kHz, receiver NF is 8 dB, antenna noise temperature is 900 K, and the bit error rate is 10^{-5}, determine the receiver sensitivity.

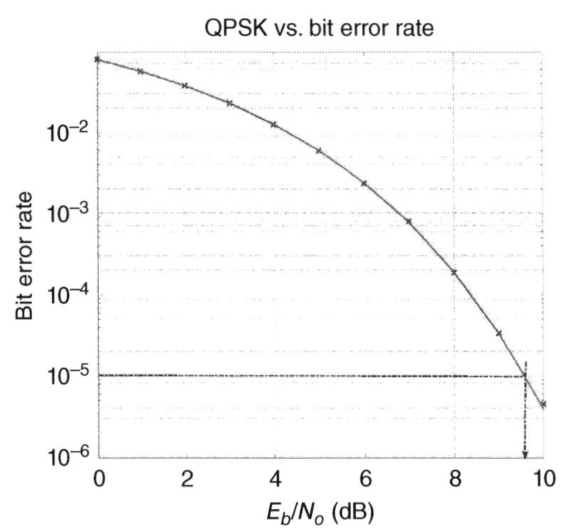

Solution
From the graph of QPSK vs. bit error rate, at bit error rate of 10^{-5} the E_b/N_o is about 9.7 dB.

The receiver sensitivity can be expressed as

$$S_{\min} = \mathrm{kB}[T_A + (F-1)T_o]\left(\frac{S_o}{N_o}\right)_{\min}, \text{ and}$$

$$\left(\frac{S_o}{N_o}\right)_{\min} = \frac{R_b}{B}\cdot\frac{E_b}{N_o} = \frac{46.6\times10^3}{30\times10^3}\times 9.3 = 14.49$$

$$\therefore S_{\min} = (1.38\times10^{-23})(30\times10^3)[900+(6.3-1)(290)](14.49) = -109.74\ \mathrm{dBm}$$

Example 6.4

For the RF receiver front-end shown below, if the input noise power from an antenna is $N_i = kT_a B$, where the subsystem bandwidth is 10 MHz and the antenna temperature is $T_a = 15$ K, determine the following:

- NF of the frontend.
- Output noise power.

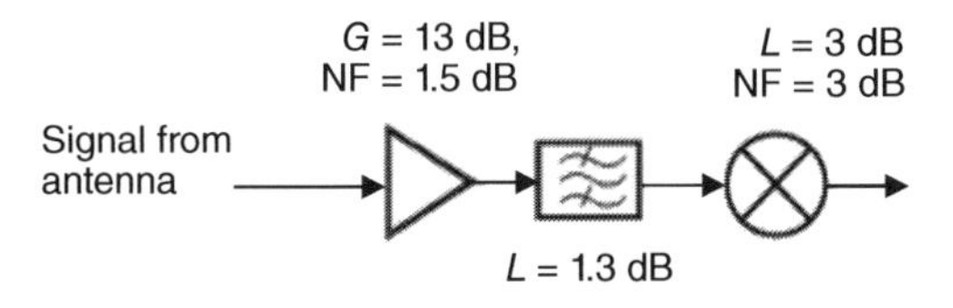

Solution

Covert the dB values to linear scale.

Use the following equations to calculate the cascaded noise factor

$$F_{acc} = F_1 + \frac{F_2-1}{G_1} + \frac{F_3-1}{G_1G_2}, \mathrm{NF} = 10\ \log(F_{acc})$$

The calculated NF = 1.75 dB

To calculate the output noise power, calculate the equivalent noise temperature of the front-end using the following equation:

$$T_e = (F-1)\,T_o$$

The calculated $T_e = 144.22$ K

Calculate the output noise using the following equation:

$$N_o = \mathrm{kGB}\,(T_a + T_e)$$

In the above equation, the cascaded gain is $G = 8.7$ dB, and the calculated output noise power is −98.585 dBm.

Example 6.5

For the RF subsystem block shown below, determine the NF and equivalent noise temperature.

G1 = 15 dB,
NF1 = 3 dB, B = 200 MHz

Demodulator

Te = 800 K

Solution

Use the following equation to calculate the NF of the demodulator

$$F_2 = 1 + \frac{T_{e2}}{T_o}$$

The calculated noise factor $F_2 = 3.76$

Calculate the cascaded noise factor and NF using the following equations

$$F_{acc} = F_1 + \frac{F_2 - 1}{G_1} \quad NF = 10 \log (F_{acc})$$

The calculate NF is NF = 3.18 dB

The equivalent noise temperature is calculated as follows:

$$T_{e1} = (F_1 - 1)\, T_o, T_e = T_{e1} + \frac{T_{e2}}{G_1}$$

The calculated noise temperature $T_e = 315.32$ k

6.4.3 Intermodulation and Spurious-Free Dynamic Range

When two or more signals separated in frequency are applied to the input of an RF block such as an amplifier operating in its nonlinear region due to high-level input signals, the device's output would contain additional frequency components called *intermodulation products*. To quantify the linear operation of an amplifier, it is essential to determine its 1-dB compression point. The 1 *dB compression point is* the point at which the output power is 1 dB below the output produced by a perfectly linear amplifier. Figure 6.4 illustrates the 1 dB compression point.

In radio receivers, the receiver's linearity is quantified based on the *third-order intercept point* (IP3). The IP3 is defined as the intersection point of the linear extension of the fundamental signal, and the third-order intermod (IM), as shown in Figure 6.4. If the input signal $V_{in}(t)$ of an amplifier is given by

$$v_i(t) = A\,[cos(2\pi f_1 t) + cos(2\pi f_2 t)],$$

then the output signal $v_o(t)$ can be expressed by a power series and is given by

$$v_o(t) = a_o + a_1 v_i(t) + a_2 v_i^2(t) + a_3 v_i^3(t) + \cdots$$

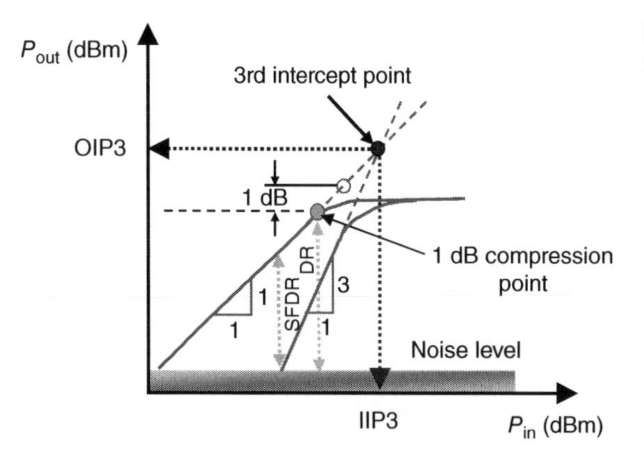

Figure 6.4 Illustration of one dB compression point.

The output signal would have frequency components that include f_1, f_2, $2f_1$, $2f_2$, $f_1 \pm f_2$, $2f_1 \pm f_2$, $2f_2 \pm f_1$, $3f_1$, and $3f_2$, where $2f_1$, $2f_2$ are the 2nd harmonics, $3f_1$, $3f_2$ are 3rd harmonics, $f_1 \pm f_2$ are the second-order intermod, and $2f_1 \pm f_2$ and $2f_2 \pm f_1$ are the third-order intermod. All harmonics beyond the fundamental components represent *harmonic distortion* (HD) that is given by

$$\mathrm{HD} = \frac{\sqrt{\sum_{i=2}^{N} v_i^2}}{v_1} \tag{6.14}$$

where v_i is the voltage of the ith frequency component. Figure 6.5 illustrates the second and third-order two-tone intermodulation products at the output of an amplifier.

In an LNA, all the second-order intermodulation products are undesired; they can easily be filtered because they are far from the fundamental frequency components f_1 and f_2. However, the third-order products cannot be easily filtered from the passband of the amplifier because they are close to the fundamental components. Such intermodulation products cause distortion to the output signal. This effect is called *third-order intermodulation distortion*. From Figure 6.4, both the linear and third-order responses exhibit compression at high input powers. Because the linear and third-order responses have different slopes, they intersect at a hypothetical point called third-*order intercept point* (IP3). The cascaded IP3 at the receiver's input is given by

$$\mathrm{IIP3} = \frac{1}{\frac{1}{\mathrm{IIP3}_1} + \frac{G_1}{\mathrm{IIP3}_2} + \frac{G_1 \cdot G_2}{\mathrm{IIP3}_3} + \frac{G_1 \cdot G_2 \cdot G_3}{\mathrm{IIP3}_4} + \cdots + \frac{G_1 \cdot G_2 \cdot G_3 \cdot \cdots \cdot G_{N-1}}{\mathrm{IIP3}_N}} \ [\mathrm{mW}], \tag{6.15}$$

where

- $\mathrm{IIP3}_1$ = input IP3 of the first stage (mW)
- $\mathrm{IIP3}_N$ = input IP3 of the last stage (mW)
- $\mathrm{G_i}$ = gain (absolute) of ith stage

The input IP3 and output IP3 are related through the gain and expressed as

$$\mathrm{OIP3} = \mathrm{IIP3} + \mathrm{G}$$

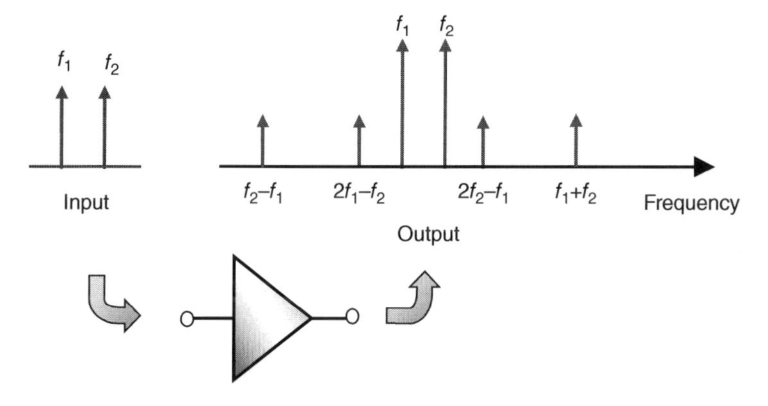

Figure 6.5 Representation of the second- and third-order two-tone intermodulation products at the output of an amplifier.

IP3 and NF conflict with each other. A practical approximation rule assumes that the IP3 is 10.6 dB greater than the 1 dB compression point. The intermodulation (IM) level at frequency $2f_1 - f_2$ is given by

$$\text{IM (dBm)} = 2\,P_{1\_out}(\text{dBm}) + P_{2\_out}(\text{dBm}) - 2\,\text{OIP3(dBm)} \tag{6.16}$$

The 1 dB compression point is a crucial parameter because it determines the amplifier's dynamic range (DR). The DR of an amplifier is the range of input powers over which the amplifier has the desirable performance. It is the range between the noise floor and the P_{1dB} compression point and given by

$$\text{DR} = P_{1\,\text{dB}} - N_o \tag{6.17}$$

where N_o is the noise at the output. Equation (6.17) indicates that higher $P_{1\,\text{dB}}$ would increase the amplifier's DR. For low-noise amplifiers, the operation is limited by the noise level and the maximum power level for which intermodulation distortion becomes unacceptable. The spurious-free dynamic range (SFDR) is given by

$$\text{SFDR} = \frac{2}{3}(\text{OIP3} - N_o) \tag{6.18}$$

If the output SNR is specified, it can be added to the N_o (noise at the output) to give the SFDR in terms of the minimum detectable signal level.

Example 6.6

For the receiver subsystem block shown blow, determine the DR and SFDR of the subsystem if the required SNR at the output is 9 dB.

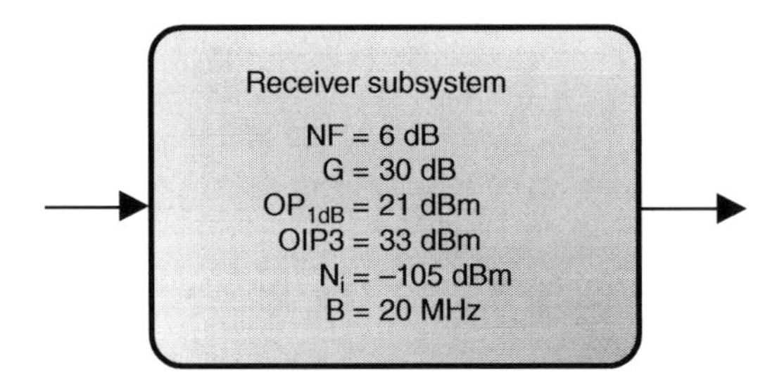

Solution

Calculate the DR using the following equations:

$$\text{DR} = P_{1\,\text{dB}} - N_o, \text{and } N_o = G(N_i + kT_eB),$$

$$T_e = (F - 1)T_o$$

the calculated $T_e = 864.5$ K, and the calculated $N_o = -66.89$ dBm.
Thus, the calculated DR is

$$\text{DR} = 21\ \text{dBm}–(-66.89\ \text{dBm}) = 87.21\ \text{dB}.$$

The SFDR is calculated as follows

$$\text{SFDR} = \frac{2}{3}(\text{IP3} - N_o)$$

$$\therefore \text{SFDR} = \frac{2}{3}(\text{IP3} - N_o - \text{SNR}), \quad \text{thus}$$

the calculated SFDR = 60.6 dB.

Example 6.7
For the amplifier shown, determine the DR. Assume the noise at the output is −99 dBm.

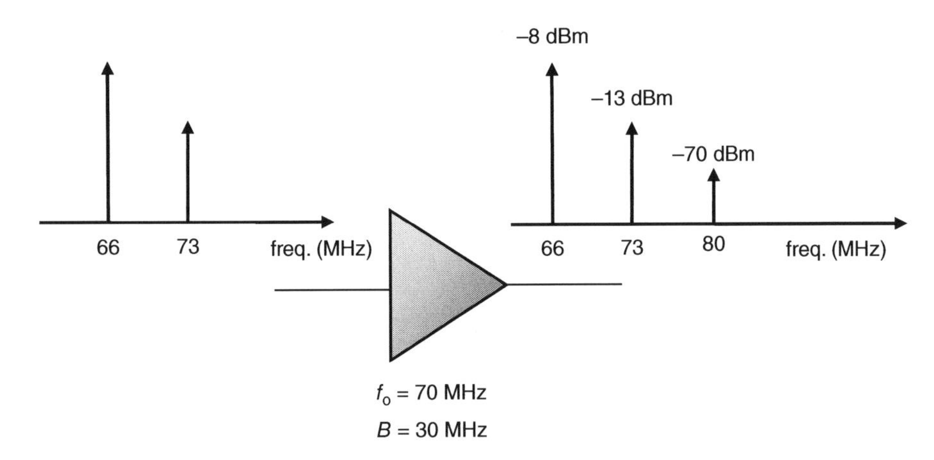

Solution
At the amplifier's output, the first signal has power $P_1 = -8$ dBm at $f_1 = 66$ MHz; the second signal has power $P_2 = -13$ dBm at $f_2 = 73$ MHz, and the intermodulation IM3 = −70 dBm at the third-order frequency component of $2f_2 - f_1 = 80$ MHz.

The OIP3 can be calculated using the following equation:

$$\text{IM (dBm)} = 2\,P_{1\_out}(\text{dBm}) + P_{2\_out}(\text{dBm}) - 2\,\text{OIP3(dBm)}$$

$$-70 = 2(-13) + (-8) - 2\,\text{OIP3}, \quad \therefore \text{OIP3} = 18\text{ dBm}$$

Knowing OIP3, the $P_{1\text{B}}$ can be determined as follow

$$P_{1\text{ dB}} = \text{OIP3–10.6 dB} = 18\text{–}10.6 = 7.4\text{ dBm}$$

Thus, the dynamic rage can be calculated as

$$\text{DR} = P_{1\text{ dB}}\text{–}N_o = 7.4\text{ dBm–(-99 dBm)} = 106.4\text{ dB}$$

Example 6.8
Determine the output IP3 of the RF subsystem block shown below.

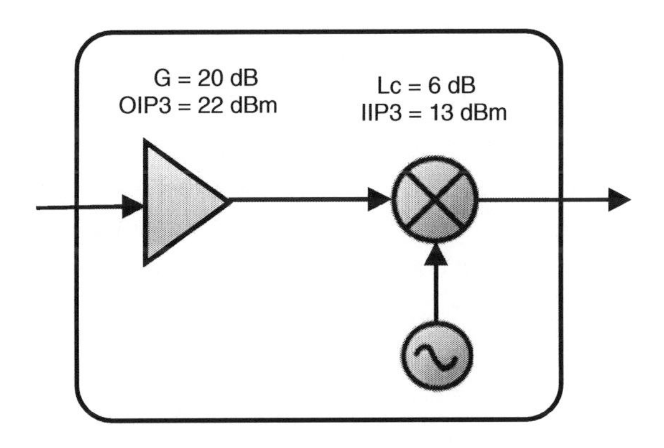

Solution

- Transfer the amplifier's OIP3 to the input, thus, IIP3 = 22 dBm − 20 dB = 2 dBm
- Covert the dBm values to milliwatts, and dB values to linear scale

$$2\,\text{dBm} = 1.59\,\text{mW}, 13\,\text{dBm} = 19.95\,\text{mW}, 20\,\text{dB} = 100$$

- Calculate the input IP3 of the subsystem using the following equation

$$\text{IIP3} = \frac{1}{\dfrac{1}{\text{IIP3}_1} + \dfrac{G_1}{\text{IIP3}_2}}\ [\text{mW}],$$

- The calculated IIP3 = −7.5166 dBm. Thus,

$$\text{OIP3} = \text{IIP3} + \text{Gain} = -7.516\,\text{dBm} + 14\,\text{dB} = 6.48\,\text{dBm}$$

Example 6.9

For the RF subsystem block shown, determine the output IP3 of the block.

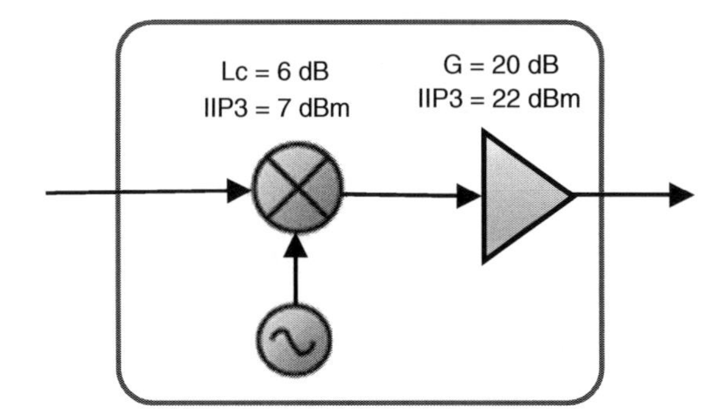

Solution

Covert the dBm values to milliwatts, and dB values to linear scale, and use the following equation to calculate the input IP3 of the block

$$\text{IIP3} = \frac{1}{\dfrac{1}{\text{IIP3}_1} + \dfrac{G_1}{\text{IIP3}_2}}\ [\text{mW}].$$

Thus, the calculated OIP3 = 6.964 dBm + 14 dB = 20.96 dBm

6.4.4 Receiver Spurious Response

Receiver spurious frequencies differ from the desired receiver frequency and can produce demodulated output in the receiver. Most receiver spurious responses are mixer spurious responses, which may or may not be further attenuated by RF selectivity in the preceding stages. Any RF frequency that satisfies the following relationship is a potential receiver spurious response:

$$\pm f_{\text{IF}} = \pm m \cdot f_{\text{RF}} \pm n \cdot f_{\text{LO}} \tag{6.19}$$

where

- f_{RF} = any incoming frequency into the RF port
- f_{LO} = LO frequency
- f_{IF} = desired IF frequency
- m = integer multiplier of RF frequency
- n = integer multiplier of LO frequency

In full-duplex superheterodyne radios, the receiver may respond to frequencies separated by the IF frequency from the transmitted frequency. Such a problem does not exist in direct-conversion receivers (i.e., Zero-IF) [7].

6.4.5 Receiver Selectivity

Receiver selectivity is the receiver's ability to detect and decode a desired signal in the presence of unwanted interfering signals, or unwanted co-channel and adjacent channel interfering signals. Usually, these unwanted signals can never be eliminated fully, but if the levels of these signals are down with respect to the wanted carrier frequency, then the receiver can function without any errors. If the levels of these interfering signals are comparable to the desired signal, then the system starts producing errors. Filters are the key receiver building blocks that contribute to the improvement of the receiver's selectivity.

6.4.6 Intermediate Frequency and Images

The choice of IF frequency is a compromise, and the goal is to achieve good selectivity. In radio receivers, the received RF signal gets mixed with the LO frequency to produce an IF signal that can be easily filtered and processed. However, if an undesired interfering signal is spaced from the RF signal by a frequency that is twice the IF frequency (i.e., *image frequency*), it will mix with the LO and get converted to a signal that falls on the desired IF signal. The image frequency is given by

$$f_i = f_s + f_{LO} \tag{6.20}$$

where,

- f_i = image frequency
- f_s = desired frequency
- f_{IF} = intermediate frequency

Figure 6.6 illustrates the problem of image frequency in superheterodyne receivers.

The image interference occurs only when the image signal reaches the mixer's input. For the receiver to have good selectivity, it is necessary to eliminate the image. The common approach to suppress the image is to use an image rejection filter placed before the mixer. Figure 6.7 illustrates the image rejection using a filter in superhet receiver.

Figure 6.7 indicates that using a high IF frequency enables significant image signal rejection, whereas a low-IF allows substantial interferences. The choice of IF frequency is influenced by the availability of filters. The image rejection filter must provide sufficient attenuation at $(\omega_{\text{interference}} + \omega_{LO})/2$. The image problem affects the receiver's sensitivity, which impacts the receiver's bit-error-rate performance.

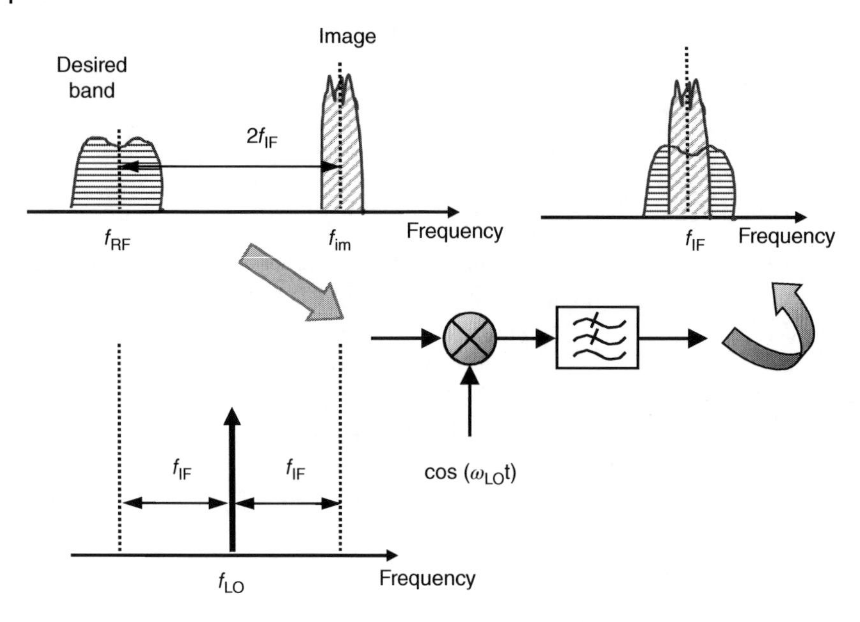

Figure 6.6 Illustration of the image frequency in superheterodyne receivers.

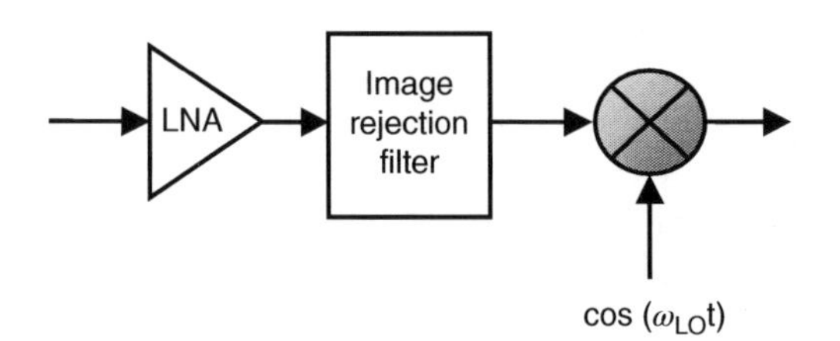

Figure 6.7 Illustration of image rejection using a filter in superheterodyne receivers.

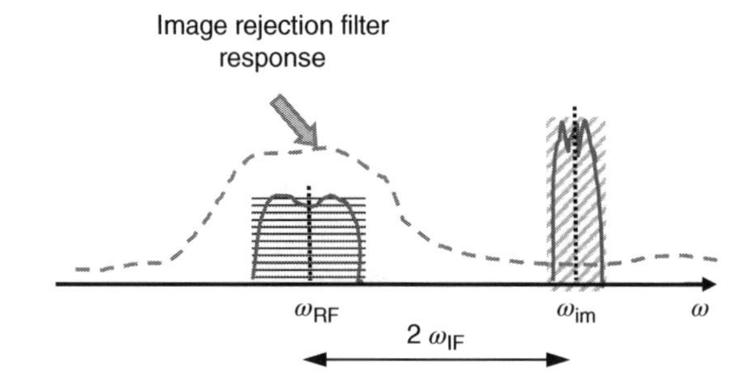

6.5 Dual-Conversion Superheterodyne Receivers

A dual-conversion superheterodyne receiver is an approach to improve the receiver's selectivity and eliminate the image problem. Figure 6.8 shows a block diagram of dual-conversion superheterodyne receiver.

In Figure 6.8, the receiver has two mixers and two LOs; therefore, it has two IF frequencies. The first mixer converts the incoming RF signal to a relatively high IF to eliminate the image signals. The second mixer converts the first IF to a much lower frequency, which enables good selectivity.

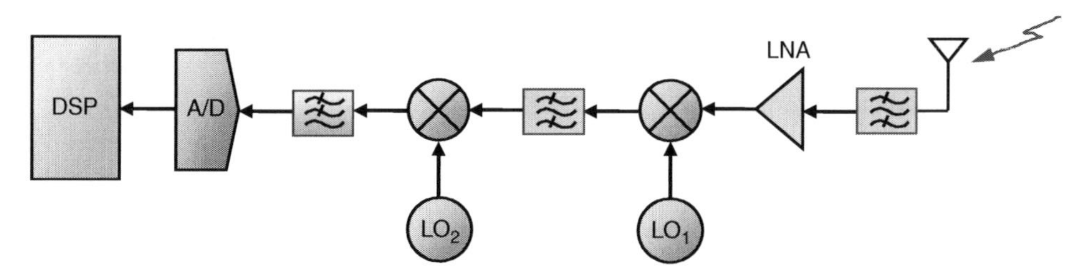

Figure 6.8 A dual-conversion superheterodyne receiver.

6.6 Direct Conversion (Zero-IF) Receiver

Direct conversion (i.e., zero-IF), also called homodyne, converts an RF signal down to baseband without any intermediate frequency stages. This conversion is achieved by setting the LO frequency equal to the RF signal frequency. In zero-IF receivers, the incoming signal gets converted directly to baseband; therefore, they perform the demodulation of the signal as part of the frequency conversion. Direct-conversion receivers do offer some advantages versus their superheterodyne counterparts. The advantage of this architecture is that it does not need a frequency plan, it does not have image frequency, lower cost, smaller circuit-board area, and reduced component count. Figure 6.9 shows a block diagram of a direct-conversion transceiver.

In Figure 6.9, the transmit BPF filter in the duplexer suppresses the transmitter's leakage into the receiver to prevent receiver desensitization. In the receiver's chain, the received signal passes through the duplexer's receive BPF filter and gets amplified by an LNA, and it is filtered further.

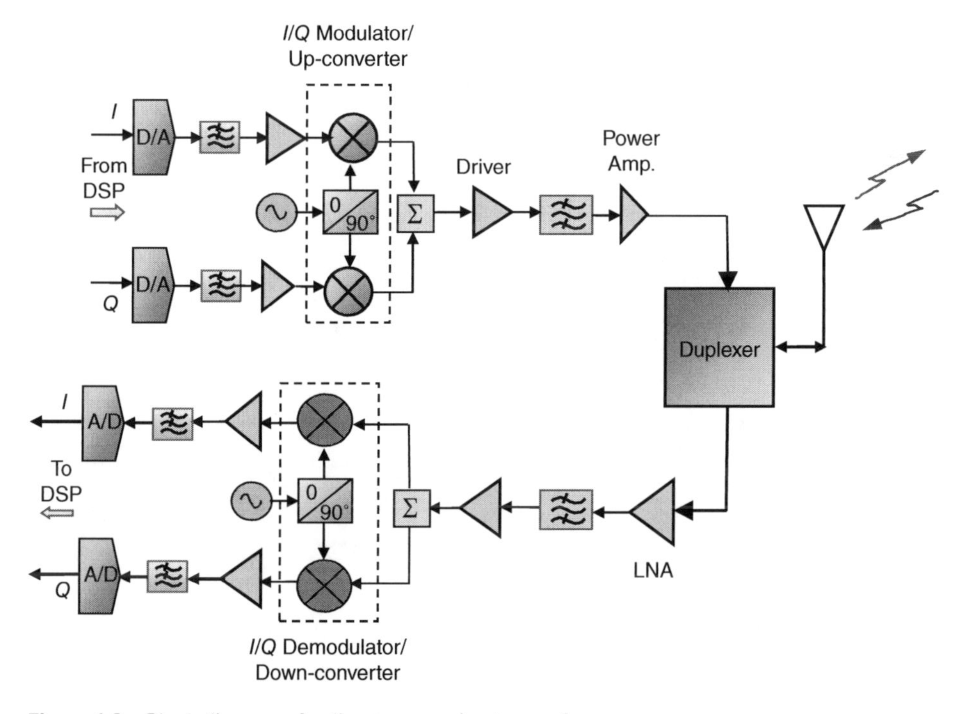

Figure 6.9 Block diagram of a direct-conversion transceiver.

The filtered RF signal is then directly down-converted to I and Q channel baseband signals by the I/Q quadrature demodulator. Finally, the I and Q signals are converted to digital signals by the analog-to-digital converters, then applied to a digital signal processing (DSP) block for further processing.

The disadvantage of the zero-IF is that the LO leakage causes self-mixing that leads to a large dc offset, which could saturate the analog-to-digital converters. The LO leakage can be minimized by maintaining high isolation between a mixer's LO and RF ports. Another way to resolve dc offset problems is by converting the input signal to a frequency that is close to but not exactly dc. This implementation is known as a low-IF receiver.

6.7 Software-Defined Radios

An SDR [5] is a radio communication system where some of the functions that are normally implemented in the analog domain are instead implemented in the digital domain, leaving only the essentials to the analog domain. The advantages of the SDR software radio are that the radio can be configured or defined by the software and uses the same hardware platform for several different radios. Figure 6.10 shows a block diagram of the SDR transceiver.

In Figure 6.10, the RF front-end includes the duplexer which shared by the transmitter and receiver chains, power amplifier (PA), RF up-converter, digital-to-analog converter; LNA, RF down-converter, analog-to-digital converter, and a frequency synthesizer which shared by the transmitter and receiver chains. The digital back-end contains digital up-converter (DUC), digital down-converter (DDC), and baseband processing blocks.

In SDRs, it is easier to have one set of hardware to handle multiple wireless technologies, and easy to add new features. Figure 6.11 shows a block diagram of an SDR receiver.

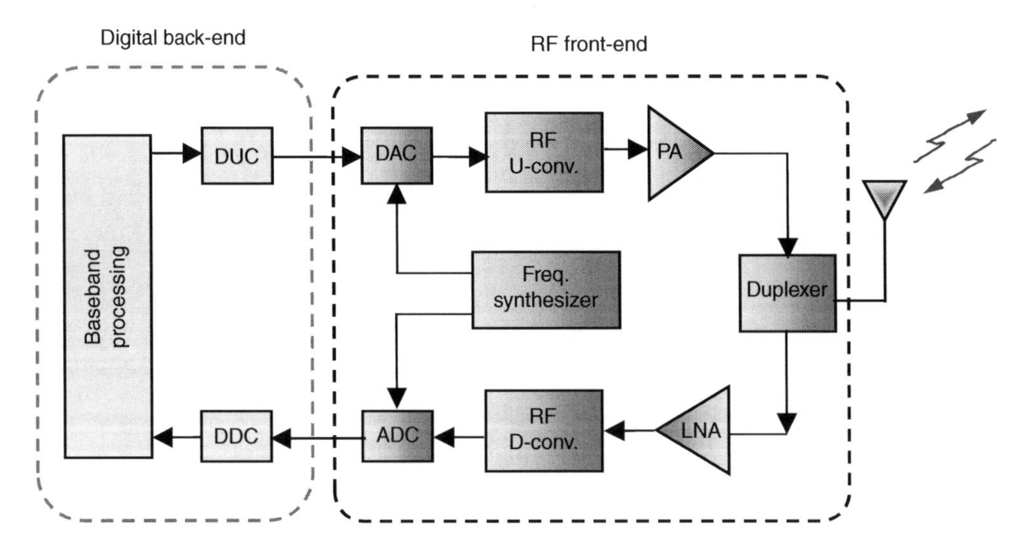

Figure 6.10 Block diagram of a software-defined radio transceiver.

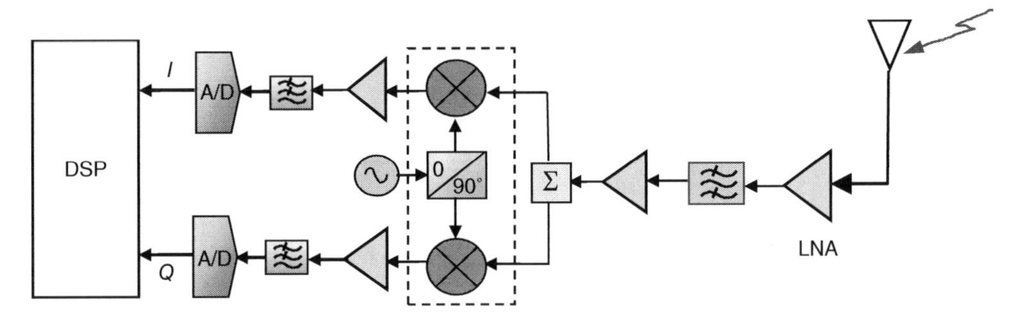

Figure 6.11 Basic structure of a software-defined radio receiver.

6.8 RF Block-Level Budget Analysis

RF block level analysis is necessary for developing RF transceivers and selecting the RF building blocks and their locations in the transmitter and receiver chains, to enable achieving the system requirements. In an RF receiver, the block-level budget analysis enables determining the cascaded NF, cascaded gain, cascaded third-order intercept point, and the number of amplification stages and their positions to achieve the balance between the NF and IP3. IP3 and NF conflict with each other, and a compromise between them would be required to meet the optimal performance. Table 6.1 shows an example of an RF block-level budget for an RF receiver frontend which requires a sensitivity of −105 dBm, and SFDR of 86 dB.

In Table 6.1, the cascaded NF does not change after the fourth RF block because of the high gain of the second block. The gain of the early stages is the parameter that influences the NF the most. The budget of Table 6.1 meets the receiver requirements with sufficient margin.

Figure 6.12 shows the Simulink [8] testbench simulation results of the RF chain; the spreadsheet calculation and Simulink simulation results are the same.

Example 6.10

Develop an RF block-level budget for a receiver frontend subsystem to meet the following requirements:

- Receiver sensitivity of −80 dBm
- Receiver bandwidth of 80 MHz
- SNR of 7 dB

Solution

First step is to determine the cascaded NF as follow:

$$S_{\min} = -174\,\text{dBm} + NF_{\text{RX}} + 10 \cdot \log(B) + \left(\frac{S}{N}\right)_o$$

Using the above equation with SNR of 7 dB, receiver sensitivity of −80 dBm, and bandwidth of 80 MHz, the calculated NF is 8.02 dB.

Table 6.1 Receiver RF block-level budget.

2110–2200 MHz frequency band								
	Receiver components							
Performance parameters	**Duplx Fltr**	**LNA**	**BPF**	**Balun**	**Chip**	**IF Balun**	**BPF**	**Hybrid**
Noise figure (dB)	2.7	0.70	2.0	0.5	3.0	0.50	1.2	4.0
Gain: Passband (dB)	–2.7	18.0	–2.0	–0.5	40.0	–0.50	–1.2	–4.0
IIP3 (dBm)	100.0	18.6	100.0	100.0	–14.0	100.0	100.0	100.0
Input power (dBm)	–60.0							
Bandwidth (MHz)	5.0							
SNR (dB)	1.0							
	Receiver components							
Cascaded parameters	**Duplx Fltr**	**LNA**	**BPF**	**Balun**	**Chip**	**IF Balun**	**Triplexer**	**Hybrid**
Noise figure (dB)	2.70	3.40	3.43	3.45	3.55	3.55	3.55	3.55
Gain: Passband (dB)	–2.70	15.30	13.30	12.80	52.80	52.30	51.10	47.10
IIP3: (dBm)	100.00	21.30	21.30	21.30	–26.80	–26.80	–26.80	–26.80
Output power (dBm)	–62.7	–44.7	–46.7	–47.2	–7.2	–7.7	–8.9	–12.9

Receiver parameter	**Calculated**
Noise figure (dB)	3.55
IIP3 (dBm)	–26.8
Gain (dB)	47.10
Sensitivity (dBm)	–110.5
SFDR (dB)	84.14

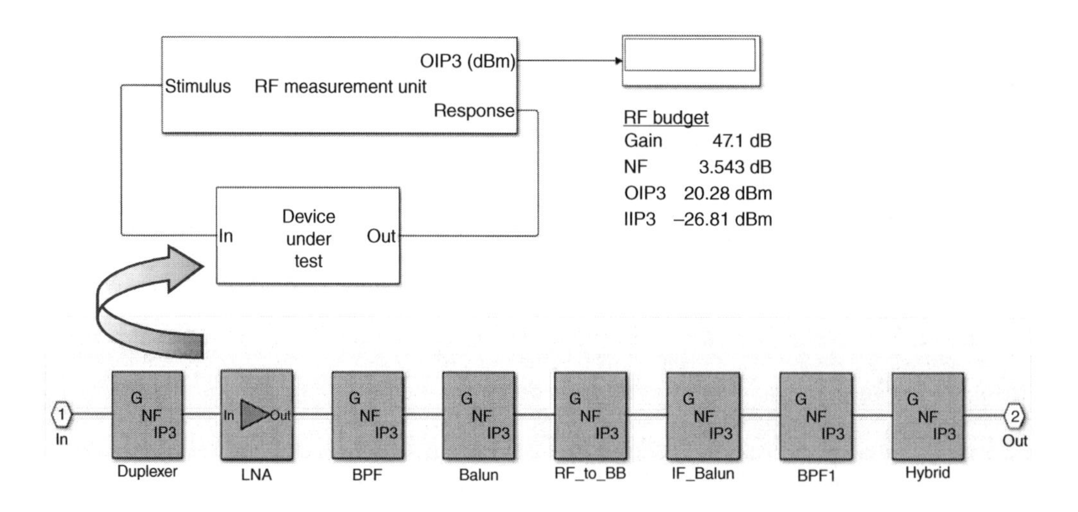

Figure 6.12 Simulink testbench of the receive chain of Table 6.1.

Figure 6.13 shows a block diagram of an RF receiver frontend subsystem block, which consists of an RF duplexer (i.e., receive bandpass filter), LNA, and down-converter mixer. The RF block-level budget of this receiver frontend is shown in Table 6.2.

Table 6.2 indicates that the selected RF blocks and their locations in the chain enable achieving the system requirement of −80 dBm receiver sensitivity and 7 dB SNR for a bandwidth of 80 MHz.

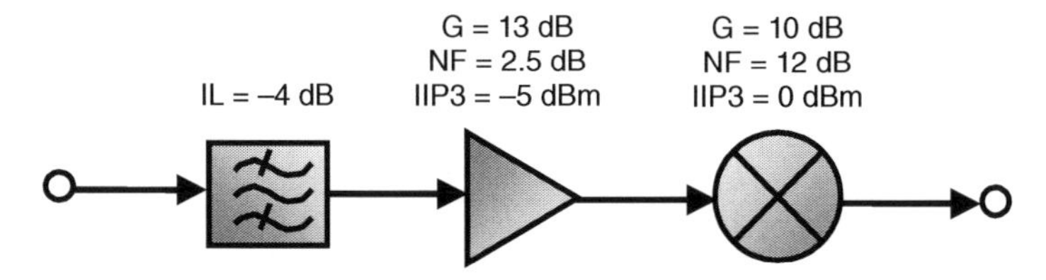

Figure 6.13 RF receiver frontend.

Table 6.2 Block-level budget of Figure 6.13.

RF block-level budget of a receiver front-end subsystem			
	Receiver building blocks		
Performance parameters	**Duplx Fltr**	**LNA**	**Mixer**
Noise figure (dB)	4.0	2.50	12.0
Gain: Passband (dB)	–4.0	13.0	10.0
IIP3 (dBm)	100.0	–5.0	0.0
Input power (dBm)	–60.0		
Bandwidth (MHz)	80.0		
SNR (dB)	7.0		
	Receiver building blocks		
Cascaded parameters	**Duplx Fltr**	**LNA**	**BPF**
Noise figure (dB)	4.00	6.50	8.02
Gain: Passband (dB)	–4.00	9.00	19.00
IIP3: (dBm)	100.00	–1.00	–9.64
Output power (dBm)	–64.0	–51.0	–41.0
Receiver parameter	**Calculated**		
Noise figure (dB)	8.02		
IIP3 (dBm)	–9.6		
Gain (dB)	19.00		
Sensitivity (dBm)	–80.0		
SFDR (dB)	81.54		

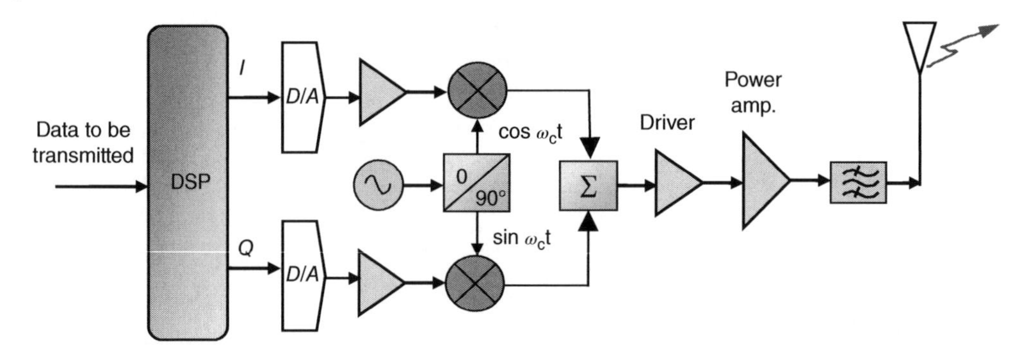

Figure 6.14 A direct-conversion transmitter.

6.9 Direct-Conversion Transmitters

Direct-conversion transmitters [7] are commonly used in cellular phone systems. Figure 6.14 shows a block diagram of direct-conversion transmitter.

In Figure 6.14, the power amplifier is needed to boost the signal to reach the receiver at the other end of the communication link. A block-level budget analysis should be performed to select the transmitter's building blocks that meet the system requirements. Table 6.3 shows an example of a transmitter RF block-level budget.

6.10 RF Transmitters System Parameters

The key performance parameters of a wireless transmitter include output power, output power DR, frequency error, EVM, occupied bandwidth (OBW), and ACLR. The following subsections describe these parameters.

6.10.1 Transmitter's Output Power

The transmitter's *output power* determines the communication range and should not exceed the requirement specification across the frequency range of operation. This parameter also influences the design specification of the transmitter's power amplifier.

6.10.2 Transmitter's Output Power Dynamic Range

The transmitter's *output power DR* is the difference between the maximum and minimum transmit power for a specific performance condition as per the standards requirements. It is limited by the transmitter's noise floor and the third-order intercept point.

6.10.3 Transmitter's Frequency Error

The *transmitter's frequency* error is the difference between the actual measured frequency and the assigned transmit frequency, it quantifies the signal quality. The frequency error is specified in parts per million (ppm), 1 ppm means $1/10^6$ part of a nominal frequency.

Table 6.3 Transmitter RF block-level budget.

		Transmitter building blocks						
Performance parameters		**Bauln**	**Gain block**	**BPF**	**Power Amp**	**Coupler**	**Isolator**	**Duplexer**
Noise figure (dB)		0.7	1.5	3.0	6.0	0.15	0.80	3.00
Gain: Passband (dB)		–0.7	22.5	–3.0	29.5	–0.15	–0.80	–3.00
Gain: Rejected band (dB)		–0.7	20.0	–35.0	20.0	–0.50	–35.00	–35.00
IIP3 (dBm)		100.0	7.5	100.0	16.5	100.00	100.00	100.00
Input noise BW (Hz)	5000000.0							
Input noise temperature (°C)	25.0							
Input power (dBm) @ the balun	–18.0							

	Transmitter building blocks						
Cascaded parameters	**Bauln**	**Gain Block**	**BPF**	**Power Amp**	**Coupler**	**Isolator**	**Duplexer**
Noise figure (dB)	0.7	2.20	2.22	2.32	2.32	2.32	2.32
Gain: Passband (dB)	–0.7	21.80	18.80	48.30	48.15	47.35	44.35
Gain: Rejected band (dB)	–0.7	19.30	–15.70	4.30	3.80	–31.20	–66.20
IIP3: In-band (dBm)	100.0	8.20	8.20	–2.67	–2.67	–2.67	–2.67

Transmitter performance parameter	**Required**
Gain (dB)	44.4
IIP3 (dBm) at the input of the gain block	2.3
Output power @ antenna port (dBm)	26.4
Out of band rejection (dB)	–66.2

6.10.4 Transmitter's Error Vector Magnitude (EVM)

EVM is one of the most important system-level parameters, it is a figure of merit for assessing the quality of digitally modulated telecommunication signals (e.g., QA and OFDM). It is the vector difference between the ideal transmit signal and the actual measured receive signal at a given time, and it is expressed as a percentage (%) or dB. Figure 6.15 illustrates the concept of EVM.

The EVM is useful because it contains information about both the amplitude and phase errors in a digital signal. It can be plotted vs. frequency and called error vector magnitude spectrum, which is helpful in finding spurs and interference that cannot be easily detected by standard power vs. frequency trances. The contributors to the EVM are the components' nonlinearity, low SNR, intersymbol interference (ISI), spurs, multipath, fading, phase noise, IQ gain imbalance, quadrature offset, and carrier feed through. Higher modulation orders require lower EVM.

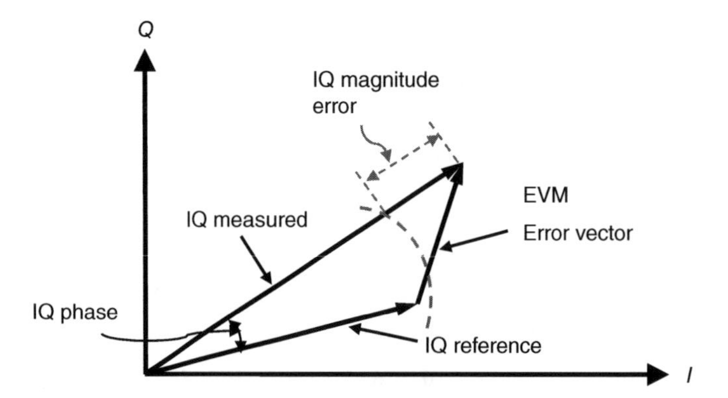

Figure 6.15 Illustration of signal's error vector magnitude.

6.10.5 Occupied Bandwidth (OBW)

Occupied bandwidth (OBW) is defined by the 3GPP (third-generation partnership project) standards as the bandwidth containing 99% of the total integrated power of the transmitted spectrum, centered on the assigned channel frequency.

6.10.6 Adjacent Channel Leakage Power Ratio (ACLR)

ACLR is used as a measure of the amount of power leaking into adjacent channels. It is defined as the ratio of the filtered mean power centered on the assigned channel frequency to the filtered mean power centered on an adjacent channel frequency.

6.10.7 Operating Band Unwanted Emissions (OBUE)

The 3GPP defines the operating band unwanted emissions as all unwanted emissions in the operating band plus a frequency range of 10 MHz above and 10 MHz below the operating band. Figure 6.16 illustrates the OBUE range.

6.10.8 Transmitter's Spurious Emission

Transmitter's *spurious emissions* are emissions at frequencies that are outside its assigned channel. These emissions are caused by unwanted transmitter effects such as harmonics emission, parasitic emission, intermodulation products, and frequency conversion products but exclude out-of-band (OOB) emissions. Regulatory agencies such as the Federal Communications Commission (FCC) in the United States and the European Telecommunications Standards Institute (ETSI) in Europe specify spurious emissions limits.

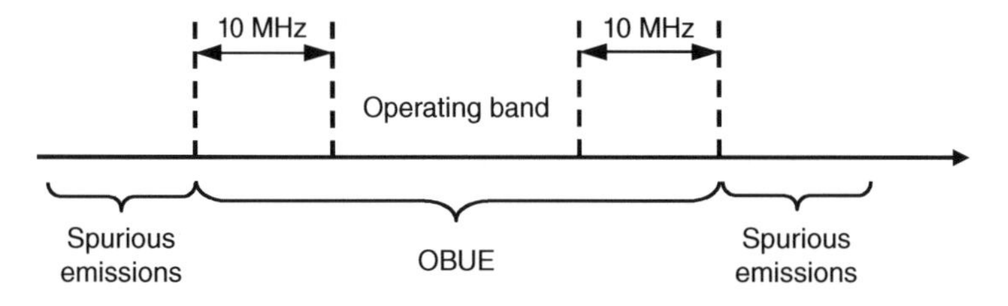

Figure 6.16 Illustration of operating band unwanted emission (OBUE).

6.10.9 Transmitter's Intermodulation

Transmitter's intermodulation products are caused by the presence of a wanted signal and an interfering signal reaching the transmitter. Such intermodulation products should be below the transmitter's spurious emission specifications. The test setups for characterizing a 5g NR base station transmitter are covered in Chapter 9.

Review Questions

6.1 Draw a block diagram of a radio transceiver and explain the function of each block.

6.2 Illustrate and describe the problem of image frequency in a full-duplex transceiver and how to mitigate such a problem.

6.3 Explain the following terms:
- receiver sensitivity,
- receiver selectivity,
- receiver intermodulation,
- receiver SFDR.

6.4 Explain the impact of receiver sensitivity on the receiver performance.

6.5 Draw a block diagram of a dual-conversion receiver and explain its advantages.

6.6 Draw a block diagram of the direct-conversion receiver and explain its advantages.

6.7 Explain the advantages and disadvantages of an SDR receiver.

6.8 Draw a block diagram of an SDR receiver and explain its advantages.

6.9 Explain the impact of the transmitter's EVM on the transmitter's performance.

6.10 Illustrate and describe the transmitter intermodulation and its impact on the system performance.

6.11 Calculate the noise temperature for a receiver that has a 7 dB NF at room temperature.

6.12 For the RF subsystem block shown below, calculate the NF and equivalent noise temperature.

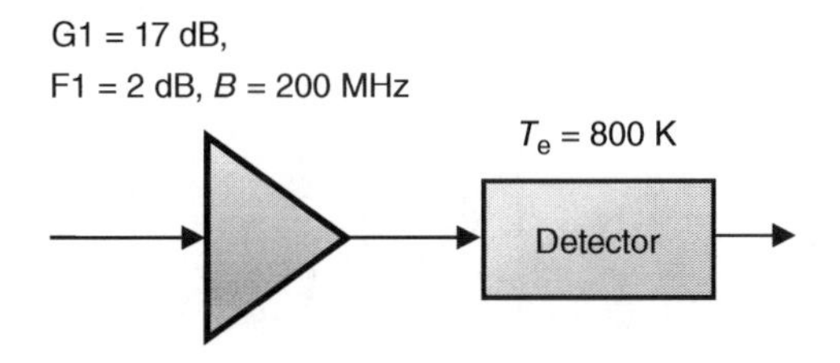

6.13 Determine the output IP3 of the RF subsystem block shown below.

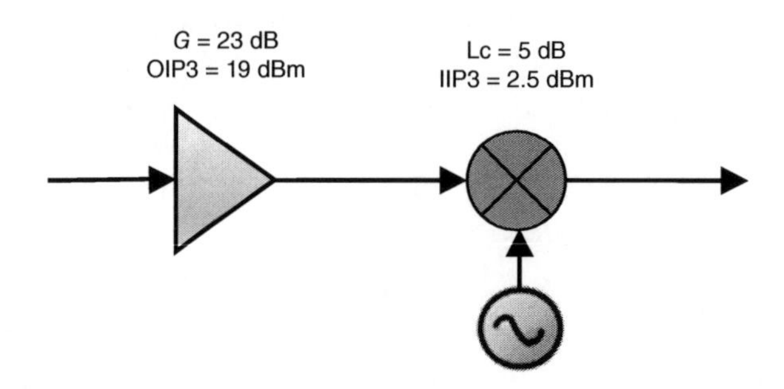

6.14 For the amplifier shown, determine the intermodulation level in dBm.

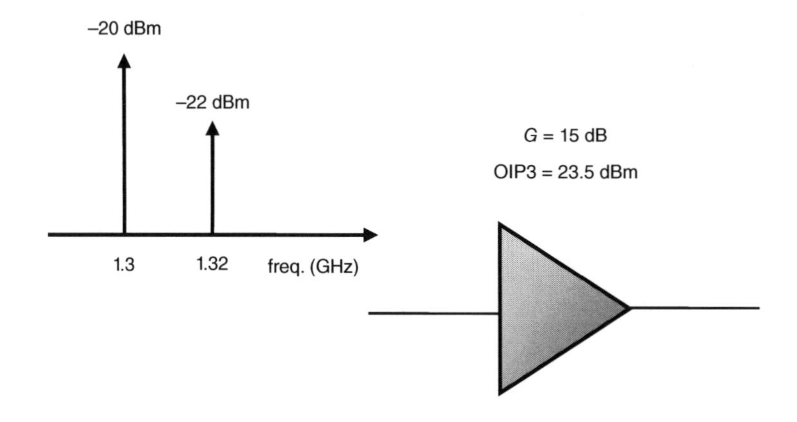

6.15 For the receiver frontend block shown, determine the DR and the SFDR if the SNR is 10 dB.

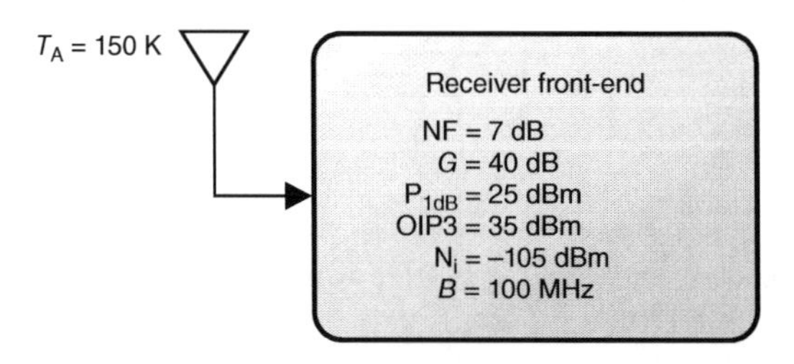

6.16 For the output spectrum of the RF subsystem block shown, determine the output DR. Assume unity gain noiseless.

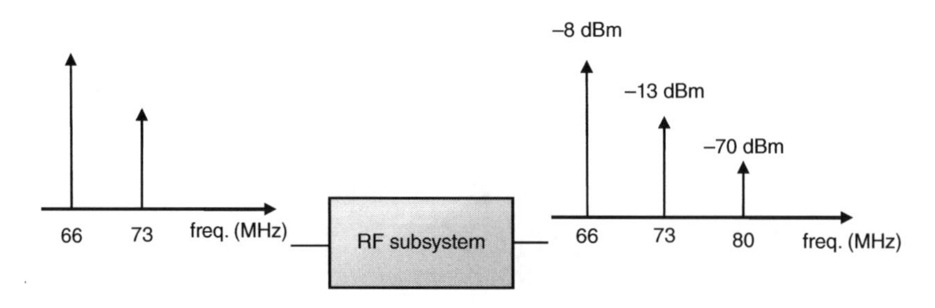

6.17 A receiver subsystem has a NF of 7 dB, a 1 dB compression point of 22 dBm, a gain of 33 dBm, and an output third-order intercept point of 35 dBm. If the subsystem is fed with a noise source with Ni = −105 dBm, and the required output SNR is 9 dB, determine SFDR. Assume the system bandwidth is 20 MHz.

6.18 For the subsystem block shown below, if the system is operating at room temperature, determine the cascaded NF, and the SNR.

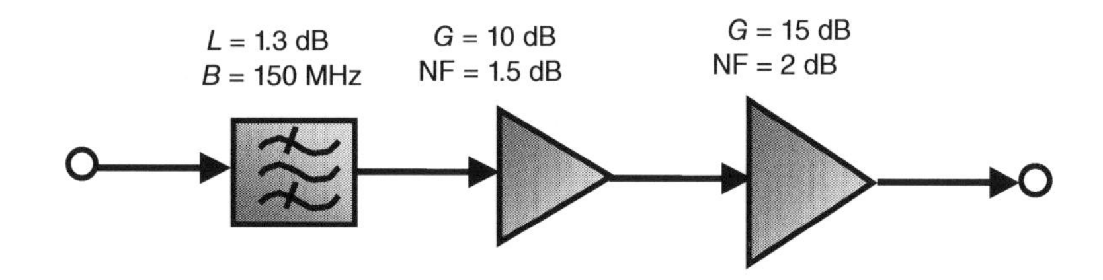

References

1 Scott R. Bullock, *Transceiver and System Design for Digital Communications*, 5th ed., The Institution of Engineering and Technology, 2017.
2 Ariel Luzzatto, Motti Haridim, *Wireless Transceiver Design: Mastering the Design of Modern Wireless Equipment and Systems*, 2nd ed., Wiley, 2016.
3 Abbas Mohammadi, Fadhel M. Ghannouchi, *RF Transceiver Design for MIMO Wireless Communications*, Springer, 2012.
4 Qizheng Gu, *RF Systems Design of Transceivers for Wireless Communications*, Springer, 2006.
5 Markus Dillinger, Kambiz Madani, Nancy Alonistioti, *Software Defined Radio: Architectures, Systems and Functions*, Wiley, 2005.
6 William F. Egan, *Practical RF System Design*, Wiley-IEEE Press, 2003.
7 A.A. Abidi, "Direct-Conversion Radio Transceivers for Digital Communications," *IEEE Journal of Solid-State Circuits*, Volume: 30, 1995.
8 Simulink – MATLAB, https://www.mathworks.com.

Suggested Readings

Alán Rodrigo Díaz-Rizo, Hassan Aboushady, Haralampos G. Stratigopoulos, "Anti-Piracy Design of RF Transceivers," *IEEE Transactions on Circuits and Systems*, Volume: 70, Issue: 1, 2023.
Armagan Dascurcu, Sohail Ahasan, et al., "A 60 GHz Phased Array Transceiver Chipset in 45 nm RF SOI Featuring Channel Aggregation Using HRM-Based Frequency Interleaving," *IEEE Radio Frequency Integrated Circuits Symposium (RFIC)*, 2022.
Christian Motz, Thomas Paireder, et al., "A Survey on Self-Interference Cancellation in Mobile LTE-A/5G FDD Transceivers," *IEEE Transactions on Circuits and Systems*, Volume: 68, Issue: 3, 2021.
Christian Motz, Thomas Paireder, Mario Huemer, "Low-Complex Digital Cancellation of Transmitter Harmonics in LTE-A/5G Transceivers," *IEEE Open Journal of the Communications Society*, Volume: 2, 2021.

Gregory M. Flewelling, “Broadband Reconfigurable Transceivers in SiGe,” *IEEE BiCMOS and Compound Semiconductor Integrated Circuits and Technology Symposium (BCICTS)*, 2020.

H.-C. Park, D. Kang, et al., “Millimeter-Wave Band CMOS RF Phased-Array Transceiver IC Designs for 5G Applications,” *IEEE International Electron Devices Meeting (IEDM)*, 2020.

Haiping Song, Xiaolei Li, et al., “Simultaneous RF Self-Interference Cancellation, Local Oscillator Generation, Frequency up- and Down-Conversion in an Integrated In-Band Full-Duplex 5G RF Transceiver Front-End,” *Journal of Lightwave Technology*, Volume: 40, Issue: 2, 2022.

Hakan Papurcu, Justin Romstadt, Steffen Hansen, Christian Krebs, Klaus Aufinger, Nils Pohl, “A Wideband Four-Channel SiGe D-Band Transceiver MMIC For TDM MIMO FMCW Radar,” *IEEE 23rd Topical Meeting on Silicon Monolithic Integrated Circuits in RF Systems,* 2023.

Jatin JM, Fouziya C, Dharmveer Pakharia, Viswanadham Ch, “A Configurable, Multi-Channel, Direct Wideband RF Sampling SWaP-C Optimized Transceiver and Processing Module for EW Systems,” *International Conference on Control, Communication and Computing (ICCC)*, 2023.

Juinn-Horng Deng, Keng-Hwa Liu, Wei-Cheng Huang, Pin-Nian Chen, Meng-Lin Ku, “Loopback Crosstalk Estimation and Compensation for MIMO Wideband Transceiver Systems: Design and Experiments,” IEEE 33rd Annual International Symposium on Personal, Indoor and Mobile Radio Communications (PIMRC), 2022.

Ming-Da Tsai, Song-Yu Yang, et al., “10.3 A 12 nm CMOS RF Transceiver Supporting 4G/5G UL MIMO,” *IEEE International Solid-State Circuits Conference - (ISSCC)*, 2020.

Mohammad Amin Karami, Kambiz Moez, “An Integrated RF-Powered Wake-Up Wireless Transceiver With −26 dbm Sensitivity,” *IEEE Internet of Things Journal*, Volume: 9, Issue: 11, 2022.

Silvester Sadjina, Christian Motz, et al., “A Survey of Self-Interference in LTE-Advanced and 5G New Radio Wireless Transceivers,” *Transactions on Microwave Theory and Techniques*, Volume: 68, Issue: 3, 2020.

Yasser M. Madany, Abdelazeem M. Abdelwahab, “Design and Analysis of RF Front-End MIMO System Based on Direct Conversion Transceiver For LTE and Wireless Communication Applications,” *International Telecommunications Conference (ITC-Egypt)*, 2022.

Yuan Wang, Heng-Tung Hsu, et al., “Design of a Compact RF Front-End Transceiver Module for 5G New-Radio Applications,” *IEEE Transactions on Instrumentation and Measurement*, Volume: 72, 2023.

Zewen Luo, Haidong Chen, Wenquan Che, Quan Xue, “Study of 28 GHz Transceiver Module Integrated with LO Source for 5G mm Wave Communication,” *IEEE MTT-S International Microwave Workshop Series on Advanced Materials and Processes for RF and THz Applications (IMWS-AMP)*, 2020.

Zhiwen Qin, Zhiming Yi, et al., “Analog Beamforming Transceiver for Millimeter-Wave Communication,” *IEEE International Students’ Conference on Electrical, Electronics and Computer Science (SCEECS),* 2022.

7

Antenna Basics and Radio Wave Propagation

This chapter covers the antenna basics including: antenna fields, radiation pattern, antenna gain, beamwidth, antenna efficiency, antenna input impedance, antenna bandwidth, antenna polarization, isotropic antenna, radiation intensity, effective radiated power, antenna types including dipole and microstrip, antenna impedance mismatch and polarization mismatch, antenna noise temperature, multielements antenna array, radio link parameters and multipath propagation, propagation pathloss, radio Fresnel zones, antenna anechoic chamber, compact antenna test range (CATR), and antenna measurements (gain, directivity, and input impedance). Review questions are provided to help the reader understand the covered topics. A list of suggested readings is provided to present the latest activities in the field of antenna design.

7.1 Chapter Objectives

On reading this chapter, the reader will be able to:

- Explain and illustrate the antenna parameters of gain, beamwidth, bandwidth, input impedance, radiation efficiency, directivity, antenna polarization, main lobe and side lobes of a radiation pattern, and the antenna's back-to-front ratio.
- Describe the difference between near and far fields of an antenna and the difference between E-plane and H-plane radiation patterns.
- Explain and illustrate different test setups for measuring antenna parameters.
- Explain and illustrate the compact antenna test range method for testing antenna radiation patterns.
- Explain the concept of a liner array antenna and its impact on the performance of a radio link.
- Compare and contrast dipole and microstrip antennas.
- Describe the Fresnel zones in a radio link.
- Explain and illustrate the delay spread of a radio channel.
- Explain the difference between frequency selective fading and flat fading.
- Calculate the free space pathloss for a given distance and operating frequency.

7.2 Introduction

Antennas [1–7] are used in wireless communication systems to radiate and receive electromagnetic (EM) waves (i.e., radio waves) to and from the free space. An antenna is a reciprocal device that can transmit and receive radio waves. The transmit and receive antennas are part of the communications radio link. The radio link is the signal path between the transmitter's output and the receiver's input, including the transmission line from the transmitter's output to the antenna, the transmit antenna, the free-space propagation path, the receive antenna, and the transmission line from the receive antenna to the receiver's input. Figure 7.1 represents the transmit and receive antenna in a radio communications link. Figure 7.2 shows the representation of the transmit and receive antennas models.

Radio waves from the transmitter to the receiver of a wireless communication link get attenuated due to the propagation path loss, diffraction, reflection, scattering, blocking, and transmission lines' losses. Radio waves are EM waves made of electric (E) and magnetic (H) fields, and they are perpendicular to each other and perpendicular to the direction of propagation. The EM waves propagate in free space with the speed of light (3×10^8 m/s). Figure 7.3 shows a representation of an EM wave.

In radio systems, the transmit antenna radiates spherical waves that are considered plane waves at a distance referred to as the far-field region. The far-field region is referred to as the ***Fraunhofer region***. The criterion for far-field operation is given by

$$R_{\text{far-field}} \geq \frac{2D^2}{\lambda} [\text{m}] \tag{7.1}$$

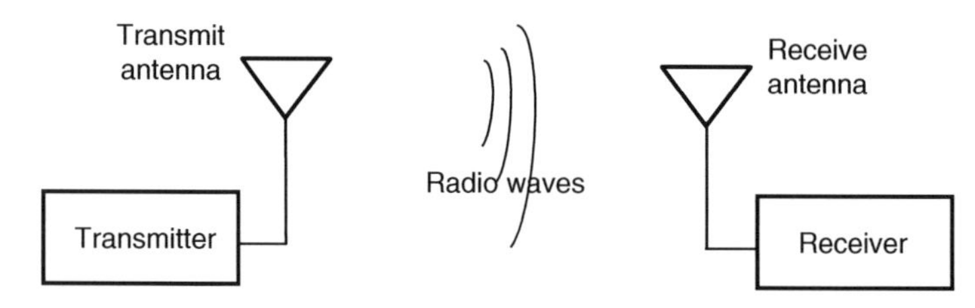

Figure 7.1 Representation of a radio communications link.

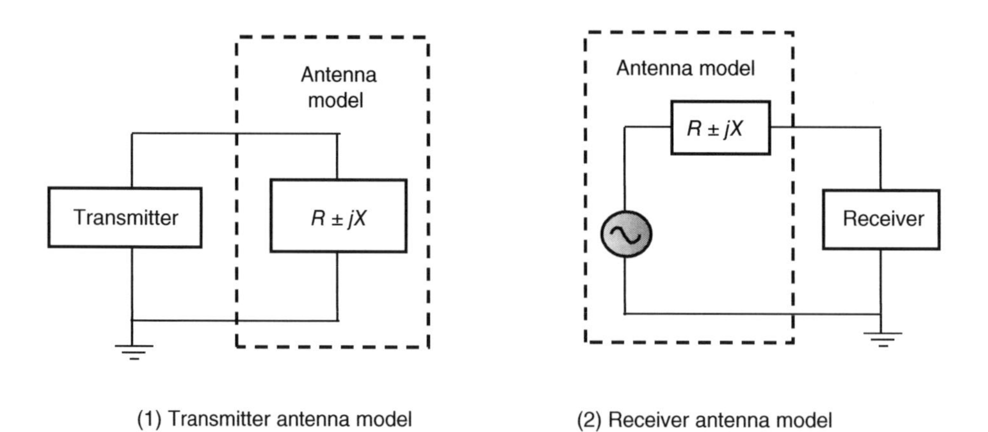

Figure 7.2 Representation of transmit and receive antenna models.

Figure 7.3 Representation of transverse electromagnetic wave.

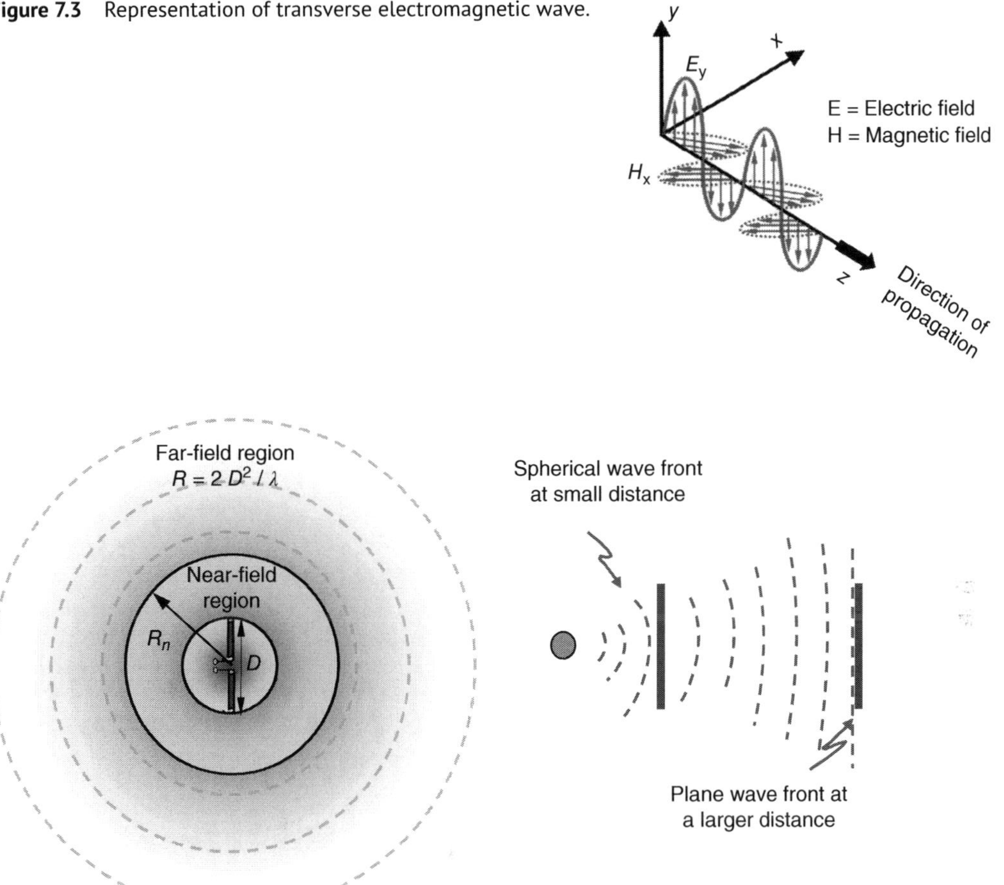

Figure 7.4 Representation of near-field and far-field regions.

where D is the maximum dimension of the antenna in meters, and λ is the free-space wavelength of the propagating signal ($\lambda = c/f$, c is the speed of light). In the far-field region, the spherical wavefront from a point source can be approximated by a uniform plane wavefront. Figure 7.4 illustrates the far and near fields. The reactive near-field region is called ***Fresnel region*** and is given by

$$R_{\text{near-field}} < 0.62\sqrt{\frac{D^3}{\lambda}} \tag{7.2}$$

In a radio system, the receiver antenna intercepts a portion of the radio waves and converts the EM waves to electrical signal to be processed in the receiver. The antenna is considered one of the most crucial building blocks in a wireless communication link because it influences the communication range, transmitter's output power, and the receiver sensitivity.

Example 7.1

For a parabolic antenna that has a diameter of 0.5 m and operates at 12.5 GHz, determine the far-field distance of the antenna.

Solution

The far-field distance is calculated as follows:

$$R_{\text{far-field}} = \frac{2D^2}{\lambda}, \qquad \lambda = \frac{c}{f} = \frac{3 \times 10^8}{12.5 \times 10^9} = 0.024\ \text{m}.$$

$$\therefore\ R_{\text{far}} = \frac{2(0.5)^2}{0.024} = 20.8\ \text{m}.$$

Example 7.2

For a dipole antenna that has a length of 25 cm and operates at 900 MHz, determine the far-field distance of the antenna.

Solution

The far-field distance is calculated as follows:

$$\lambda = \frac{c}{f} = \frac{3 \times 10^8}{900 \times 10^6} = 0.3\ \text{m}.$$

$$\therefore R_{\text{far-field}} = \frac{2D^2}{\lambda} = \frac{2(0.25)^2}{0.3} = 0.375\ \text{m},$$

7.3 Antenna Fields

Antenna fields are typically represented in a spherical coordinate system. Figure 7.5 shows electric field (i.e., E) components in a spherical coordinate system.

The radiated electric field of an antenna that is located at the origin of a spherical coordinate is expressed as

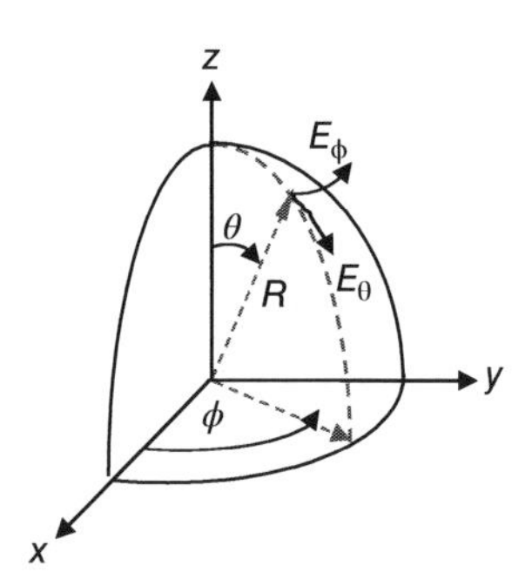

Figure 7.5 Electric field components in a spherical coordinate system.

$$\overline{E}(r,\theta,\varphi) = [\hat{\theta}\, F_\theta(\theta,\varphi) + \hat{\emptyset} F_\emptyset(\theta,\varphi)]\, \frac{e^{-jk_o r}}{r} \qquad V/m \tag{7.3}$$

where

$\overline{E}$ = electric field

$F_\theta(\theta, \varphi), F_\emptyset(\theta, \varphi)$ = radiation pattern functions

$\hat{\theta}, \hat{\emptyset}$ = unit vectors in spherical coordinate system

r = distance from the origin of the sphere coordinates to the field point

$k_o = 2\pi/\lambda$ in free space

λ = wavelength

$c = 3 \times 10^8$ m/s, speed of light

Equation (7.3) indicates that the E field propagates with a phase variation of $e^{-jk_o r}$ and an amplitude variation of $1/r$. For an antenna to radiate, it must be excited by a time-varying source. The electric field may be polarized in either $\hat{\theta}$ or $\hat{\emptyset}$ direction but not in the radial direction. Thus, this signal is defined as transverse electromagnetic wave (TEM). For a TEM wave, the electric and magnetic fields are given by

$$H_\varphi = \frac{E_\theta}{\eta_o} \tag{7.4}$$

$$H_\theta = \frac{E_\phi}{\eta_o} \tag{7.5}$$

where η_o is the free space wave impedance (120 π or 377 Ω). The Poynting vector which represents the power flow of an EM fields is given by

$$\overline{S} = \overline{E} \times \overline{H}^{*} \quad \text{w/m}^2 \tag{7.6}$$

and the time-average Poynting vector is given by

$$\overline{S}_{\text{avg}} = \frac{1}{2}\ Re[\overline{E} \times \overline{H}^{*}] \quad \text{w/m}^2 \tag{7.7}$$

7.4 Antenna Radiation Pattern and Parameters

The radiation pattern of an antenna represents the variation of the field intensity of the antenna as a function of angles relative to a reference axis. It is usually measured in the far-field in two orthogonal planes called vertical and horizontal (*sometimes called E-Plane because it has an E field and H-Plane because it has an H field*) planes. These planes are related to the orientation in which the antenna is located. Figure 7.6 shows the elevation and azimuth radiation patterns of a 900 MHz

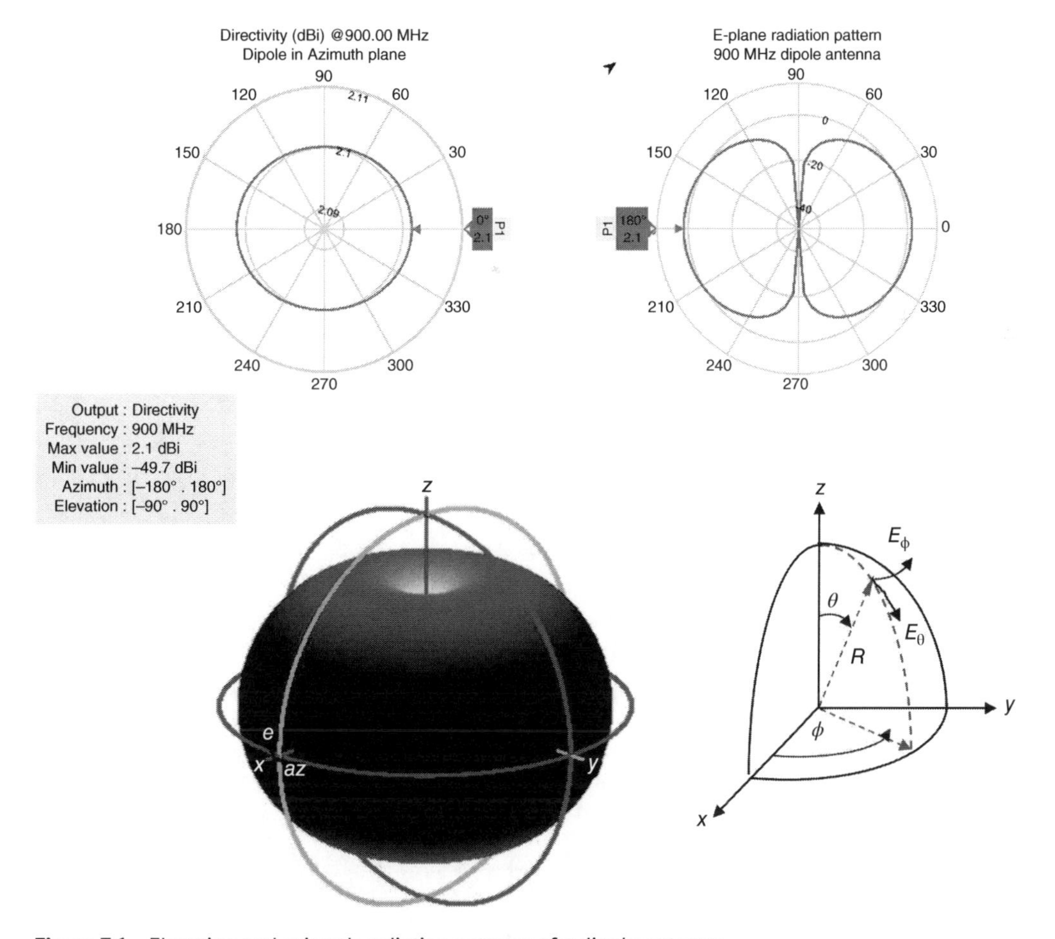

Figure 7.6 Elevation and azimuth radiation patterns of a dipole antenna.

dipole antenna and its coordinate system. In Figure 7.6, the *xz*-cut is the E-plane, and the *xy*-cut is the H-plane.

The far-field radiation pattern parameters of an antenna are illustrated in Figure 7.7. The lobes of an antenna's radiation pattern are illustrated in Figure 7.8.

In an ideal antenna, 100% of the radiated power is concentrated in the main lobe, and no other lobes exist. However, in real antennas, additional lobes (i.e., side lobes and back lobes) exist due

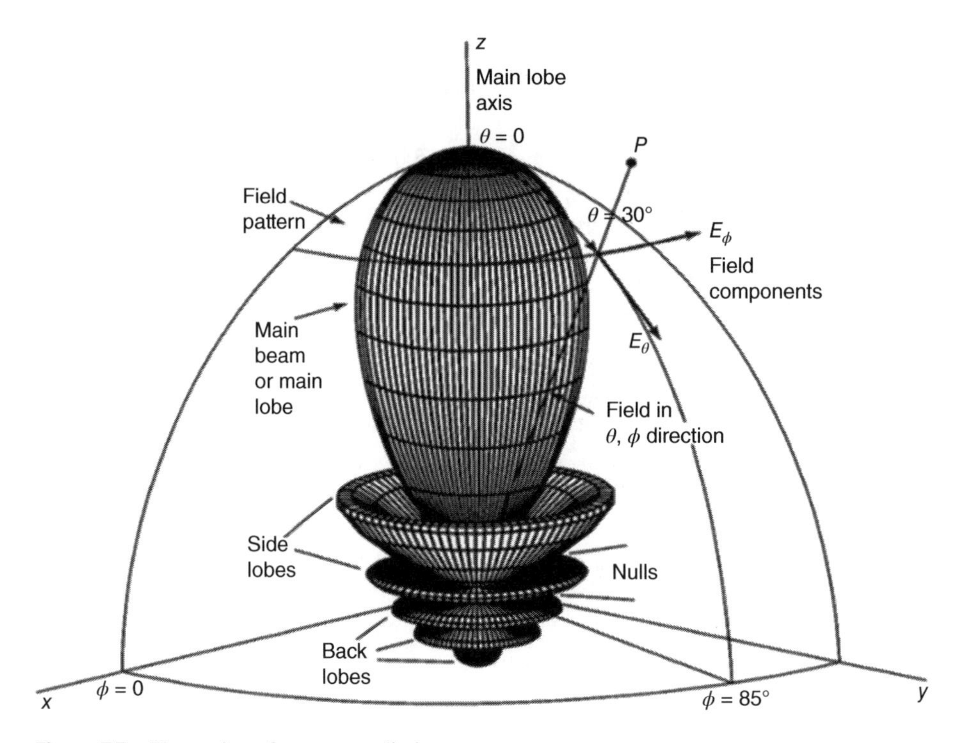

Figure 7.7 Illustration of antenna radiation pattern parameters.

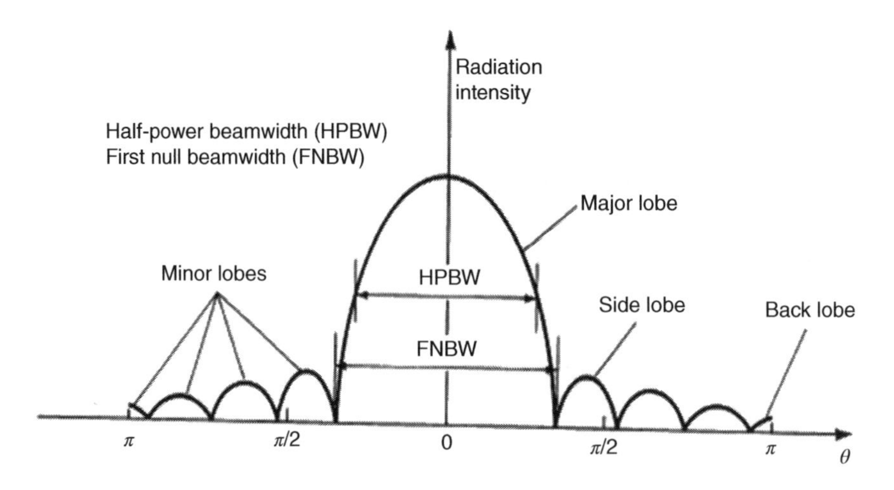

Figure 7.8 Lobes of an antenna radiation pattern.

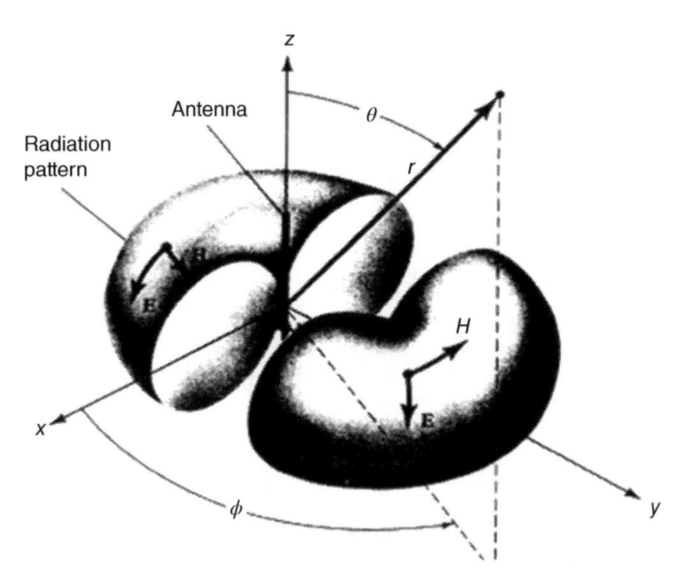

Figure 7.9 Radiation patterns of an omnidirectional antenna.

to the antenna's manufacturing process and materials. These side lobes represent unwanted signals that could cause harmful interference to other systems. In regions close to the antenna, the fields change rapidly with the distance and contain both radiating energy and reactive energy. The near-field region is referred to as the Fresnel region.

7.4.1 Antenna Gain

Antenna gain is a passive phenomenon; thus, the antenna does not add power but redistributes the power to provide more radiated power in a specific direction than would be transmitted by an isotropic antenna (isotropic antenna radiated equally in all directions). When the gain of an antenna goes up, the beamwidth becomes narrow, and as a result, the radiation pattern becomes more directive. If a reference antenna for estimating the antenna gain is an isotropic antenna, the gain will be expressed in dBi (i.e., decibels relative to isotropic), but if the reference is a dipole antenna, the gain will be expressed in terms of dBd (decibels relative to dipole, 0 dBd = 2.15 dBi). Figure 7.9 shows a radiation pattern of an omnidirectional antenna.

For an isotropic antenna (i.e., direction independent), no loss mechanisms, therefore, the input power and the radiated power are the same. However, for a real antenna, there will be losses that reduce its radiation efficiency η_{rad}. Thus, the antenna gain can be expressed as

$$G = \eta_{\text{rad}} D \tag{7.8}$$

where D is called the antenna's directivity, which is a measure of its ability to direct RF energy in a desired direction. It is defined as the ratio of the maximum radiation intensity $U_{\max}$ to the average radiation intensity U_{av} and given by

$$D = \frac{U_{\max}}{U_{\text{av}}} \tag{7.9}$$

For an isotropic antenna, the average radiation intensity U_{av} is given by

$$U_{\text{av}} = \frac{P_t}{4\pi r^2} \quad [\text{W/m}^2], \tag{7.10}$$

For a directive antenna, the radiation density is expressed as

$$U_{av} = \frac{G_t P_t}{4\pi r^2} \quad [\text{W/m}^2] \tag{7.11}$$

where r is the radius of an imaginary sphere enclosing the source, and $4\pi r^2$ is the area of that sphere at any given distance, P_t is the power at the antenna input, and G_t is the antenna gain.

There are several commercial software Antenna design tools [8–13] that enable characterizing the antenna and shorten the design cycle and time to market.

Example 7.3

For the link shown below, determine the gain of the receive antenna.

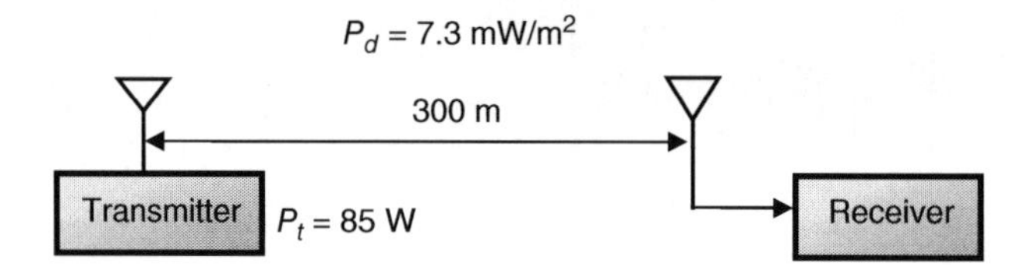

Solution

The receiver's antenna gain can be calculated as follows:

$$P_{av} = \frac{G P_t}{4\pi r^2} \quad [\text{W/m}^2],$$

$$\therefore\ G = 7.3 \times 10^{-3} \times 4\pi \times (300)^2/85 = 99.79 \quad \text{or} \quad 19.99\ \text{dBi}$$

7.4.2 Antenna Beamwidth

The antenna beamwidth is defined as the angle between the half-power points of the main lobe of the radiation pattern and is referred to as HPBW (i.e., half-power beamwidth), as shown in Figure 7.8.

7.4.3 Antenna Efficiency

The antenna efficiency (η_{rad}) is the ratio of the total radiated power to the antenna's input power and is given by

$$\eta_{rad} = \frac{P_{rad}}{P_{in}} = \frac{P_{in} - P_{loss}}{P_{in}} \tag{7.12}$$

where P_{rad} is the radiated power by the antenna, P_{in} is the transmitter's output power fed to the antenna's input, and P_{loss} is the power lost in the antenna due to the antenna materials.

7.4.4 Antenna Aperture Efficiency

Aperture antennas are antennas that have well-defined aperture area from which the radiation occurs. Such antennas include reflector antennas, horn antennas, and array antennas. The maximum directivity of an aperture antenna is given by

$$D_{max} = \frac{4\pi A}{\lambda^2}. \tag{7.13}$$

where A is the area of the aperture, and λ is the operating wavelength of the antenna. In practice, the maximum directivity is impacted by the amplitude and phase characteristics of the aperture. The ratio of the maximum directivity to the actual directivity is defined as the *aperture efficiency*. Thus, the directivity of an aperture antenna is expressed as

$$D = \eta_{ap} \frac{4\pi A}{\lambda^2}. \tag{7.14}$$

Example 7.4

For a reflector antenna that has a diameter of 0.457 m and operating frequency of 12.4 GHz, determine the directivity if the antenna aperture efficiency is 65%.

Solution

The directivity can be calculated as follows:

$$D = \eta_{ap} \frac{4\pi A}{\lambda^2}, \qquad \lambda = \frac{c}{f} = \frac{3 \times 10^8}{30 \times 10^6} = 10\,\text{m}.$$

$$A = \pi(\text{diameter}/2)^2 = \pi(0.457/2)^2 = 0.164\,\text{m}^2.$$

$$\therefore\ D = 0.65\,\frac{4\pi(0.164)}{(10)^2} = 2289 = 33.6\,\text{dB}$$

Example 7.5

For an antenna that has input power of 40 W and an efficiency of 98% and assuming the maximum radiation density is 20 mW/m^2 at 100 m from the antenna, determine the directivity and the gain relative to an isotropic antenna.

Solution

The directivity is given by

$$D = \frac{U_{max}}{U_{av}}, \quad \text{and} \quad \eta_{rad} = \frac{P_{rad}}{P_{tx}}$$

$$\therefore\ P_{rad} = 0.98\,(40) = 39.2\,\text{W}.$$

The average radiated power density is

$$U_{av} = \frac{P_{tr}}{4\pi\, r^2} = \frac{39.2}{4\pi\,(100)^2} = 3.12 \times 10^{-4}\ \frac{\text{W}}{\text{m}^2}, \text{thus}$$

$$\therefore\ D = \frac{20 \times 10^{-3}}{3.12 \times 10^{-4}} = 64.1 \quad \text{or} \quad 18\,\text{dB}$$

$$G = \eta_{rad} \cdot D = 0.98\,(64.1) = 62.83 \text{ or } 17.98\,\text{dBi}$$

Example 7.6

For a transmitter that puts a 50 W RF signal into an antenna that has gain of 2.15 dB, determine the radiated power density of the signal at 48.3 km.

Solution

The radiated power density is given by

$$U_o = \frac{P_{tr}}{4\pi\, r^2} = \frac{G \times P_{in}}{4\pi\, r^2} = \frac{1.64 \times 50}{4\pi\,(48300)^2} = 2.8 \times 10^{-9} = 1.7\,\text{nW/m}^2.$$

Example 7.7

For an antenna which has a 10 W signal at its input and has efficiency of 90% determine the radiated power.

Solution

The radiated power can be calculated as follows:

$$\eta_{\text{rad}} = \frac{P_{\text{rad}}}{P_{\text{in}}}, \ \therefore \ P_{\text{rad}} = \eta_{\text{rad}} \times P_{\text{in}} = 0.9\,(10) = 9\ \text{W}$$

7.4.5 Antenna Input Impedance

Antennas are considered one-port networks and can be described by their S_{11} parameters or their reflection coefficients Γ. Knowing Γ, the input impedance of an antenna can be calculated from the following equation

$$Z_{\text{in}} = Z_0 \frac{1+\Gamma}{1-\Gamma} \tag{7.15}$$

where Z_0 is the characteristic impedance of the transmission line that is connected to the antenna. The antenna impedance is a complex quantity given by

$$Z_{\text{in}} = R_{\text{ANT}} + X_{\text{ANT}} \tag{7.16}$$

where R_{ANT} is the resistive part of the antenna impedance (it includes the radiation resistance and the losses of the antenna), and X_{ANT} is the reactive part of the antenna impedance. The impedance matching between the antenna's input and the RF system should ensure maximum power transfer.

In a radio transmitter, if the antenna is not matched, the transmitter's output power will be reduced and that impacts the communication range. In a receiver, if the antenna is not matched, the SNR will be reduced and as a result the receiver sensitivity will be degraded, which increases the system bit-error rate.

7.4.6 Antenna Bandwidth

The bandwidth of an antenna is the range of frequencies over which the antenna parameters such as input impedance, gain, directivity, beamwidth, side-lobe levels, etc., must meet the system specifications.

7.4.7 Antenna Polarization

Antenna polarization describes the orientation of the electric field (E-plane) with respect to the plane of propagation. Receive antennas intercept only the electric field components which align with their polarization. The three commonly used polarizations are vertical polarization, horizontal polarization, and circular polarization. Figure 7.10 shows a vertically polarized antenna.

If the transmitting and receiving antennas have different polarizations, portion of the signal will be lost, and if the transmitting and receiving polarization angles differ by 90 degrees, no signal will be received. A circularly polarized signal can be received by any linearly polarized antenna. Figure 6.9 shows liner and circular polarizations of a microstrip patch antenna.

The antenna polarization depends on the feeding of the antenna, in Figure 7.11, the vertically polarized patch antenna has the electric field E_y perpendicular to the plane of the propagation,

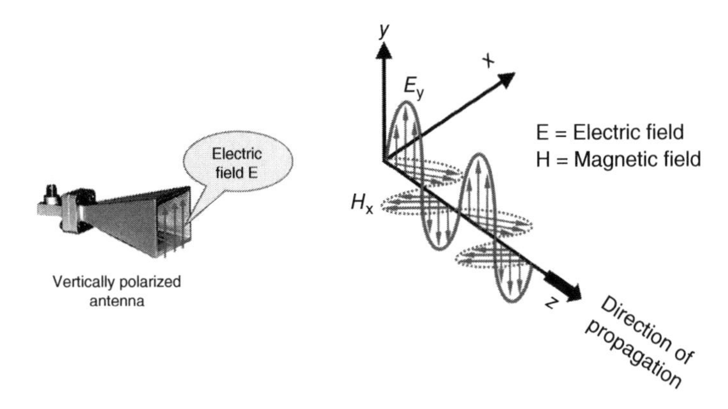

Figure 7.10 Vertically polarized antenna.

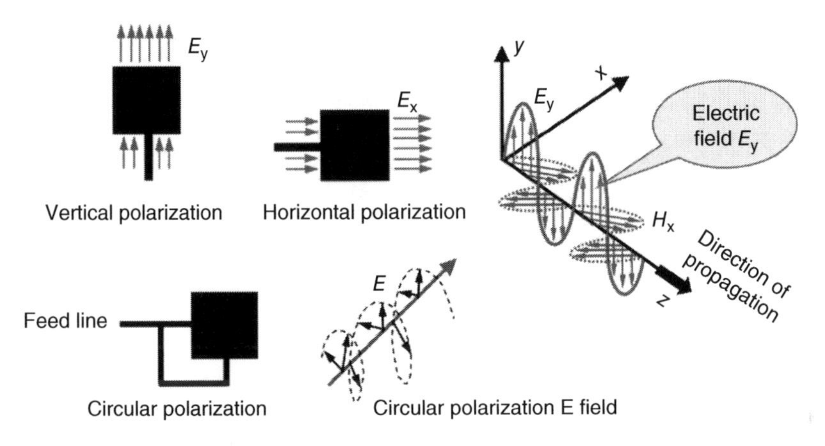

Figure 7.11 Linear and circular polarizations of microstrip antenna.

whereas the horizontally polarized patch has the electric field parallel to the plane of propagation. Feeding the patch at two different points and a 90-degree phase shift between the currents at each feed point would provide circular polarization. In circular polarization, the transmitted signal can be detected by the receive antenna regardless of its polarization. Circular polarization can be right-hand circular (RHC) or left-hand circular (LHC) polarization, and it depends on the rotation of the wave.

7.5 Isotropic Antenna

An isotropic antenna is a hypothetical point-source radiator that radiates equally in all directions (directivity = 1 = 0 dB). Figure 7.12 shows a representation of an isotropic point-source antenna.

Assuming the transmit power of a point-source radiator is $\boldsymbol{P_t}$, the power density $\boldsymbol{P_d}$ at a distance $\boldsymbol{r}$ from the source can be written as

$$P_d = \frac{P_t}{4\pi\, r^2} \quad (W/m^2) \tag{7.17}$$

Equation (7.17) indicates that the power density varies inversely as the square of the distance (r) from the point-source radiator. All practical antennas are characterized with respect to a reference isotropic antenna.

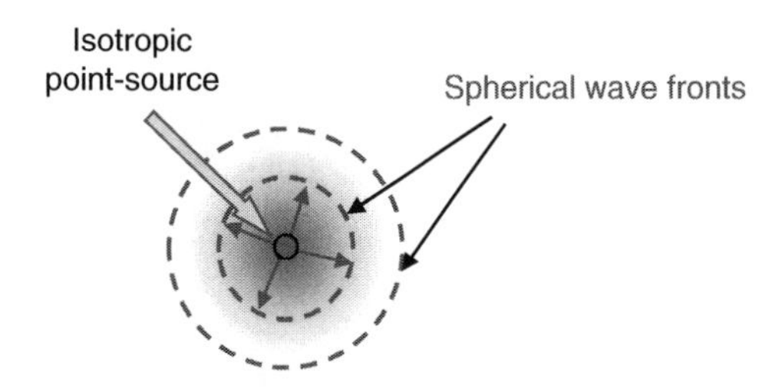

Figure 7.12 Representation of an isotropic point-source antenna.

7.6 Fields Due to Short Antenna

The simplest type of a wire antenna is a short, very thin wire of length ***L*** and carrying a current ***I*** **sin(ωt)**. When the current flows in the antenna, it generates a magnetic field ***H*** in its surrounding environment, and this magnetic field has an associated electric field ***E***. Both the electric and magnetic fields create EM waves according to Maxwell's equations [2–4]. Figure 7.13 shows a short antenna placed along the *z*-axis at the center of a spherical coordinate.

In Figure 7.13, the electric and magnetic field components at a point ***P*** that is located at a distance ***R*** from the antenna can be written as

$$E_\theta = j\eta_o \frac{kILe^{-jkr}}{4\pi r}\left[1 + \frac{1}{jkr} - \frac{1}{(kr)^2}\right] \cdot \sin(\theta) \tag{7.18}$$

$$E_r = \eta_o \frac{kILe^{-jkr}}{2\pi r^2}\left[1 + \frac{1}{jkr}\right] \cdot \cos(\theta) \tag{7.19}$$

$$E_\emptyset = 0 \tag{7.20}$$

where E_θ, E_ϕ, and E_r are the electric field components, and $\boldsymbol{\eta_o}$ is the free-space wave impedance ($\eta_o = 120\pi$), ***I*** is the electric current, and $\boldsymbol{k} = 2\pi/\lambda$. In the far-field region, r^2 and r^3 can be neglected; thus, the fields can be written as

$$E_\theta = j\eta_o \frac{kILe^{-jkr}}{4\pi r} \cdot \sin(\theta) \tag{7.21}$$

$$H_\emptyset = j\frac{kILe^{-jkr}}{4\pi r} \cdot \sin(\theta) \tag{7.22}$$

$$E_r = 0 \tag{7.23}$$

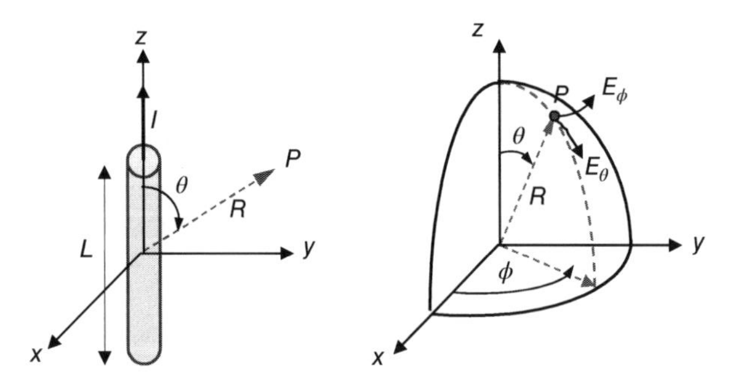

Figure 7.13 Short linear antenna in a spherical coordinate.

Figure 7.14 Transverse electric (*E*) and magnetic (*H*) fields.

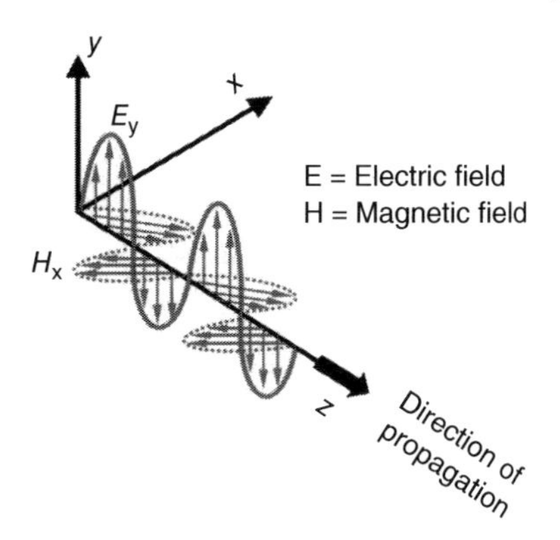

The only nonzero fields are E_θ and H_ϕ, this means the total electric and magnetic fields are transverse to each other. The free-space wave impedance η_o can be written as

$$\eta_o = \frac{E_\theta}{H_\phi} \tag{7.24}$$

where

$$H_\phi = \frac{E_\theta}{\eta_o}, H_\theta = \frac{E_\phi}{\eta_o} \tag{7.25}$$

The electric field intensity E is expressed as

$$E = \sqrt{E_\theta^2 + E_\phi^2} \tag{7.26}$$

Figure 7.14 illustrates the E and H fields, and the direction of propagation of a TEM wave.

7.7 Received Power and Electric Field Strength

The transmitted EM waves spread out as they travel through the free space. Thus, a receiving antenna of an effective area (A_{eff}) parallel to the wavefronts will collect power, P_r, which is given by

$$P_r = P \cdot A_e \cdot \eta_{\text{rad}} \tag{7.27}$$

where η_{rad} is the antenna efficiency, P is the radiated power ($P = E^2/\eta_o$), and A_e is the effective area that also called effective aperture, as shown in Figure 7.15.

The antenna's effective area is given by

$$A_e = \frac{\lambda^2}{4\pi} G_r \tag{7.28}$$

For an isotropic antenna, η_{rad} is one and the received power P_r at a distance $\boldsymbol{d}$ from the transmitter can be written as

$$P_r = \frac{P_t \cdot G_t}{4\pi d^2} \cdot \left(\frac{\lambda^2 G_r}{4\pi}\right) = P_t \cdot G_t \cdot G_r \cdot \left(\frac{\lambda}{4\pi d}\right)^2 \tag{7.29}$$

where P_t is the transmitted power, G_t and G_r are the gains of the transmit and receive antennas, respectively, and λ is the wavelength. Equation (7.29) can be written as

$$\frac{P_r}{P_t} = \frac{G_t G_r}{\left(\frac{4\pi d}{\lambda}\right)^2} \tag{7.30}$$

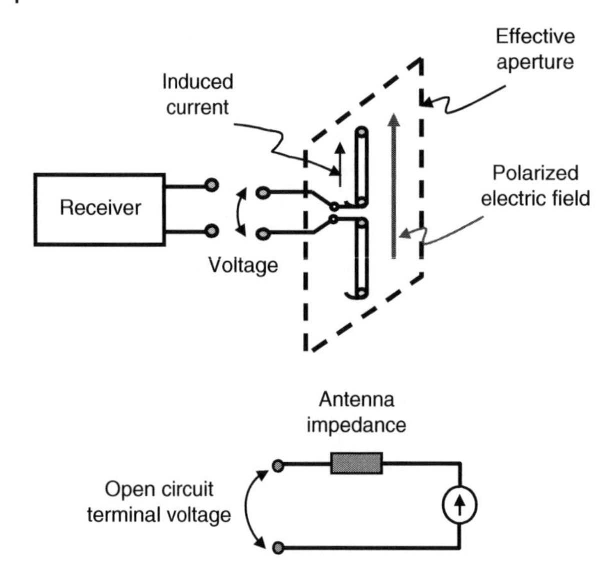

Figure 7.15 Illustration of antenna aperture.

Equation (7.30) can be expressed as

$$P_r(\text{dBm}) = P_t(\text{dBm}) + G_t(\text{dB}) + G_r(\text{dB}) - \text{PL(dB)} \tag{7.31}$$

where PL is the propagation path loss, thus,

$$\text{PL (dB)} = 10 \times \log\left(\frac{4\pi d}{\lambda}\right)^2 \tag{7.32}$$

Example 7.8

Determine the effective area of a half-wavelength lossless dipole antenna that operates at a frequency of 30 MHz.

Solution

The effective area can be calculated as follows:

$$A_e = \frac{\lambda^2}{4\pi} G_r, \qquad \lambda = \frac{c}{f} = \frac{3 \times 10^8}{30 \times 10^6} = 10\,\text{m}.$$

A half-wave dipole has an antenna gain of 1.64 (i.e., 2.15 dBi). Thus, the effective area is

$$A_e = \frac{(10)^2}{4\pi}(1.64) = 13.05\,\text{m}^2$$

Example 7.9

Calculate the received power at a receiver if the transmitter is located 20 km from the receiver; the antenna gains of the transmitter and the receiver are 30 dBi and 35 dBi, respectively, the transmitter's power is 37 dBm, and the operating frequency is 4 GHz.

Solution

The received power can be calculated as follows:

$$\text{PL (dB)} = 10 \times \log\left(\frac{4\pi d}{\lambda}\right)^2 = 20\,\log\left(4\pi \times 20 \times 10^3 \frac{4 \times 10^9}{3 \times 10^8}\right) = 130.5\,\text{dB}$$

$$P_r(\text{dBm}) = P_t(\text{dBm}) + G_t(\text{dB}) + G_r(\text{dB}) - \text{PL}$$

$$\therefore\ P_r(\text{dBm}) = 37 + 30 + 35 - 130.5 = -28.5\,\text{dBm}$$

Example 7.10
For the radio link shown below, determine the following:

free-space path loss
antenna gain
received power in dBm

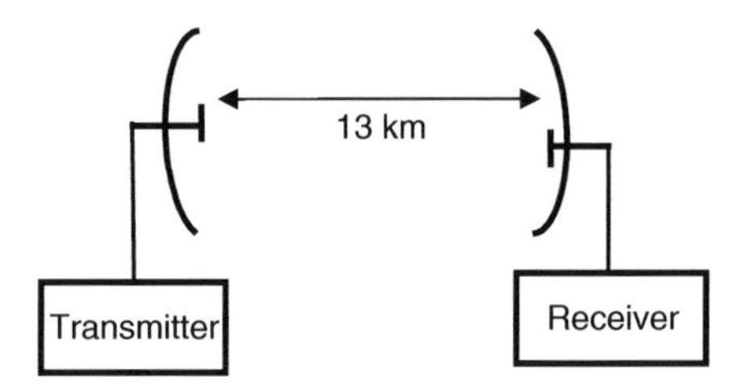

Solutions
The free space path loss is given by

$$\mathrm{PL(dB)} = 10 \times \log \left(\frac{4\pi d}{\lambda} \right)^2$$

$$\therefore \ \mathrm{PL} = 20 \ \log \left(\frac{4\pi \, (13 \times 10^3)}{3 \times 10^8 / 3 \times 10^9} \right) = \mathbf{124.26 \ dB}.$$

The antenna gain can be calculated as follows:

$$G_r = \frac{4\pi}{\lambda^2} A_e, \quad A_e = \pi(1)^2, \quad \lambda = 0.1 \text{ m}. \quad \text{Thus,} \qquad G_r = \frac{4\pi}{(0.1)^2}(\pi) = \mathbf{35.96 \ dB}$$

The received power can be calculated as

$$P_r(\mathrm{dBm}) = P_t(\mathrm{dBm}) + G_t(\mathrm{dB}) + G_r(\mathrm{dB}) - \mathrm{PL(dB)},$$

$$\text{and} \quad P_t(\mathrm{dBm}) = 10 \ \log(2) + 30 = 33 \ \mathrm{dBm}$$

$$\therefore \ P_r(\mathrm{dBm}) = 33 + 35.96 + 35.96 - 124.26 = \mathbf{-19.32 \ \ dBm}$$

7.8 Effective Radiated Power

The effective radiated power (ERP) is the actual power radiated by the antenna in a given direction and is given by

$$\mathrm{ERP} = P_t(\mathrm{dBm}) + G_t(\mathrm{dB}) \tag{7.33}$$

where $\boldsymbol{P_t}$ is the transmitter's output power, and $\boldsymbol{G_t}$ is the antenna gain. Figure 7.16 illustrates the concept of ERP.

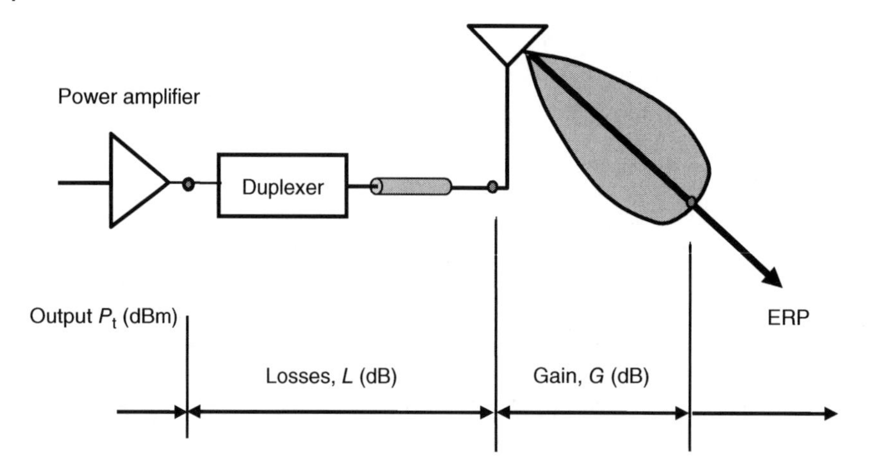

Figure 7.16 Illustration of effective radiated power ERP.

Knowing the transmit power, distance between the transmit and receive antennas, the operating frequency, and the antenna gains of the transmitter and the receiver, and the carrier-to-noise ratio at the input of the receiver can be calculated.

7.9 Antenna Types

Antennas can be classified in terms of their structures such as wire antennas (dipole, loop, and helix), aperture antennas (e.g., horn, slot), and microstrip antennas. They also can be defined in terms of their gain such as high gain (dish antenna), medium gain (horn horn), and low gain (a single microstrip patch). Antennas are also classified in terms of their radiation patterns such as omni-directional (dipole), pencil beam (dish), or fan beam (an array). In addition, antennas can be classified in terms of their bandwidth such as wideband antennas (e.g., helix, spiral, and log periodic) and narrowband (patch and slot). Figure 7.17 shows some antenna types.

7.9.1 Dipole Antenna

Dipole and monopole antennas are commonly used in wireless communication systems because of their radiation characteristics that are suited for many applications. Figure 7.18 shows a dipole antenna and its equivalent monopole antenna.

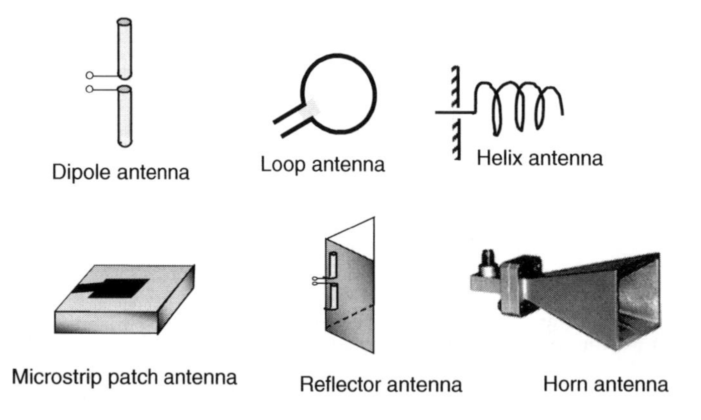

Figure 7.17 Various antenna types.

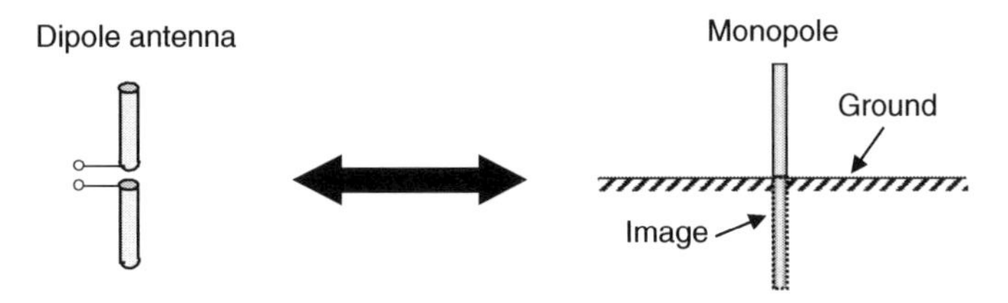

Figure 7.18 Dipole antenna and its equivalent monopole.

From Figure 7.18, using ground plane with a monopole causes an image of the monopole; the monopole element; and its image form a dipole antenna. A $\lambda/2$ dipole antenna has a beamwidth of 78 degrees, gain of 1.64 (i.e., 2.15 dBi), and input impedance about 73 + j43 ohms.

Figure 7.19 shows the simulated directivity of the azimuth (i.e., H-plane) and elevation (i.e., E-plane) radiation patterns of a 900 MHz dipole antenna. The simulation was performed using MatLab antenna toolbox [8]. The 3D radiation pattern and the antenna structure are shown in Figure 7.20.

From Figure 7.19, the dipole antenna has a horizontal radiation pattern (H-plane) that is an omni-directional, whereas the elevation radiation pattern (E-plane) is figure-eight shape. Figure 7.21 shows the simulated antenna's input impedance, and S_{11}.

The simulation results show S_{11} of −15.4 dB at 900 MHz, which indicates good matching.

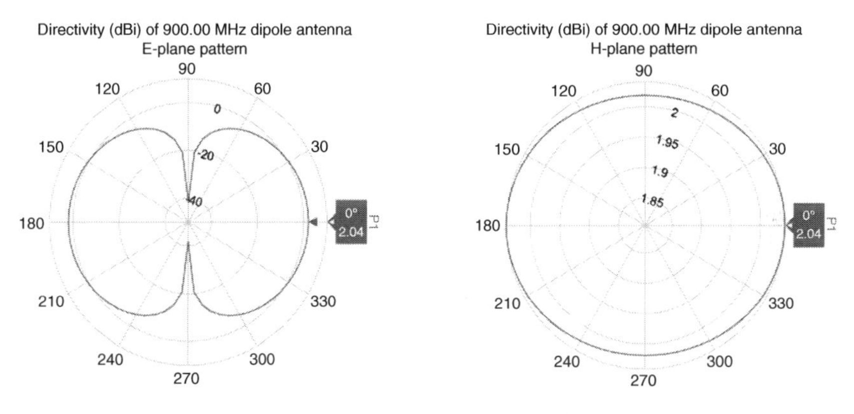

Figure 7.19 Elevation and azimuth radiation patterns of 900 MHz dipole.

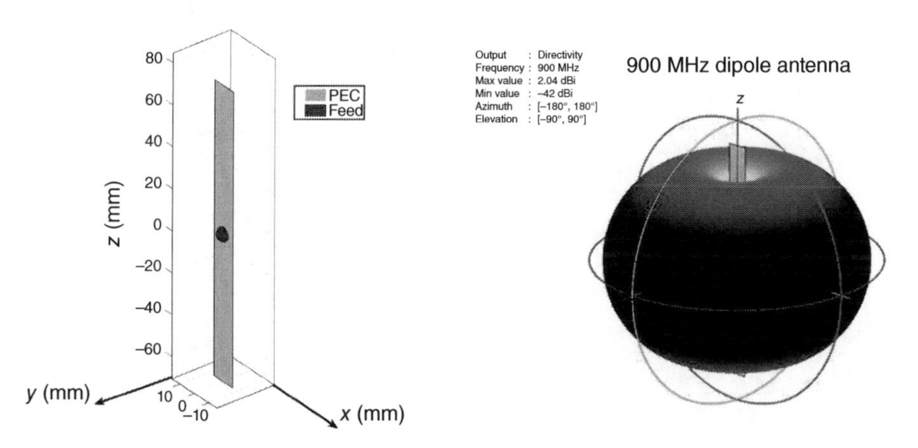

Figure 7.20 900 MHz dipole and its 3D radiation pattern.

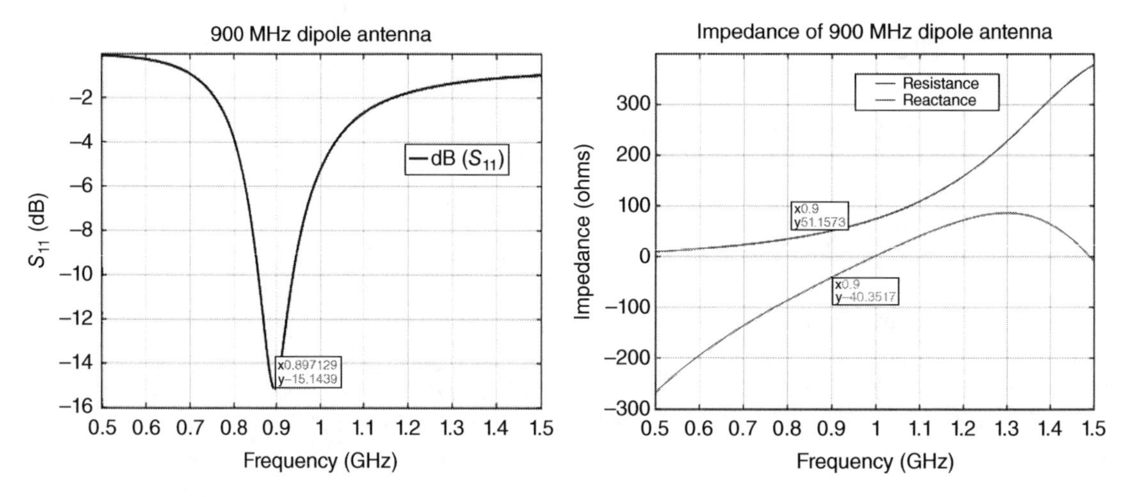

Figure 7.21 Simulated input impedance and S_{11} of a 900 MHz dipole antenna.

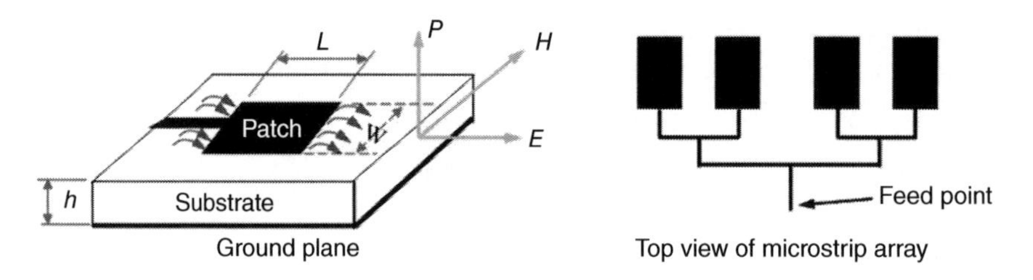

Figure 7.22 Rectangular patch antenna and the top view of a microstrip linear array.

7.9.2 Microstrip Antenna

A microstrip antenna [6], also known as a patch antenna, is the most used antenna in RF and mmWave (millimeter wave) applications, and it is the basic antenna element in a microstrip antenna array. A microstrip antenna consists of a metallic patch, which could be circular or rectangular, and is etched on a substrate; the other side of the substrate is covered with a ground plane. Figure 7.22 shows a rectangular patch antenna and the top view of a microstrip linear array.

In Figure 7.22, ***L*** is the patch's length, ***W*** is the patch's width, ***h*** is the substrate's height, ***E*** is the electric field, ***H*** is the magnetic field, and P is the direction of the radiated power. The patch's dimensions, the substrate's relative dielectric constant, ε_r ($\varepsilon_r = \varepsilon/\varepsilon_o$, ε and ε_o are the permittivities of the substrate and the free space, respectively), and the position of the feed pint have significant influence on the antenna radiation pattern, and the antenna input impedance.

The width and length of a microstrip patch [6] are given by

$$W = \frac{c}{f_o}\sqrt{\frac{2}{\varepsilon_r + 1}} \tag{7.34}$$

$$L = L_{\text{eff}} - 2\Delta L \tag{7.35}$$

where L_{eff} and ΔL are the effective and extension lengths and given by

$$L = \frac{c}{2\, f_o \sqrt{\varepsilon_e}}, \tag{7.36}$$

$$\Delta L = 0.412 \times h \times \frac{(\varepsilon_e + 0.3)\left(\frac{W}{h} + 0.26\right)}{(\varepsilon_e + 0.253)\left(\frac{W}{h} + 0.8\right)} \tag{7.37}$$

where the effective dielectric constant ε_e is given by

$$\varepsilon_e = \frac{\varepsilon_r + 1}{2} + \frac{\varepsilon_r - 1}{2} \times \sqrt{\left(1 + \left(\frac{12\text{h}}{W}\right)\right)} \tag{7.38}$$

The frequency of operation of a patch antenna is approximately given by

$$f_r = \frac{c}{2L\sqrt{\varepsilon_r}} = \frac{c}{2L\sqrt{\varepsilon_r}} \tag{7.39}$$

Equation (7.36) indicates that the microstrip antenna should have a length equal to one half of a wavelength within the dielectric (i.e., substrate) medium. The width W of the microstrip antenna controls its input impedance.

Figure 7.23 shows a 53.5 GHz microstrip patch antenna on Rogers RO4350 substrate and was simulated using MatLab [8], and the simulated 3D radiation pattern of the antenna structure is shown Figure 7.24. The antenna's E-plane radiation pattern is shown in Figure 7.25, and the simulated S_{11} is shown in Figure 7.26

The simulation results of Figure 7.26 show excellent matching at 53.94 GHz.

7.10 Antenna Impedance Mismatch

In a transmitter or receiver subsystem, the antenna input impedance must be matched to the subsystem to ensure maximum power transfer. The antenna mismatch, η_{mismatch}, is expressed as

$$\eta_{\text{mismatch}} = 1 - |\Gamma|^2 \tag{7.40}$$

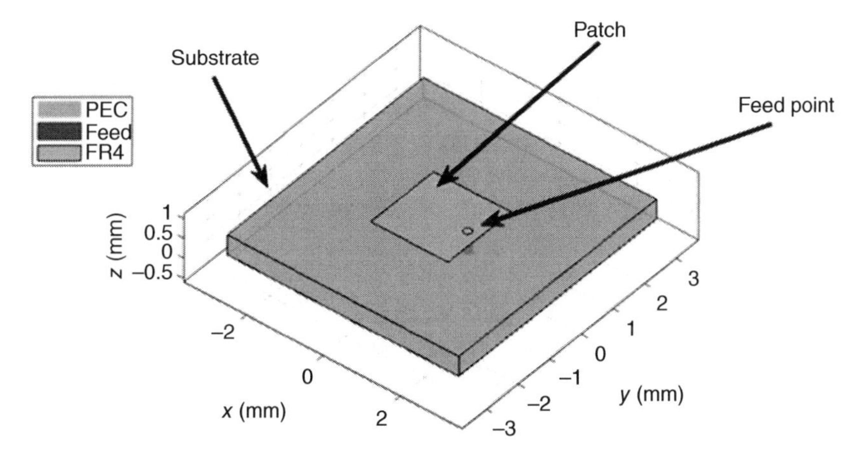

Figure 7.23 53.5 GHz microstrip patch antenna on Rogers RO4350 substrate.

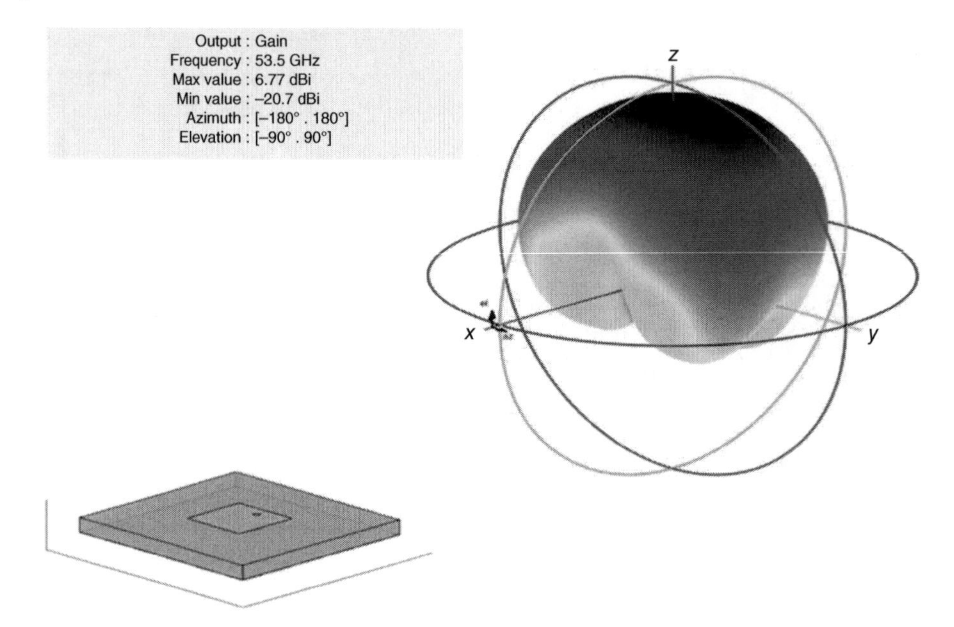

Figure 7.24 3D Radiation pattern of a 53.5 GHz microstrip patch antenna on Rogers RO4350 substrate.

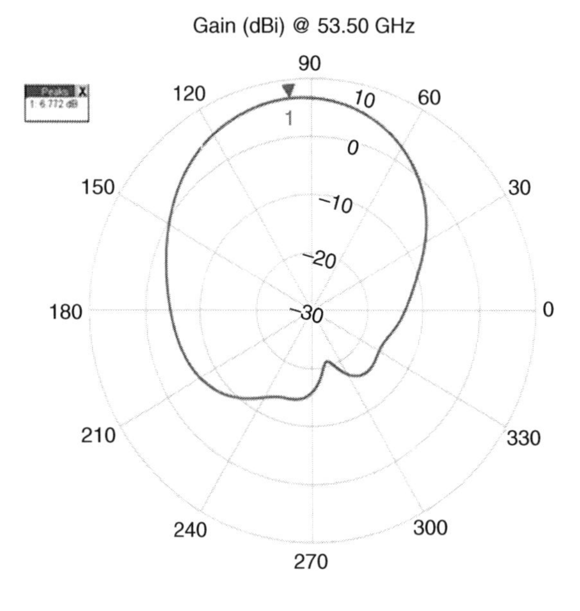

Figure 7.25 E-plane radiation pattern of a 53.5 GHz microstrip patch antenna on Rogers RO4350 substrate.

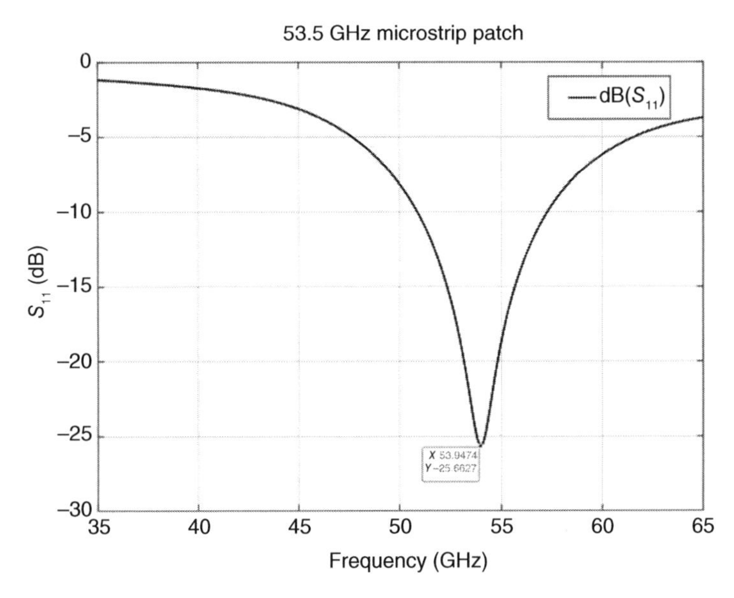

Figure 7.26 Simulated S_{11} of a 53.5 GHz microstrip patch antenna on Rogers RO4350 substrate.

$$\Gamma = \frac{Z_L - Z_{\text{ant}}}{Z_L + Z_{\text{ant}}} = \frac{\text{VSWR} - 1}{\text{VSWR} + 1} \tag{7.41}$$

where $\boldsymbol{Z}_{\text{L}}$ is the impedance of the transmission line connected to the antenna, and Z_{ant} is the antenna impedance.

7.11 Antenna Polarization Mismatch

Maximum transmission between the transmit and the receive antennas requires that the polarization of the two antennas is in the same direction. The polarization loss factor (PLF) defines the power that not captured from the incident wave by the antenna and is given by

$$\text{PLF} = \frac{P_r}{P_i} \tag{7.42}$$

where P_r is the received power and P_i is the power at the antenna's input. The values of PLF are between 0 and 1

$$0 \leq \text{PLF} \leq 1$$

7.12 Antenna Noise Temperature

Antenna temperature, $\boldsymbol{T}_{\mathrm{a}}$, is a parameter that describes how much noise power delivered to the receiver's input by the antenna. The noise power is given by

$$P_{\mathrm{TA}} = KT_{\mathrm{A}}B \tag{7.43}$$

where

$K =$ Boltzmann's constant 1.38×10^{-23} [J/K]
$B =$ bandwidth in Hz
$T_{\mathrm{A}} =$ antenna temperature

and

$$T_{\mathrm{A}} = \frac{1}{4\pi}\int_0^{2\pi}\int_0^{\pi} R(\theta,\varphi)\, T(\theta,\varphi) \sin\theta \, d\theta d\varphi \tag{7.44}$$

$R(\theta, \varphi) =$ radiation pattern of the antenna
$T(\theta, \varphi) =$ temperature distribution in spherical coordinate

7.12.1 Antenna Gain-to-Noise-Temperature (*G/T*)

Antenna gain-to-noise-temperature (G/T) is a figure of merit in the characterization of antenna performance, where G is the antenna gain in dB at the receive frequency, and T is the equivalent noise temperature of the receiving antenna in kelvins. ***G/T*** is expressed as

$$G/T\,(\mathrm{dB}) = 10 \log \frac{G}{T_{\mathrm{A}}}\ [\mathrm{dB}/K]. \tag{7.45}$$

The reason G/T parameter is important because the signal-to-noise ratio (SNR) at the receiver's input is proportional to that parameter. The received power at the receiver is defined by the Friis equation which is given by

$$P_r = \frac{G_t G_r P_t \lambda^2}{(4\pi R)^2}\ [\mathrm{W}]. \tag{7.46}$$

where G_t, G_r are the gains of the transmit and receive antennas, P_t is the transmitter's power, λ is the wavelength of the signal, and R is the distance between the transmit and receiver antennas. Thus, the signal power delivered by the receive antenna to a matched receiver's input can be written as

$$S_i = G_t G_r P_t \frac{\lambda^2}{(4\pi R)^2}\ [\mathrm{W}], \quad \text{and} \qquad N_i = k\, T_{\mathrm{A}} B$$

where N_i is the noise at the receiver's input, K is the Boltzmann's constant, and B is the radio system bandwidth. Thus,

$$\frac{S_i}{N_i} = \left(\frac{G_r}{T_{\mathrm{A}}}\right) \frac{G_t P_t\, \lambda^2}{\mathrm{kB}\,(4\pi R)^2} [\mathrm{W}]. \tag{7.47}$$

Equation (7.47) indicates that the $\frac{S_i}{N_i}$ is proportional to the $\left(\frac{G_r}{T_{\mathrm{A}}}\right)$ of the receiver's antenna. All the parameters in Eq. (7.47) except the $\left(\frac{G_r}{T_{\mathrm{A}}}\right)$ are fixed by the transmitters' design and its location from the receiver. Thus, for a fixed transmitter, the receiver's performance can be optimized by increasing the receiver's $\left(\frac{G_r}{T_{\mathrm{A}}}\right)$.

Example 7.11
For the communication link shown, determine the transmit power if the required signal-to-noise ratio at the receiver SNR = 0 dB.

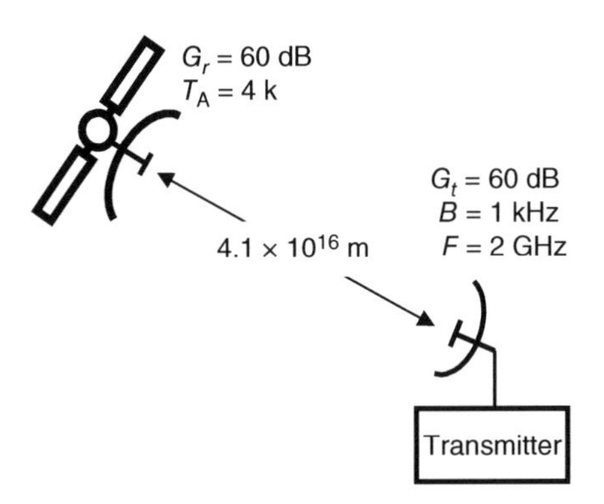

Solution
First, convert the dB values to linear scale.

$$\therefore G_r = G_t = 10^{(60\ \mathrm{dB}/10)} = 10^6,$$

$$\frac{S_i}{N_i} = \frac{G_r G_t P_t\, \lambda^2}{kTB\,(4\pi R)^2}$$

$$P_t = \frac{kTB\,(4\pi R)^2 (SNR)}{G_r G_t \lambda^2}$$

$$\lambda = \frac{c}{f} = \frac{3\times 10^8}{2\times 10^9} = 0.15$$

$$\therefore\ P_t = \frac{(1.38\times 10^{-23})\times 4\times 1000\,(4\pi\times 4.1\times 10^{16})^2(1)}{10^6\times 10^6(0.15)^2} \cong 650\ \mathrm{k\,W}.$$

7.13 Multielements Antenna (Array)

An antenna array is a directive antenna that consists of several radiating elements spaced and phased so that their contributions add in the desired direction to give higher gain and directivity and cancel in other directions. High-gain antennas enable extending the communication range or relaxing the transmitter's power specifications. However, practically some of the fields in the undesired direction will not be cancelled fully, and as a result, there will be some side and back lobes. If the antenna elements are equally spaced along a straight line, the array will be called a linear array. Antenna arrays could be one or two dimensional. The spacing between the elements as well as the phase of the current in each element influences the array's radiation pattern and gain. The higher the number of the elements, the higher the gain, and the narrower the beamwidth of the pattern. A directive narrow beam can be easily blocked, and as a result, the system's performance can be degraded significantly. To overcome such a problem, an adaptive antenna array would be required to search and beamform to the available signal path. Beamforming technique is discussed in Chapter 8. Figure 7.27 shows an N-element microstrip antenna array.

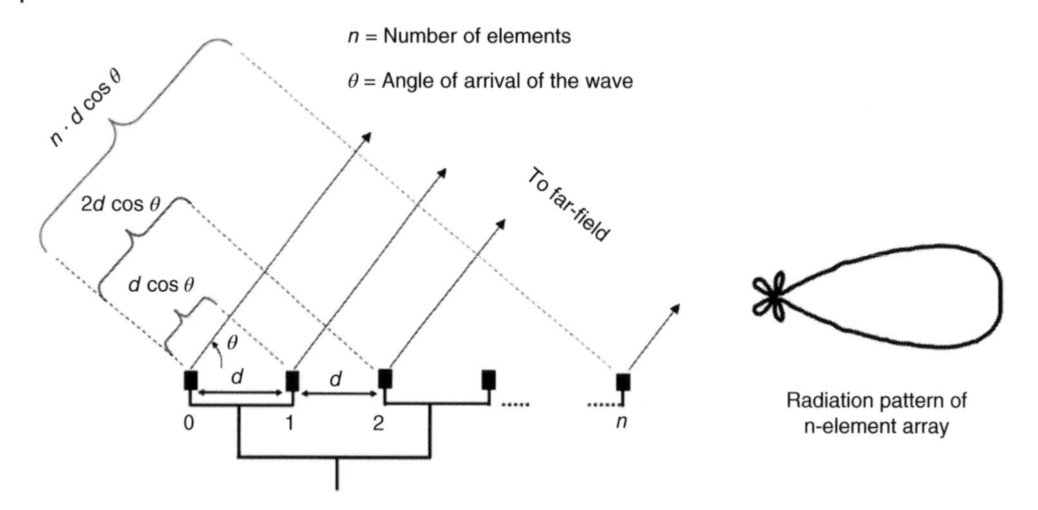

Figure 7.27 N-element microstrip antenna array.

In Figure 7.27, $\boldsymbol{\theta}$ is the angle of arrival of the radio wave, $\boldsymbol{n}$ is the number of antenna elements, and $\boldsymbol{d}$ is the spacing between adjacent elements. For an n-element antenna array with equi-space distance between the elements, the total electric field of the antenna is given by

$$E_{\text{total}} = E_{\text{o}}\,[1 + e^{j\psi} + e^{j2\psi} + e^{j3\psi} + \cdots + e^{jn\psi}] \tag{7.48}$$

where

$$\psi = \beta d \cos(\theta) + \delta, \tag{7.49}$$

$$\beta = \frac{2\pi}{\lambda}$$

$\boldsymbol{\psi}$ is the phase difference that is created by the path length difference, and δ is the progress phase shift between the elements (i.e., the phase by which the current in any element leads the current in the preceding element), and $\boldsymbol{E}_{\text{o}}$ is the field of the element. From Eq. (7.49), the relative phase between the elements will depend on the angel of arrival, the phase of the current in each element, and the space between the elements. Equation (7.48) can be written as

$$E_{\text{total}} = E_{\text{o}} \left| \frac{1 - e^{jn\psi}}{1 - e^{j\psi}} \right| = E_{\text{o}} \left| \frac{\sin(n\psi/2)}{\sin(\psi/2)} \right| \tag{7.50}$$

A conventional method to model the radiation pattern of the entire array is to multiply the pattern of one element by the *array factor*, which is the radiation pattern that would result if the antenna elements were isotropic (i.e., point sources). Thus, Eq. (6.50) can be expressed as

$$E_{\text{total}} = E_{\text{element}} \times \text{AF} \tag{7.51}$$

Equation (7.50) is referred to as pattern multiplication, where $\boldsymbol{E}_{\text{element}}$ is the amplitude of the element's field, and $\boldsymbol{AF}$ is the array factor and is given by

$$\text{AF} = \frac{\sin(n\psi/2)}{\sin(\psi/2)} \tag{7.52}$$

The array factor does not depend on the type of the elements used in the array, but rather depends on their spacing and the phase difference between the radiation from the elements. Figure 7.28 shows simulated radiation patterns of 7-element and 13-element antenna arrays.

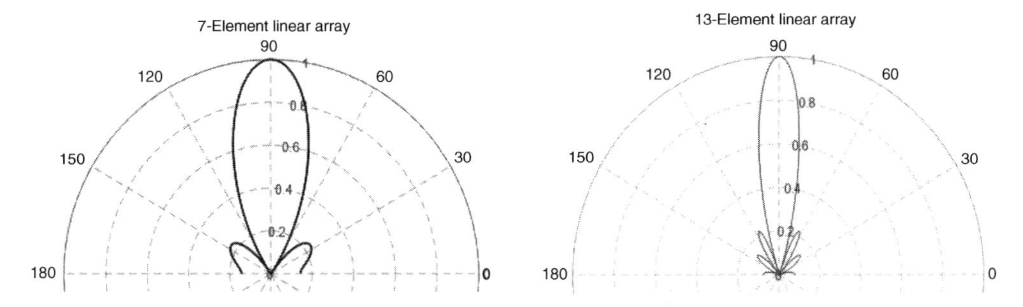

Figure 7.28 Simulated array factor for 7-element and 13-element linear arrays.

Figure 7.28 shows that increasing the number of elements from 7 elements to 13 elements produced a narrower antenna beam, which helps in reducing the interference with other collocated radio systems.

7.14 Multipath Propagation

In radio communication links, the radio link is the RF path from the transmitter's output to the receiver's input, including the transmission lines' losses, propagation path loss, and the antenna gains. The radio waves from the transmitter to the receiver are subject to a variety of propagation effects that impact their amplitude, phase, or frequency. These effects are

- reflection from the ground or large object
- diffraction from the edges of objects
- scattering from foliage or small objects
- attenuation from rain or the atmosphere
- Doppler from moving objects

The propagation effects could reduce the signal level at the receiver, which limits the communications range and degrades the transmission data rate. The propagation environment between the transmitter and the receiver of any radio channel dictates the system specifications. Figure 7.29 shows an illustration of multipath propagation.

In Figure 7.29, when the transmitted signal reaches the receiver through multipath, multiple copies of the transmitted signal may arrive at the receiver with different phases, which could degrade the signal-to-noise ratio (SNR) significantly. If the signals arrive at the receiver with equal amplitude and out-of-phase, they cancel each other resulting in a deep fading and loss of the signal. Therefore, it is essential to provide a sufficient link margin to overcome the signal degradation due to multipath effects. When the received multipath components of the transmitted signal experience the same amount of attenuation, this results in a *flat fading*. Fading is the variation of

Figure 7.29 illustration of multipath propagation.

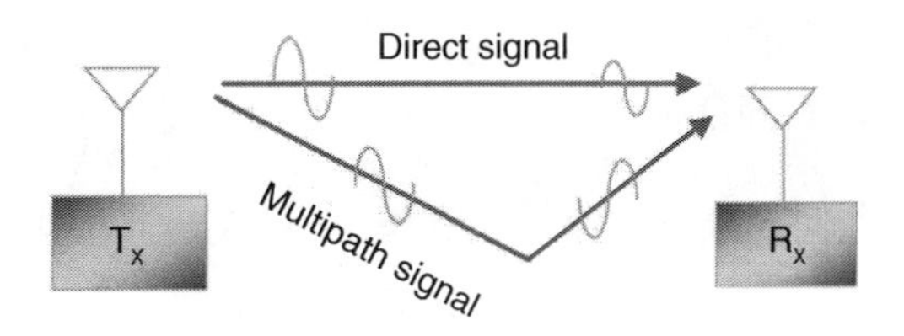

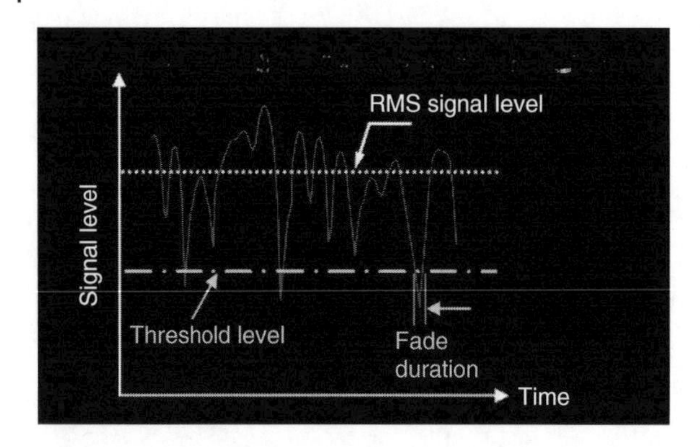

Figure 7.30 Illustration of flat fading.

the received signal with time. When the fading is independent of frequency, it is referred to as *flat fading*. Figure 7.30 illustrates flat fading.

If the signal level goes below a specified threshold level, the signal will be considered in fading condition. Flat fading is a narrowband fading that could cause a substantial reduction in the SNR, which results in significant degradation in the system's BER. Severe fading could result in a signal reduction of more than 30 dB and requires more transmit power to maintain the required BER performance.

Shadow fading occurs when the line-of-sight (LOS) path is blocked by an obstruction such as a building or hill and is referred to as a *slow fading*. When the transmitted signal arrives to the receiver through multipath propagation, the time between the first signal arrives and the last copy of the signal is defined as the ***delay spread***. Figure 7.31 illustrates the delay spread of a radio link.

When the multipath signals of the first transmitted pulse overlap with the second transmitted pulse (i.e., intersymbol interference, ISI), the receiver would not be able to distinguish between the first and the second pulses and that increases the BER (bit-error rate). Delay spread is crucial in broadband and determines how fast the data can be transmitted over the channel.

A ***frequency selective fading*** occurs when different components of the transmitted signal experience different fading. In this fading type, the spectral components of the transmitted signal are not all affected by the same amount. The fading effects can be mitigated by using antenna space diversity (multiple receive antennas); space separation of half-wavelength between the antennas is sufficient to obtain two uncorrelated signals. Signal-processing techniques are used to choose the best antenna output or to combine all the antennas outputs coherently.

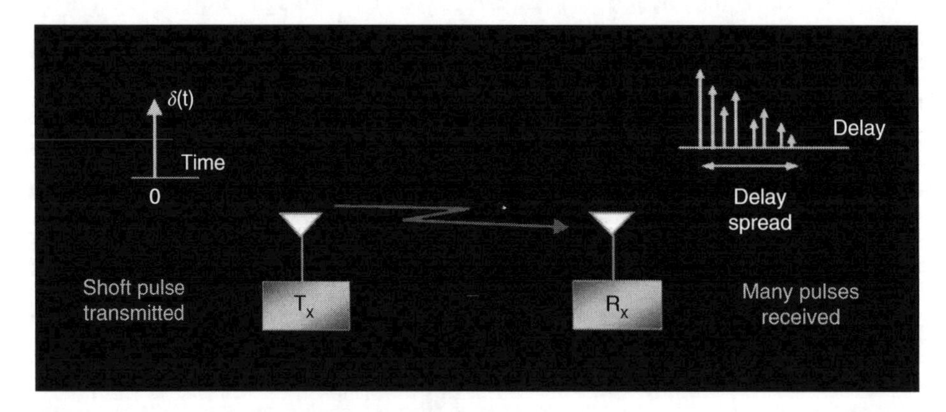

Figure 7.31 Illustration of delay spread.

OFDM (*Orthogonal Frequency Division Multiplexing*) is another approach to mitigate the impact of frequency-selective fading. Frequency-hopping spread spectrum (FHSS) is also used to mitigate the effect of frequency-selective fading.

7.14.1 Radio Wave Propagation Path Loss

In wireless communications links, propagation path loss is the basic parameter used in the link budget to determine the required signal strength at the receiver input to achieve a specific BER. The path loss over a distance ***d*** between the transmitter and the receiver is given by

$$\mathrm{PL}(d) = \mathrm{PL}(d_o) + 10\,n\log_{10}\left(\frac{d}{d_o}\right) \tag{7.53}$$

where

n = the path loss exponent
d_o = reference distance from the transmitter
$\mathrm{PL}(d_o)$ = path loss at a close-in reference distance d_o

$$\mathrm{PL}(d_o) = 20\log_{10}\left(\frac{4\pi d_o}{\lambda}\right)$$

λ = wavelength of the transmitted signal

The path loss at a distance d from the transmitter depends on the frequency of operation, the distance between the transmitter and the receiver, and the exponential loss $\boldsymbol{n}$, which depends on the objects in the propagation environment and could reflect or diffract the signal. In an indoor environment with wall partitions, the walls would significantly impact signals of mmWave frequencies. Figure 7.32 shows simulated path loss in free space and in an indoor environment as a function of frequency.

Figure 7.33 illustrates the path loss between the transmitter and the receiver antennas in a radio link.

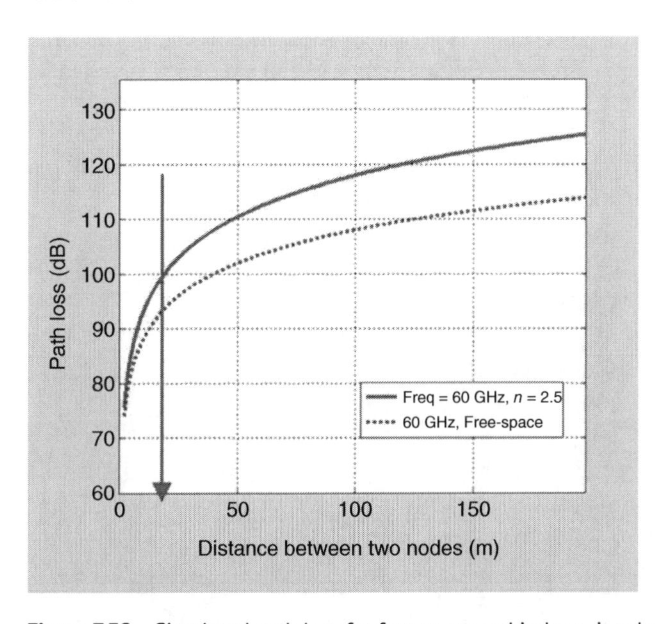

Figure 7.32 Simulated path loss for free space and indoor signals.

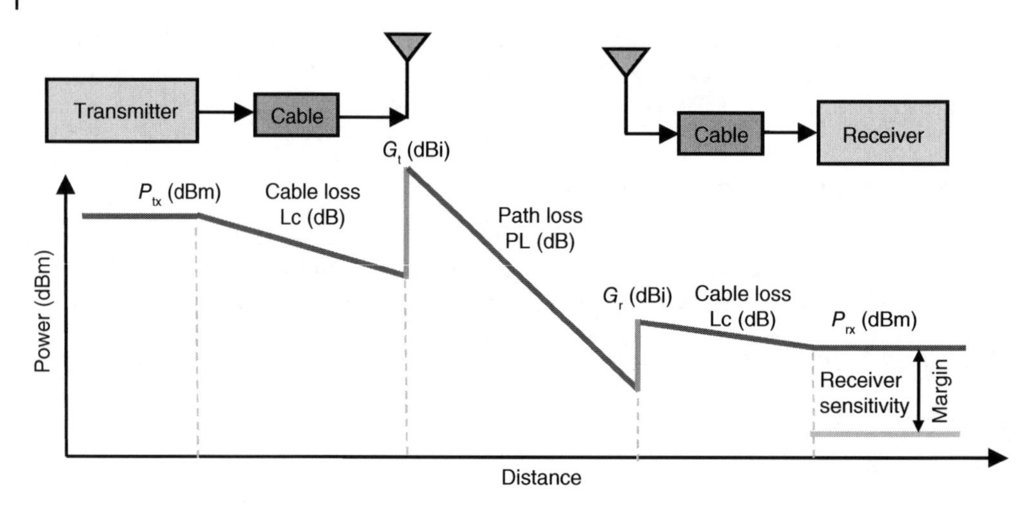

Figure 7.33 Representation of propagation path loss vs. distance.

In Figure 7.33, the power at the transmitter's output is P_{tx} (dBm), which is attenuated by the cable's insertion loss L_c (dB). The antenna's gain G_{tx} (dBi) increases the power level to meet the system specifications. The transmitted signal gets attenuated by the propagation path loss PL, which is proportional to the frequency and the distance between the transmit and receive antennas. The gain of the receive antenna increases the received signal level, which gets attenuated by the cable's insertion loss L_c (dB). The difference between the received signal at the receiver's input and the receiver sensitivity represents the link margin. The link margin considers other losses, such as polarization mismatch and fading. Figure 7.34 shows a test setup for measuring the propagation path loss exponent (n).

The propagation path loss can be measured using an RF signal generator to generate a signal with a specific frequency, two antennas that operate at the frequency of interest, and a spectrum analyzer. Measuring the signal level at a reference distance around the transmitter and knowing the transmit power, the path loss can be estimated as the difference between the transmit and the receive signal levels. The same approach can be repeated for different distances around the transmitter, and the

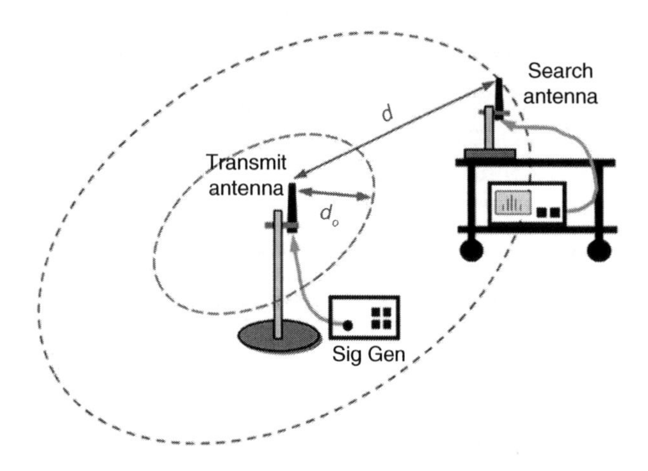

Figure 7.34 Measuring propagation path loss exponent.

average path loss for these distances can be estimated. Knowing the path losses and the distances between the transmit and receive antennas, the propagation loss exponent $\boldsymbol{n}$ can be determined using Eq. (7.53).

7.14.2 Radio Fresnel Zones

In radio communication links, a Fresnel zone is one of a few concentric ellipsoids, which define volumes in the radiation pattern of a circular aperture. Fresnel zones result because of the diffracted signals. The cross section of the first (innermost) Fresnel zone is circular. Subsequent Fresnel zones are annular (doughnut-shaped) in cross section and concentric with the first. Figure 7.35 shows an illustration of the radio Fresnel zones.

Radio waves will travel in a straight line from the transmitter to the receiver if they are unobstructed. However, if there are reflective surfaces along the path, such as bodies of water or smooth terrain, the radio waves reflecting from these surfaces may arrive either out of phase or in phase with the signals that travel directly to the receiver. Waves that reflect from surfaces within an even Fresnel zone are out of phase with the direct-path wave and reduce the power of the received signal, whereas waves that reflect in an odd Fresnel zone are in phase with the direct-path wave and can enhance the power of the received signal. Fresnel zone calculations enable determining where an obstacle will cause in phase or out of phase reflections between the transmitter and the receiver.

Obstacles in the first Fresnel zone will create signals with a path-length phase shift of 0–180 degrees, in the second zone, they will be 180–360 degrees out of phase, and so on. Thus, *even numbered zones have the maximum phase cancelling effect,* and odd numbered zones may add to the signal power. To maximize received signal level, the obstruction must be removed from the signal's line of sight (LOS). The strongest signals are on the direct line and always lie in the first Fresnel zone. The Fresnel zone radius is given by

$$\text{Radius} = \sqrt{\frac{n\lambda d_1 d_2}{d_1 + d_2}} \tag{7.54}$$

where

d_1 = distance between the transmitter and the obstacle
d_2 = distance between the obstacle and the receiver
λ = the signal's wavelength
n = the Fresnel zone number

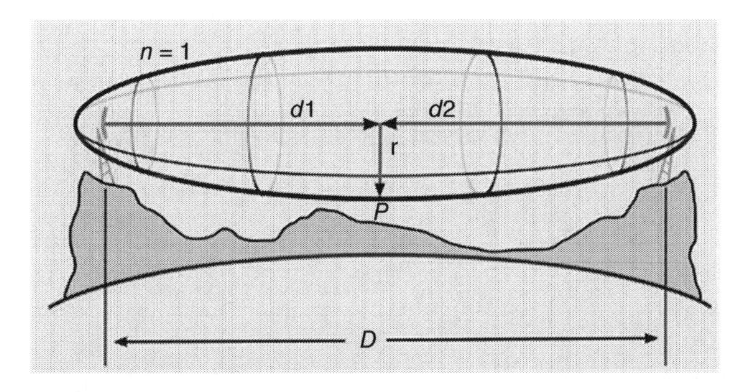

Figure 7.35 Radio Fresnel zones.

7.15 Antenna Characterization

7.15.1 RF Antenna Anechoic Chamber

An anechoic chamber is a shielded room, which has radio wave absorbing material applied to the walls, ceiling, and floor to stop reflections of radio waves inside the chamber. Also, it prevents any external RF interference signals from entering the chamber. Figure 7.36 shows a photograph of an antenna anechoic chamber.

In Figure 7.36, the interior surfaces of the chamber are covered with radiation absorbent material (RAM) that are shaped to absorb incident radio waves from any direction. Figure 7.37 shows a pyramidal radiation absorbent material.

Anechoic chambers are used for testing antennas, radars, and for performing measurements of antenna radiation patterns and EM interference. The chamber size ranges from small sizes as that of a microwave oven to ones as large as an aircraft hangar. The performance quality of the chamber is determined by its lowest operating frequency. At lower radiated frequencies, far-field measurement can require a large and expensive chamber; however, this issue can be mitigated by using compact antenna test range method.

7.15.2 Compact Antenna Test Range (CATR)

A *Compact Antenna Test Range* (CATR) [14–18] is an antenna measurement method, which generates nearly plane waves in a short distance compared to that required for far-field distance

Figure 7.36 Photograph of an antenna anechoic chamber [14].

Figure 7.37 A pyramidal radiation absorbent material (RAM). Source: Bryan Tong Minh / Wikimedia Commons / CC BY 2.0.

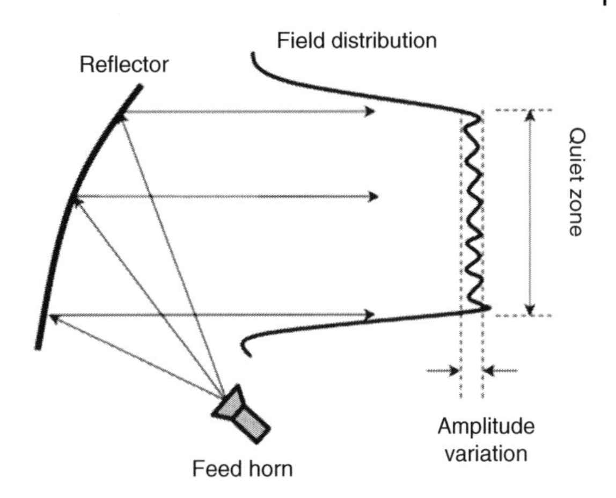

Figure 7.38 Illustration of a single-reflector CATR system and its quiet zone.

(i.e., $>2D^2/\lambda$). In this method, a horn antenna (feed) that transmit an RF signal is located in the focal of a parabolic reflector, which is used to reshape the launching field of the horn to a local area with ideal uniform phase and amplitude distribution, this area is referred to as quiet zone (QZ) and used for performing antenna far-field measurements. Figure 7.38 shows an Illustration of a single-reflector CATR system and its quiet zone.

The CATR method is becoming very popular for RF conforming testing of mmWave 5G NR (new radio) devices. For mmWave 5G NR radios, testing must be done over the air since there are no RF connectors for performing conducted testing because the antenna is part of the transceiver. Figure 7.39 shows a practical CATR system. The CATR system is the most economical option for accurate OAT measurements. It has much lower path loss compared to direct far-field chambers.

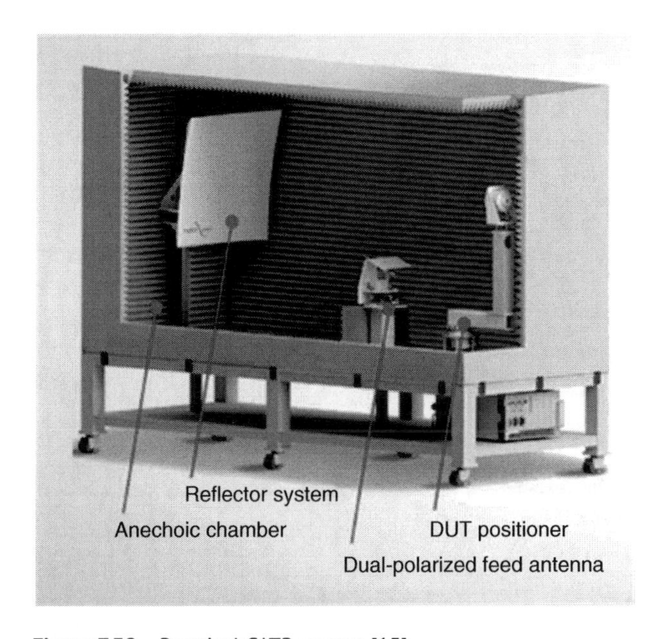

Figure 7.39 Practical CATR system [15].

7.16 Antenna Measurements

7.16.1 Antenna Gain Measurements

7.16.1.1 Absolute Gain Method

In absolute antenna gain measurement method, two identical antennas are used, and the gain is calculated using the free-space Friis' transmission equation which is given by

$$\frac{P_r}{P_t} = \left(\frac{\lambda}{4\pi R}\right)^2 G_t G_r \tag{7.55}$$

where P_r, P_t, G_r, G_t, λ, and R are the transmit power, received power, transmit antenna gain, receive antenna gain, wavelength, and the distance between the transmit and receive antennas, respectively. Figure 7.40 shows a test setup for measuring the gain of an antenna.

In Figure 7.40, one of the antennas act as the test antenna (transmit antenna) and the other acts as the antenna under test (receive antenna). Since the two antennas are the same, their gains are equal (i.e., $G_t = G_r = G$), thus, Eq. (6.55) can be written as

$$20\log_{10} G = 20\log_{10}\left(\frac{4\pi R}{\lambda}\right) + 10\log_{10}\left(\frac{P_r}{P_t}\right)$$

$$\therefore\ G_{\text{dB}} = \frac{1}{2}\left[20\log_{10}\left(\frac{4\pi R}{\lambda}\right) + S_{21}\right] \tag{7.56}$$

Equation (7.56) can be used to calculate the antenna gain. In Eq. (6.56), the first term is the free-space path loss, and the second term can be obtained by measuring S_{21} (forward transmission) using a VNA (vector network analyzer).

7.16.1.2 Gain-Transfer Method

In the gain-transfer method, the gain of the antenna under test, G_{test}, is determined by measuring the power from the antenna under test and measuring the power from an antenna with known

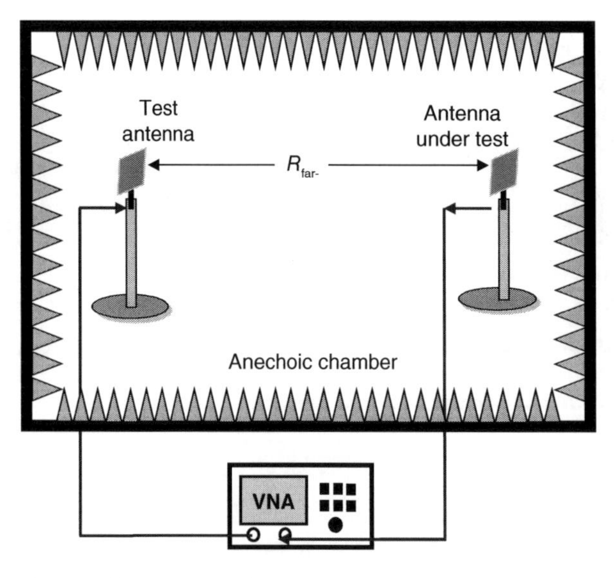

Figure 7.40 Test setup for measuring the antenna gain.

gain (standard gain antenna, G_s). The gain of the antenna under test can be calculated as

$$(G_{\text{test}})_{\text{dB}} = (G_s)_{\text{dB}} + 10\log_{10}\left(\frac{P_{\text{test}}}{P_s}\right) \tag{7.57}$$

In this test method, the antenna under test must be in the far-field region.

7.16.2 Antenna Radiation Pattern and Directivity Measurement

Antenna directivity can be estimated by measuring the E-plane and H-plane radiation patterns in an anechoic chamber. From the measured E-plane and H-plane, the beamwidths (i.e., $\theta_{3\text{dB}}$, $\phi_{3\text{dB}}$) of the two planes can be determined, and the antenna's solid angel Ω_A can be calculated as

$$\Omega_A = \Theta_{3\text{dB}} \cdot \varphi_{3\text{dB}} \tag{7.58}$$

knowing the antenna's solid angle, the antenna's directivity can be approximated as

$$D(\theta, \varphi) = \frac{4\pi}{\Omega_A} \tag{7.59}$$

Figure 7.41 shows a simplified diagram of the test setup for measuring the antenna's radiation patterns.

7.16.3 Antenna Input Impedance Measurement

The input impedance of an antenna can be measured using one-port measurement of a vector network analyzer to measure S_{11}, or the reflection coefficient. The input impedance is calculated as

$$Z_i = Z_o \frac{1+\Gamma}{1-\Gamma} \tag{7.60}$$

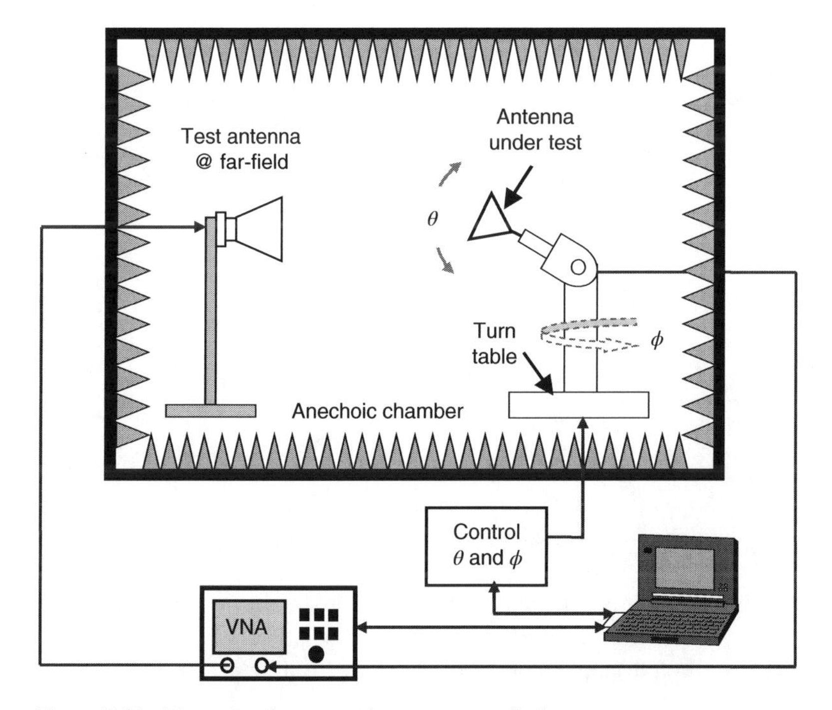

Figure 7.41 Test setup for measuring antenna radiation patterns.

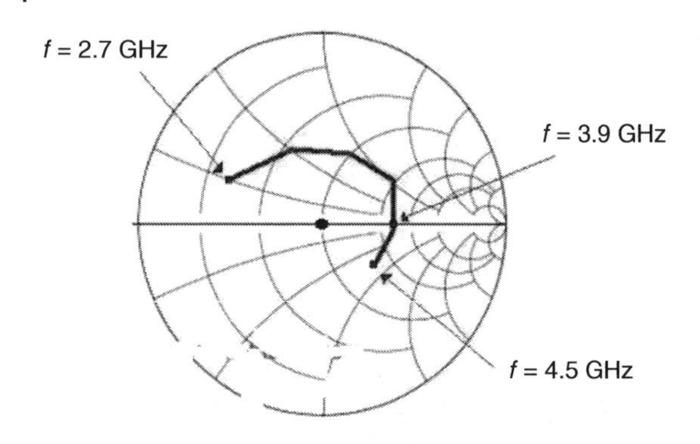

Figure 7.42 Impedance measurement vs. frequency.

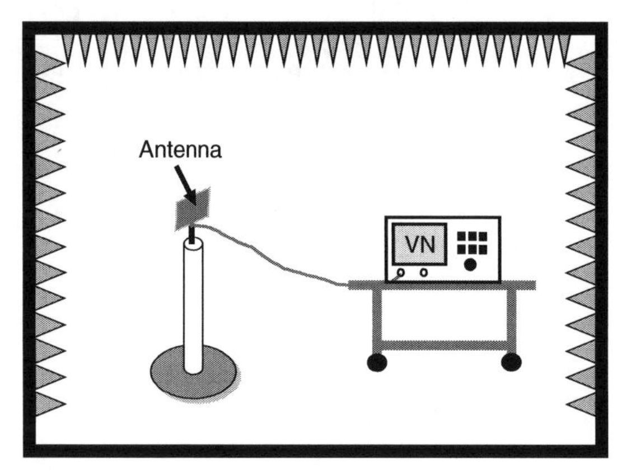

Figure 7.43 Test setup for measuring antenna input impedance.

Figure 7.42 shows an example of measured impedance vs. frequency on Smith chart of a vector network analyzer, and Figure 7.43 shows the test setup for measuring the input impedance of an antenna.

For measuring the mutual coupling between individual antenna elements in an antenna array, the inputs of the elements must be terminated except the two elements to be tested.

Review Questions

7.1 Explain and illustrate the following antenna parameters.

7.2 Gain, beamwidth, bandwidth, input impedance, radiation efficiency, directivity, main lobe, side lobe, and back-to-front ratio.

7.3 Explain the difference between E-plane and H-plane radiation patterns.

7.4 Describe the difference between the near and far fields of an antenna.

7.5 Explain why an anechoic chamber is required for antennas characterization.

7.6 Explain and illustrate the absolute gain method for measuring the antenna gain.

7.7 Explain and illustrate the gain-transfer method for measuring the gain of an antenna.

7.8 Draw and explain a test method for measuring the radiation pattern of an antenna.

7.9 Explain and illustrate a test setup for measuring the input impedance of an antenna.

7.10 Explain the difference between an antenna dipole and a monopole antenna.

7.11 List the advantages and applications of a rectangular microstrip antenna.

7.12 What is the impact of antenna polarization on the performance of a radio receiver?

7.13 Explain the impact of antenna mismatch on the performance of a radio receiver.

7.14 Define the antenna noise temperature and explain its impact on the performance of a radio receiver.

7.15 List the advantages of antenna arrays and explain the impact of array parameters on radio link performance.

7.16 Write an equation for an antenna array factor.

7.17 Explain and illustrate the delay spread of a radio channel.

7.18 What is the impact of the delay spread on a radio link performance?

7.19 Explain the difference between frequency selective fading and flat fading.

7.20 Describe a test setup for measuring the propagation path loss of a radio channel.

7.21 Describe the Fresnel zones in a radio link.

References

1 Mohammod Ali, *Reconfigurable Antenna Design and Analysis*, Artech House, 2021.
2 Balanis Constantine, *Antenna Theory, Analysis and Design*, 4th ed., Wiley, 2016.
3 Stutzman Warren, Thiele, G., *Antenna Theory and Design*, 3rd ed., Wiley, 2012.
4 John Daniel Kraus, Ronald J. Marhefka, *Antennas for all Applications*, McGraw-Hill Inc., 2002.
5 John Volakis, *Antenna Engineering Handbook*, McGraw-Hill, 2018.
6 Ramesh Garg, *Microstrip Antenna Design Handbook*, Artech House, 2001.
7 Binod Kumar Kanaujia, Deepak Gangwar, Jugul Kishor, Surendra Kumar Gupta, *Printed Antennas: Theory and Design*, CRC Press, 2020.
8 MatLab, antenna tool, box, https://www.mathworks.com

9 Advanced Design Simulator (ADS), https://www.keysight.com
10 Electromagnetic Professional (EMPro), https://www.keysight.com
11 Ansys HFSS (High Frequency Simulation Software), http://www.ansys.com
12 CST Studio, https://www.3ds.com
13 EMPIRE XPU, https://empire.de
14 Joseph Timothy, Vladimair Omelianv, Slawomir Koziel, Adrian Bekasiewicz, "Low-Cost Antenna Positioning System Designed With Axiomatic Design," *MATEC Web of Conferences* Volume: 127, 2017.
15 NSI-MI Technologies *Portable Millimeter Wave CATR for Testing 5G Antenna Systems*, NSI-MI Technologies, https://www.nsi-mi.com.
16 Vitawat Sittakul, Sarinya Pasakawee, "Design of Millimeter-Wave Antenna for Compact Antenna Test Range (CATR)," *International Symposium on Antennas and Propagation*, IEEE, 2022.
17 Michael D. Foegelle, "Validation of CATRs for 5G mmWave OTA Testing Applications," *15th European Conference on Antennas and Propagation (EuCAP)*, 2021.
18 Corbett Rowell, Benoit Derat, Adrian Cardalda-Garcia, "Multiple CATR Reflector System for 5G Radio Resource Management Measurements," *15th European Conference on Antennas and Propagation (EuCAP)*, IEEE, 2021.

Suggested Readings

Aamir Rashid, Syed Shahid Shah, et al., "Mutual Coupling Reduction in MIMO Antenna for 5G Application by Self-Decoupled Method," *2nd International Conference on Vision Towards Emerging Trends in Communication and Networking Technologies (ViTECoN)*, 2023.

Aidi Ren, Haoran Yu, et al., "A Broadband MIMO Antenna Based on Multimodes for 5G Smartphone Applications," *IEEE Antennas and Wireless Propagation Letters*, Volume: 22, Issue: 7, 2023.

Anouk Hubrechsen, Steven J. Verwer, et al., "Pushing the Boundaries of Antenna-Efficiency Measurements Towards 6G in a mm-Wave Reverberation Chamber," *IEEE Conference on Antenna Measurements & Applications (CAMA)*, 2021.

Basem Aqlan, Hamsakutty Vettikalladi, et al., "A Low-Cost Sub-Terahertz Circularly Polarized Antenna for 6G Wireless Communications," *IEEE International Symposium on Antennas and Propagation and USNC-URSI Radio Science Meeting (APS/URSI)*, 2021.

Chao You, Yejun He, et al., "Design of Slot-Coupled Broadband 5G mmWave Base Station Antenna Based on Double-Layer Patch," *IEEE MTT-S International Microwave Workshop Series on Advanced Materials and Processes for RF and THz Applications (IMWS-AMP)*, 2022.

Chow-Yen-Desmond Sim, Jeng-Jr Lo, Zhi Ning Chen, "Design of a Broadband Millimeter-Wave Array Antenna for 5G Applications," *IEEE Antennas and Wireless Propagation Letters*, Volume: 22, Issue: 5, 2023.

Fang-Fang Fan, Qing-Lin Chen, et al., "A Wideband Compact Printed Dipole Antenna Array with SICL Feeding Network for 5G Application," *IEEE Antennas and Wireless Propagation Letters*, Volume: 22, Issue: 2, 2023.

Farzad Karami, Halim Boutayeb, Larbi Talbi, "IoT-Based Transceiver Antenna System for 5G Future High-Speed Train Communications," *17th European Conference on Antennas and Propagation (EuCAP)*, 2023.

Jinghui Xue, Yuke Guo, Luyu Zhao, "Microstrip Antenna Array Based on LTCC for 5G Millimeter-Wave Application," *International Applied Computational Electromagnetics Society Symposium (ACES-China)*, 2022.

Josheena Gnanathickam, Gajula Thanusha, Nesasudha Moses, "Design and Development of Microstrip Patch Antenna for 5G Application," *International Conference on Computer Communication and Informatics (ICCCI)*, 2023.

K. Jeyavarshini, H. Jeevapriya, et al., "Design of Compact Y-Shape Antenna for 5G Smartphones," *International Conference on Networking and Communications (ICNWC)*, 2023.

Kuan-Hsun Wu, Hsi-Tseng Chou, et al., "Considerations of SiP Based Antenna in Package/Module (AiP/AiM) Design at Sub-Terahertz Frequencies for Potential B5G/6G Applications," *IEEE 71st Electronic Components and Technology Conference (ECTC)*, 2021.

M. Dabbous, A. Ngom, et al., "Reconfigurable Antenna Array for 5G Small Cells," *17th European Conference on Antennas and Propagation (EuCAP)*, 2023.

M. Rabbani, J. Churm, et al., "Enhanced Data Throughput Using 26 GHz Band Beam-Steered Antenna for 5G Systems," *16th European Conference on Antennas and Propagation (EuCAP)*, 2022.

Mahdi Nouri, Hamid Behroozi, et al., "A Learning-Based Dipole Yagi-Uda Antenna and Phased Array Antenna for mmWave Precoding and V2V Communication in 5G Systems," *IEEE Transactions on Vehicular Technology*, Volume: 72, Issue: 3, 2023.

Md Fahim Foysal, Saheel Mahmud, A. K. M. Baki, "A Novel High Gain Array Antenna Design for Autonomous Vehicles of 6G Wireless Systems," *International Conference on Green Energy, Computing and Sustainable Technology (GECOST)*, 2021.

Ming Wu, Bing Zhang, Yanping Zhou, Kama Huang, "A Double-Fold 7×8 Butler Matrix-Fed Multibeam Antenna with a Boresight Beam for 5G Applications," *IEEE Antennas and Wireless Propagation Letters*, Volume: 21, Issue: 3, 2022.

Ming-An Chung, Bing-Ruei Chuang, "Design a Broadband U-Shaped Microstrip Patch Antenna on Silicon-Based Technology for 6G Terahertz (THz) Future Cellular Communication Applications," *10th International Conference on Internet of Everything, Microwave Engineering, Communication and Networks (IEMECON)*, 2021.

Mojtaba Sohrabi, Ronny Hahnel, et al., "5G mmWave Dual-Polarized Stacked Patch Antenna," *51st European Microwave Conference (EuMC)*, 2022.

Nonchanutt Chudpooti, Nattapong Duangrit, et al., "THz Photo-Polymeric Lens Antennas for Potential 6G Beamsteering Frontend," *International Symposium on Antennas and Propagation (ISAP)*, 2021.

S. Y. Zhu, "Silicon-Based High Gain Terahertz Metasurface Antennas Working at 1 THz," *IEEE International Workshop on Electromagnetics: Applications and Student Innovation Competition (iWEM)*, 2021.

Sagar Juneja, Rajendra Pratap, Rajnish Sharma, "Design of a Highly Directive, Wideband and Compact End Fire Antenna Array for 5G Applications," *IEEE International Conference of Electron Devices Society Kolkata Chapter (EDKCON)*, 2022.

Sheng-Chi Hsieh, Hong-Sheng Huang, et al., "Design of Dual-Band (28/39GHz) Antenna-in-Package with Broad Bandwidth for 5G Millimeter-Wave Application," *IEEE 24th Electronics Packaging Technology Conference (EPTC)*, 2022.

Shu-Chuan Chen, Chih-Kuo Lee, Sheng-Min Li, "Compact Multi-Input Multi-Output Loop Antenna System for 5G Laptops," *International Symposium on Antennas and Propagation (ISAP)*, 2021.

Yixiang Fang, Ying Liu, et al., "Reconfigurable Structure Reutilization Low-SAR MIMO Antenna for 4G/5G Full-Screen Metal-Frame Smartphone Operation," *IEEE Antennas and Wireless Propagation Letters*, Volume: 22, Issue: 5, 2023.

Yuehui Gao, Junlin Wang, Xin Wang, Zhanshuo Sun, "Extremely Low-Profile Dual-Band Antenna Based on Single-Layer Square Microstrip Patch for 5G Mobile Application," *IEEE Antennas and Wireless Propagation Letters*, Volume: 22, Issue: 7, 2023.

Zeeshan Siddiqui, Marko Sonkki, et al., "Dual-Band Dual-Polarized Planar Antenna for 5G Millimeter-Wave Antenna-in-Package Applications," *IEEE Transactions on Antennas and Propagation*, Volume: 71, Issue: 4, 2023.

8

Introduction to MIMO and Beamforming Technology

This chapter discusses the basic concept of multiple-input multiple-output (MIMO) and beamforming technologies, focusing on the RF architecture of beamforming subsystems. MIMIOs and beamforming are very crucial for wireless systems such as 5G NR (Fifth-Generation New Radio-access) communication systems to enable achieving high data rate and ultra-low latency transmission. A brief overview of 5G technology including the technology evolution, 5G frequency bands, 5G frame structure, numerology and subcarrier spacing, and 5G resource grid is presented. This overview helps the reader to understand the RF specifications of the conformance testing that is discussed in Chapter 9, to characterize the RF performance of 5G NR base station transmitters and receivers. Massive MIMO, MIMO channel, and beamforming types including analog, digital, and hybrid beamforming are introduced and explained. Review questions are provided to help the readers understand the chapter's contents. A list of references and suggested readings are provided to present the recent research and development activates in MIMO and beamforming technologies.

8.1 Chapter Objectives

After completing this chapter, the reader will be able to:

- List the advantages of 5G NR technology, and explain the application areas of this technology.
- Explain the advantages and challenges of using mmWave frequencies for 5G NR technology applications.
- Describe the frame structure and resource grid of 5G NR transmission.
- Explain the advantages of using MIMO technology and write an equation for a radio channel matrix of a 2×2 MIMO system.
- Compare and contrast analog, digital, and hybrid beamforming types.

8.2 Overview of 5G NR Technology and Beyond

The fifth-generation new radio-access (5G NR) [1–3] is a wireless broadband cellular technology with new standards defined by the 3GPP (third generation partnership project) organization [1]. The 5G is not a revolutionary new standard but an evolutionary standard based on previous wireless mobile networks such as G1, G2, G3, and G4. The first generation G1 was deployed in the 1980s and delivered analog voice; the second generation G2 was deployed in the early 1990s and introduced

Essentials of RF Front-end Design and Testing: A Practical Guide for Wireless Systems, First Edition. Ibrahim A. Haroun.

digital voice; the third generation G3 was deployed in the early 2000s and brought mobile data; the fourth generation 4G was deployed in the 2010s and introduced mobile broadband. All these previous generations led to the 5G technology.

The *International Telecommunication Union* (ITU) has defined three main application areas for the enhanced capabilities of 5G; these areas are Enhanced Mobile Broadband (*eMBB*), Ultra Reliable Low Latency Communications (*URLLC*), and Massive Machine Type Communications (*mMTC*). The *eMBB* was deployed in 2020.

The eMBB uses 5G as a progression from 4G Mobile Broadband services, with 10–20 Gbps peak, 100 Mbps whenever needed, 1000× more traffic, support for high mobility (500 km/h), and network energy saving by 100 times. This will benefit areas of higher traffic, such as stadiums, cities, and concert venues. The URLLC is intended for ultra responsive, less than 1 ms air interface latency, ultra reliable and available (99.9999%), and low to medium data rate (50 kbps–10 Mbps). The mMTC is intended for high density of devices, low data rate (1–100 kbps), and 10 years battery. Figure 8.1 summarizes the 5G application areas.

In 5G NR systems, latency is the time delay between initiating action and the time the action takes place. In applications such as the advanced driver assistance system (ADAS), the delay time is significantly vital for safety in new cars and the target latency is 1 ms or less. The mmWave frequencies from 24 GHz to 100 GHz are used in 5G NR are intended to support large bandwidths and very high data rates. The challenge of high propagation path losses at mmWave frequencies [4] can be compensated for by using MIMO and beamforming techniques.

The future 6G technology [5] is expected to be deployed in 2030 and expected to provide a peak data rate of around 100 Gbps. In this technology, the latency is expected to be reduced from ms to μs order. The device densities in internet of things (IoT) connectivity are expected to be much denser than the 5G. Energy efficiency and spectral efficiency are also expected to be better than 5G. More artificial intelligence (AI) and machine learning applications will be implemented in 6G. In addition, mobile handsets would not be just used for communication, but they would become personal digital assistance to people. Also, security is expected to be much better than the 5G. The device-to-device communication in 6G technology would be a hybrid of optical and wireless communications.

eMBB = *enhanced mobile broadband*

- very high data rates
- very high traffic capacity

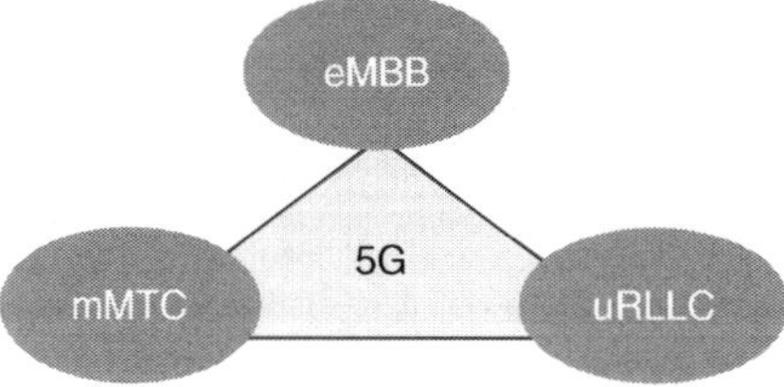

mMTC = *massive machine-type communications*

- massive number of devices
- very low device cost
- very low energy consumption

uRLLC = *ultra-reliable low latency*

- very low latency
- ultra-high reliability and availability

Figure 8.1 5G application areas.

One of the cornerstones of the 6G wireless systems is the use of the terahertz (THz) [6] frequency bands (0.1–10 THz) because of the large amount of available bandwidth. In addition, THz frequencies enable achieving high channel capacity and very high-resolution sensing. However, at THz frequencies, the propagation path loss is very high, but using high-gain antennas and ultra-massive MIMO (UM-MIMO) antenna systems can compensate for such losses.

8.3 5G NR Frequency Ranges

The frequency bands for 5G NR are separated into two different frequency ranges called frequency range 1 (FR1) and FR2. The FR1 includes sub-6 GHz frequency bands, some of which are traditionally used by previous standards but have been extended to cover potential new spectrum offerings from 410 MHz to 7125 MHz. The maximum channel bandwidth defined for FR1 is 100 MHz. The frequency range 2 (FR2) includes a frequency range from 24.25 GHz to 71.0 GHz; the minimum channel bandwidth is 50 MHz, and the maximum bandwidth is 400 MHz. Table 8.1 lists the 5G NR frequency ranges.

Most 5G NR are deployed in the mid-band of FR1, whereas FR2 deals with mmWave frequencies. At high frequencies in the mmWave range, the propagation path loss [4] is high compared with frequencies below 7 GHz. The propagation path loss is expressed as,

$$PL_{dB} = 32.44 + 20\log_{10}(f_{MHz}) + 20\log_{10}(d_{km}) \quad (8.1)$$

where f is the operating frequency in MHz, and d_{km} is the distance between the transmit and receive antennas in km. Equation (8.1) indicates that the higher the frequency, the higher the propagation path loss. Figure 8.2 shows the simulated free space propagation path loss as a function of frequency in a 1 km link (i.e., the distance between the transmitter and the receiver).

From Figure 8.2, the path loss at 4.5 GHz (i.e., in the FR1 band) is 105.5 dB, whereas at 24.25 GHz (i.e., in the FR2 band), the path loss increased to 120.1 dB, which indicates a high path loss at mmWave frequency. However, at mmWave the wavelength is small, resulting in a small-size antenna that enables having a compact array antenna with high gain. The antenna gain and its directivity go hand in hand; the greater the gain, the more directive the antenna. Antenna directivity is crucial to the channel capacity of 5G NR systems. The antenna's high-gain compensates for the channel propagation pathless. At high frequencies, the effect of Doppler shift impacts the mobility of the user equipment (UE) and the base station; therefore, the 5G mmWave technology is more efficient for fixed wireless access (FWA) applications.

There are many individual frequency bands in the 5G New Radio technology; they are numbered ***n1*** to ***n261***, some of these bands are assigned for frequency-division multiplexing (FDD) transmission, and some are assigned for time-division multiplexing (TDD) transmission. For sub-7 GHz 5G technology, the frequency subcarriers are 15-, 30-, and 60-kHz, while for mmWave the permitted subcarriers are 60-, 120-, and 240-kHz. In the 5G mmWave systems, the transmission mode is TDD.

Table 8.1 5G frequency ranges.

Frequency range designation	Frequency range
FR1	450 MHz–7125 MHz
FR2	24.25 GHz–71.0 GHz

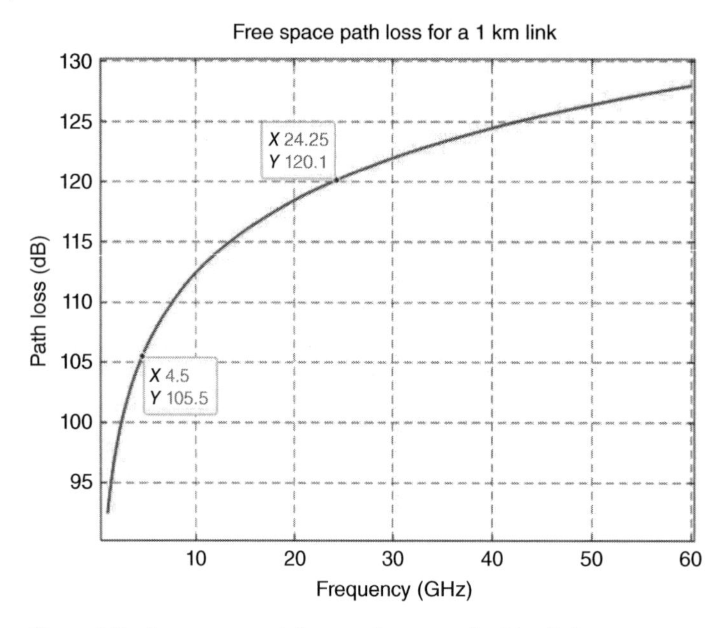

Figure 8.2 Free space path loss vs. frequency for 1 km link.

The FR1 is suitable for high mobility applications, whereas FR2 is suitable for high-capacity and high data rate scenarios.

8.4 5G NR Radio Frame Structure

The time-domain radio frame in 5G technology is a unit of 10 ms that is divided into 10 subframes, and each subframe is 1 ms. Figure 8.3 shows a radio frame structure of 5G NR transmission. In Figure 8.3, for subcarrier spacing (SCS) of 15 kHz, the subframe has only one slot (*a slot is 14 OFDM*

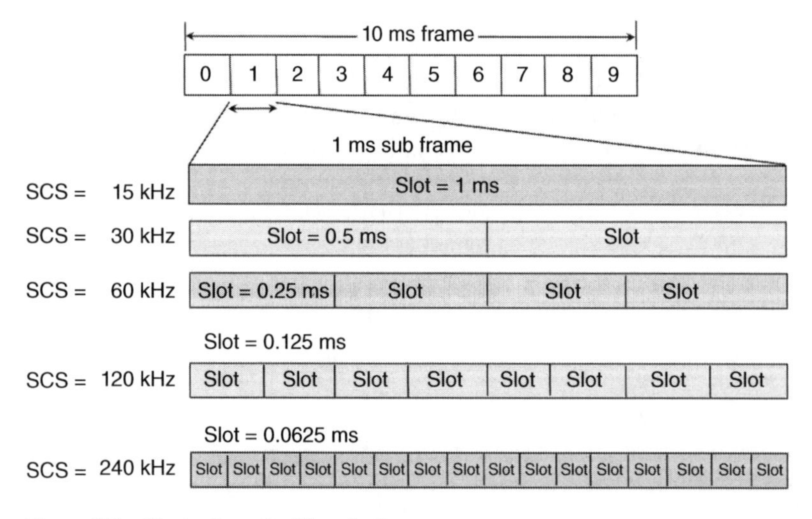

Figure 8.3 Illustration of a 5G radio frame.

"*orthogonal-frequency-division multiplexing*" *symbols*), and the slot's time is 1 ms. For SCS of 30 kHz the subframe has two slots, and each slot's time is 0.5 ms, and for SCS of 60 kHz the subframe has four slots with a slot time of 0.25 ms. For SCS of 120 kHz, the subframe has eight slots, and the slot time is 0.125 ms. Thus, the slot time is flexible in 5G NR systems and results in reducing the latency between the transmitter and the receiver of the communication link, and increasing the data rate transmission.

8.5 5G NR Numerology

In 5G radios, numerology refers to the configuration of waveform parameters. Different numerologies are considered as OFDM-based subframes having different parameters such as subcarrier spacing, symbol time, CP (cyclic prefix) size, etc. The subcarrier spacing is expressed as:

$$\text{SCS} = 15 \times 2^{\mu}\ \text{kHz} \tag{8.2}$$

where μ takes positive values, and in the future, it might take negative values. The following are channel spacing subcarriers for different numerologies:

- $\mu = 0$, SCS = 15 kHz
- $\mu = 1$, SCS = 30 kHz
- $\mu = 2$, SCS = 60 kHz
- $\mu = 3$, SCS = 120 kHz
- $\mu = 4$, SCS = 240 kHz

In 5G technology, the number of resource blocks (a resource block "RB" is a block of 12 subcarriers) changes, but the RB stays the same. The RB in the frequency domain consists of 12 subcarriers; thus, for numerology 0 where the SCS is 15 kHz, the RB is 180 kHz. For numerology 1, the resource block is 12 × 30 kHz (360 kHz), and for numerology 2 the resource block bandwidth is 12 × 60 kHz (720 kHz). Similarly, the resource block bandwidths for numerology 3 and 4 are 1440 kHz and 2880 kHz, respectively.

Numerology 0, 1, and 2 are used in the FR1 band, whereas numerology 3 and 4 are used in the FR2 band. Table 8.2 lists the subcarrier spacing associated with different numerologies in 5G NR technology.

From Table 8.2, depending on which numerology is used, the subframe will have one or more slots. Higher numerology results in a greater number of slots, and every slot has 14 symbols.

Table 8.2 SCS associated numerology in 5G numerology.

Numerology	SCS (kHz)	Slot duration (ms)	Number of slots in 1 ms subframe	Number of symbols in 1 ms subframe
0	15	1	1	14
1	30	0.5	2	28
2	60	0.25	4	56
3	120	0.125	8	112
4	240	0.0625	16	224

8.6 5G NR Resource Grid

The 5G NR resource grid is characterized by one subframe in time domain, and full carrier bandwidth in frequency domain, and Figure 8.3 illustrates a resource grid in 5G NR transmission. In Figure 8.3, the vertical axis represents the subcarriers (i.e., frequency domain) and the horizontal axis represents the symbols in one subframe (i.e., time domain).

In Figure 8.4, the resource element (RE) is one subcarrier and one OFDM symbol, and the resource block is 12 subcarriers in the frequency domain. For sub-6 GHz, the subcarrier spacing is 15 kHz, 30 kHz, and 60 kHz, while for mmWave frequencies, the subcarrier spacing is 60 kHz, 120 kHz, and 240 kHz. Thus, the 5G NR resource grid has flexible numerologies (i.e., SCS, symbol time, CP size). Table 8.3 shows the bandwidths for FR1 band, and Table 8.4 shows the bandwidths for the FR2 band.

The frequency range FR2 supports wider channel bandwidths than that of FR1. Wider bandwidths increase the radio channel capacity, which is given by

$$C = B \cdot \log_2(1 + S/N) \tag{8.3}$$

where C is the channel capacity in bits/sec, B is the channel bandwidth in Hz, and S/N is the signal-to-noise ratio in linear scale. As an example, for a channel bandwidth of 50 MHz and SNR of 20 dB, the channel capacity is

$$C = 50 \times 10^6 \times \log_2(1 + 100) \cong 332.9 \text{ Mbps}$$

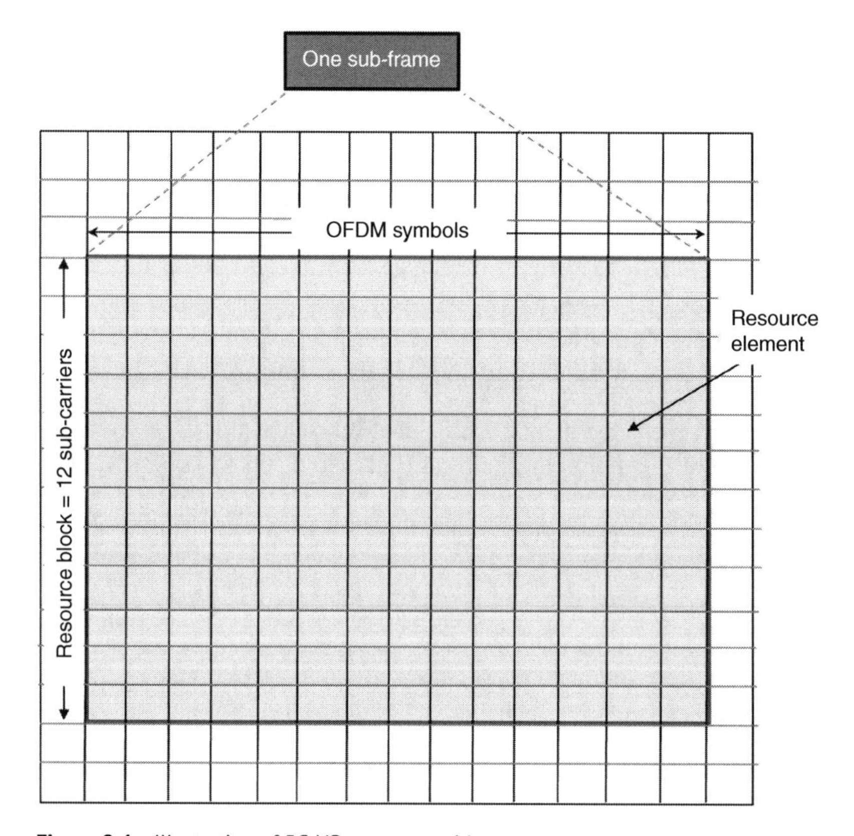

Figure 8.4 Illustration of 5G NR resource grid.

Table 8.3 5G NR bandwidths for FR1.

	FR1												
	Channel bandwidth (MHz)												
	5	10	15	20	25	30	40	50	60	70	80	90	100
SCS (KHz)	N_{RB}	N_{RB}	N_{RB}	N_{RB}	N_{RB}	N_{RB}	N_{RB}	N_{RB}	N_{RB}	N_{RB}	N_{RB}	N_{RB}	N_{RB}
15	25	52	79	106	133	160	216	270	na	na	na	na	na
30	11	24	38	51	65	78	106	133	162	189	217	245	273
60	na	11	18	24	31	38	51	65	79	93	107	121	135

Table 8.4 5G NR bandwidths for FR2.

	FR2			
	Channel bandwidth (MHz)			
	50	100	200	400
SCS (KHz)	N_{RB}	N_{RB}	N_{RB}	N_{RB}
60	66	132	264	na
120	32	66	132	264

8.7 Massive MIMO for 5G Systems

Massive MIMO [7–11] technology uses a large number of antenna elements at the transmitter and the receiver ends of a wireless communication link to increase the link throughput and capacity. The channel capacity of a MIMIO system is expressed as

$$C = M \cdot N \cdot B \log_2 (1 + S/N) \tag{8.4}$$

where M is the number of transmit antennas, N is the number of receive antennas, B is the bandwidth in Hz, and S/N is the signal-to-noise ratio in linear scale. Thus, increasing the number of transmit and receive antennas increases the system's capacity, which enables supporting a large number of users. MIMO technology uses spatial diversity and spatial multiplexing to transmit independent data signals that are referred to as streams. In multi-user MIMO (MU-MIMO), the transmitter simultaneously sends different streams to different users using the same time and frequency resources (i.e., sharing resources), which increases the network capacity and spectral efficiency. Spectral efficiency can be enhanced by adding more antennas to provide more streams.

At mmWave frequencies, the wavelength is in the millimeters range (wavelength = $(3 \times 10^8$ m/s$)/$frequency in Hz), resulting in a small-size antenna element. A small-size antenna element permits the development of a large-size array antenna (i.e., a large number of antenna elements). A large-size array antenna provides high gain and directivity, which contribute to good coverage and interference mitigation. For 5G NR, the 3GPP has specified 32 × 32 MIMO (32 transmit antennas, and 32 receive antennas). Figure 8.5 illustrates an antenna panel of 64 radiating elements for a massive MIMO application; in Figure 8.5, UE1 and UE2 are the user equipment 1 and 2.

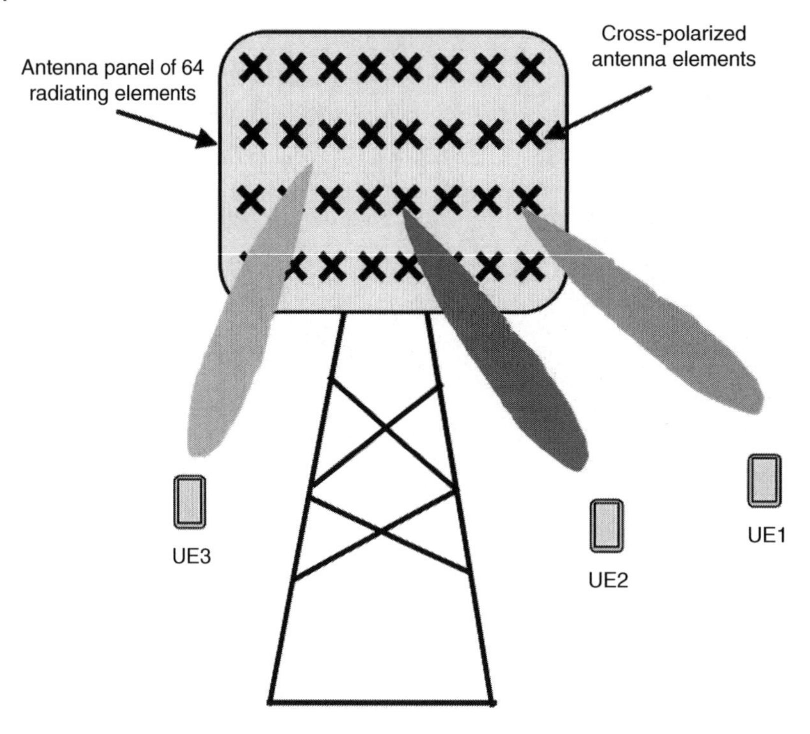

Figure 8.5 Illustration of an antenna panel of 64 elements for a 32T32R MIM system.

In a wireless channel, various reflected signals from objects such as buildings, cars, mountains, etc., arrive at the receive antennas with different time delay, angle of arrival, and attenuation. Thus, each antenna element receives a different version of the signal that was transmitted; such signals can be processed and combined to improve the signal-to-noise ratio at the receiver. This technique is referred to as spatial diversity. In a Full Dimension Multiple-Input Multiple-Output (FD-MIMO) [10, 11], the antenna system forms beams in both horizontal and vertical directions, to cover anywhere in 3D spaces. Figure 8.6 [8] illustrates the basic concept of an FD-MIMO.

8.7.1 Simplified Mathematical Model of a MIMO Channel

Massive MIMO systems can be modeled as multiple transmit and receive antennas with a fading channel between each transmit and receive antenna; thus, the system is a collection of a large number of fading channels between each transmit and receive antenna. Figure 8.6 shows an illustration of a channel matrix of a MIMO system (Figure 8.7).

In Figure 8.6, the transmit antennas x_1, x_2, ..., x_t are represented as a vector called transmit vector and is given by:

$$x = \begin{bmatrix} x_1 \\ x_2 \\ \cdot \\ \cdot \\ \cdot \\ \cdot \\ x_t \end{bmatrix} \tag{8.5}$$

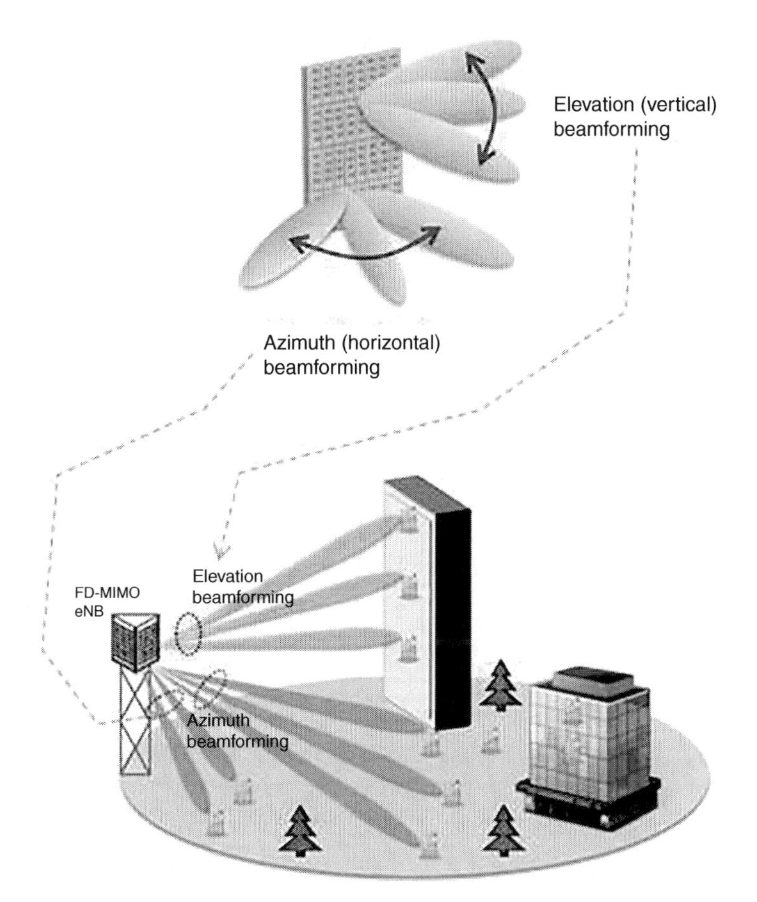

Figure 8.6 Illustration of a full dimension MIMO system [8].

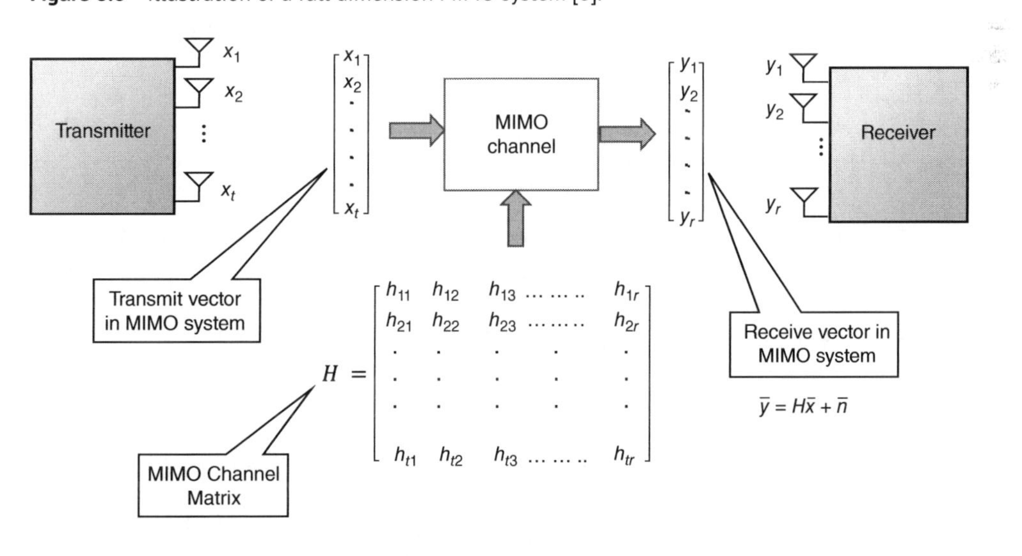

Figure 8.7 Simplified channel matrix model in MIMO system.

At the receiver, the receive antennas are stacked in a "**y**" dimensional vector called receive vector and is given by:

$$y = \begin{bmatrix} y_1 \\ y_2 \\ . \\ . \\ . \\ . \\ y_r \end{bmatrix} \quad (8.6)$$

The transmit vector passes through the channel as an input vector and shows at the channel's output as a receive vector. The receive vector, y, represents the transmit vector multiplied by the channel matrix plus the Additive White Gaussian Noise (AWGN); y is given by:

$$\bar{y} = H\bar{x} + \bar{n} \quad (8.7)$$

where n is the noise, and H is the channel matrix, which is expressed as

$$H = \begin{bmatrix} h_{11} & h_{12} & h_{13} & \cdots\cdots & h_{1r} \\ h_{21} & h_{22} & h_{23} & \cdots\cdots & h_{2r} \\ . & . & . & . & . \\ . & . & . & . & . \\ . & . & . & . & . \\ h_{t1} & h_{t2} & h_{t3} & \cdots\cdots & h_{tr} \end{bmatrix} \quad (8.8)$$

In Eq. (8.8), the channel coefficient, h_{ij}, represents the propagation path between transmit antenna ith and receive antenna jth. The transmitted information can be extracted from the receiver by multiplying the receive vector by the inverse matrix of the MIMO channel.

8.8 Beamforming Technology

Beamforming [12] is a technique that enables focusing the antenna beam (i.e., the main lobe of the antenna's radiation pattern) in a specific direction rather than broadcasting over a wide area. Beamforming is used with phased array antennas, which can change the shape and the direction of the radiation pattern without physically moving the antenna. In a phased array antenna, the individual antenna elements are placed in such a way that their signals sum up and provide higher gain in a desired direction. This is achieved by controlling the phase difference between the antenna elements; the phase shift is calculated to provide constructive interference in the desired direction, while destructive interference may occur in other directions. The phase shift from one element to the next to achieve the desired angle is given by

$$\Delta\varphi = \left(\frac{2\pi}{\lambda}\right) \cdot d \cdot \sin\theta \quad (8.9)$$

where $\boldsymbol{d}$ is the distance between the radiating elements and $\boldsymbol{\theta}$ is the beam steering angle. Figure 8.8 shows an Illustration of an electronically steered phase array antenna.

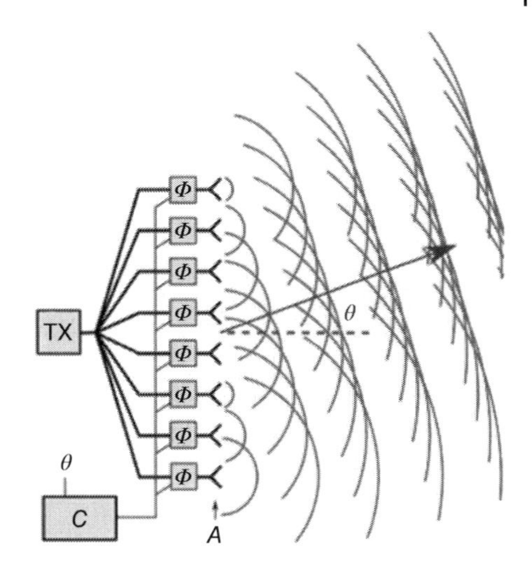

Figure 8.8 Illustration of electronically steered phase array antenna.

In Figure 8.8, the signal from the transmitter (TX) is sent to the antenna elements (A) via a phase shifter (ϕ), which can be controlled by the digital signal processing of the transmitter (C). Thus, the radio waves from separate antenna elements can be combined to increase the power radiated in desired directions, and to be suppressed in undesired directions. The phased array antenna is well suited for mmWave frequency applications because the higher the frequency, the smaller the antennas, and therefore, a large number of elements can be used in a small area.

The mmWave frequencies in 5G NR wireless systems enable the implementation of a large number of antenna elements, which support the creation of 3D beamforming (i.e., vertical and horizontal beams toward the UE). The large-size antenna array increases the capacity of the communication channel, which is essential, particularly in areas with high-rise buildings. Also, it provides high antenna gain that extends the communication range or allows relaxing the transmitter's power specifications. However, the propagation path loss at mmWave frequencies is high, but such a loss can be compensated for by using beamforming. The gain of a multi-element antenna array is given by

$$G\,(\mathrm{dB}) = G_{\mathrm{element}}(\mathrm{dB}) + 10\log_{10}(N) \tag{8.10}$$

where N is the number of antenna elements in the array. Increasing the number of elements in an antenna array result in a high gain narrow beam radiation pattern, which can be directed in a specific direction. Figure 8.9 shows simulated radiation patterns of a single radiating element (dipole antenna), 7-element linear array, and 13-element linear array.

From Figure 8.9, the radiation pattern of a single-element dipole antenna is the same in all directions (i.e., not directive). By having a 7-element linear array with the elements spaced 0.25 λ, the antenna beam becomes narrow and directive (i.e., higher gain), and increasing the number of elements in the antenna array to 13 elements results in producing a narrower beam. Therefore, the antenna beams can be focused in a specific direction by changing the progressive phase between neighboring antenna elements. Also, more than one beam can be generated in different directions by dividing the radiating elements into subarrays. The radiation patterns in Figure 8.10 illustrate the concept of beam steering by controlling the phase between neighboring radiating elements of an antenna array.

Figure 8.10 demonstrates that by changing the phase between the neighboring elements of the antenna array, the antenna beam can be directed in a specific direction.

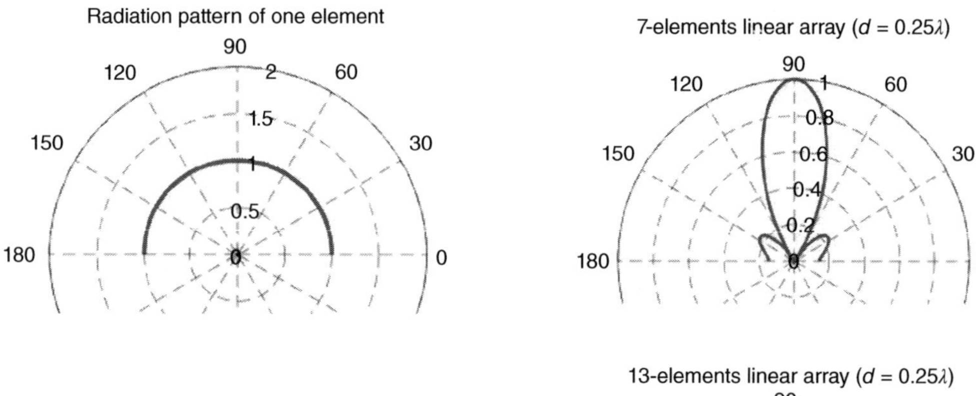

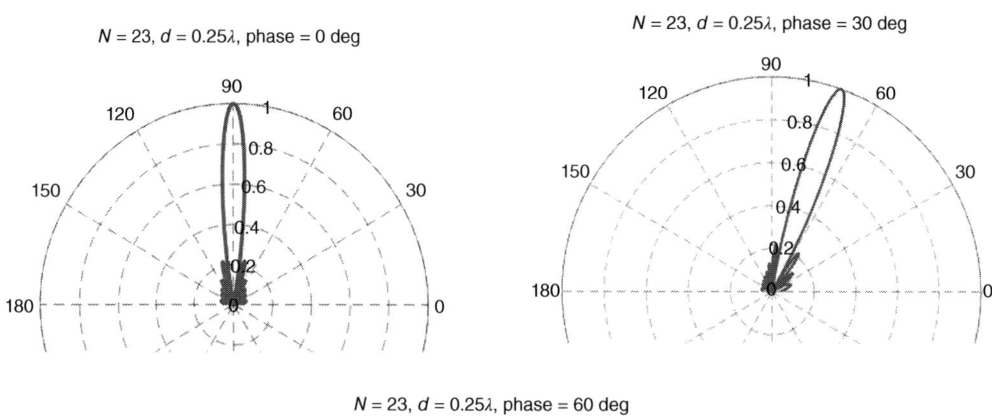

Figure 8.9 Normalized radiation patterns of single radiating element, 7-element, and 13-element linear arrays.

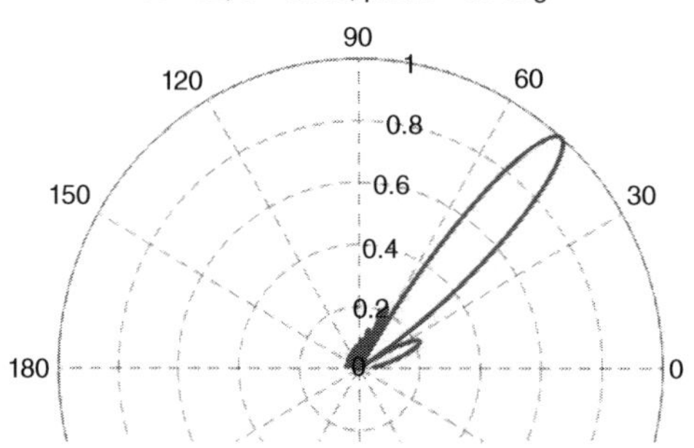

Figure 8.10 Illustration of antenna array beam steering by controlling the progressive phase between neighboring radiating elements.

There are three basic architectures for antenna beamforming, which include analog beamforming, digital beamforming, and hybrid beamforming. These beamforming types are analog beamforming, digital beamforming, and hybrid beamforming.

8.8.1 Analog Beamforming

In analog beamforming, only a single RF chain in the transmitter is used for feeding all the antenna elements, as shown in Figure 8.11.

In Figure 8.11, the output of the RF chain is split among multiple antenna elements using a power splitter, and each antenna signal path has an analog phase shifter. The phase shifters adjust the phase of the signals in each antenna element to direct the beam of the antenna array in the desired direction. In this type of beamforming, one beam is sent at a time. The advantage of analog beamforming includes low cost, low power consumption, and increased reliability. By distributing the phase shifters, if one RF path fails the system still functions. The disadvantage is less flexibility and the difficulty of reconfiguring.

8.8.2 Digital Beamforming

In digital beamforming, each antenna has a dedicated RF chain; the phases and amplitudes of the signals in the antenna elements are controlled by the transmitter's digital baseband processing. This type of beamforming provides full flexibility to create a superposition of beams and vary the power over the antenna elements. However, it suffers high power consumption and signaling overheads. Also, each antenna requires an RF chain, which increases the system's cost. Figure 8.12 shows digital beamforming in a wireless transmitter. In digital beamforming, multiple beams can be sent simultaneously.

8.8.3 Hybrid Beamforming

Hybrid beamforming combines aspects of analog and digital beamforming. In this technique, each subantenna array has a single RF power splitter and an RF chain, and each antenna element in

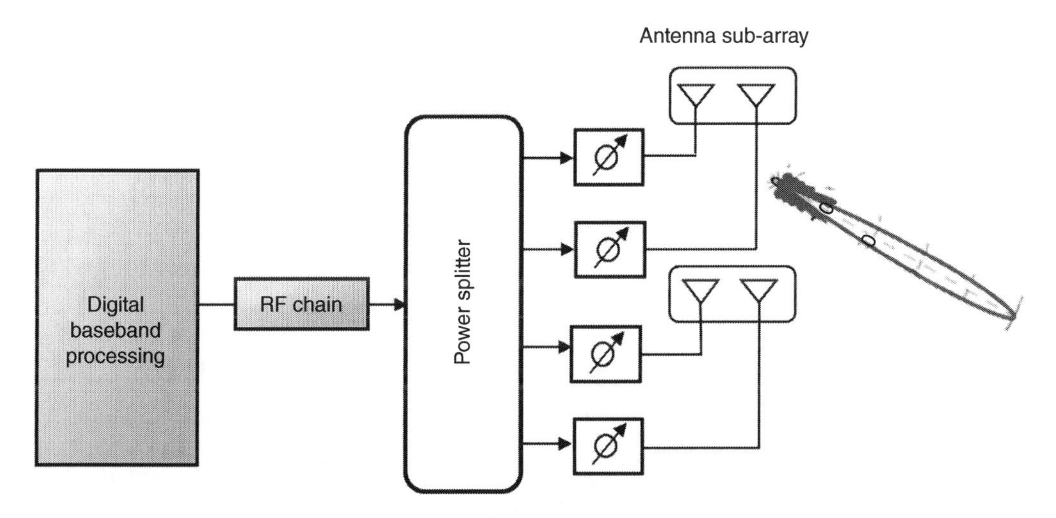

Figure 8.11 Illustration of an analog beamforming in a wireless transmitter.

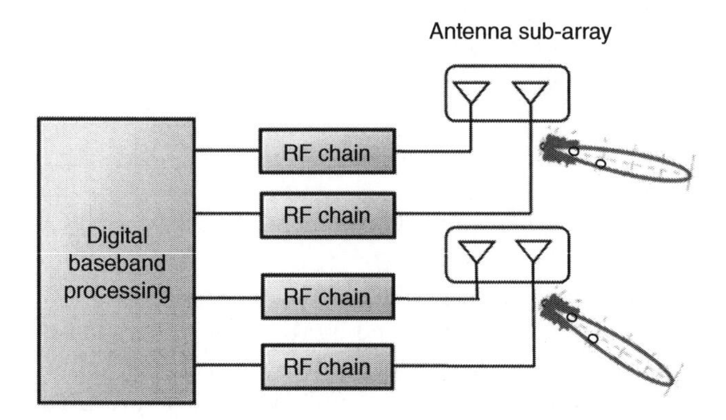

Figure 8.12 Illustration of digital beamforming in a wireless transmitter.

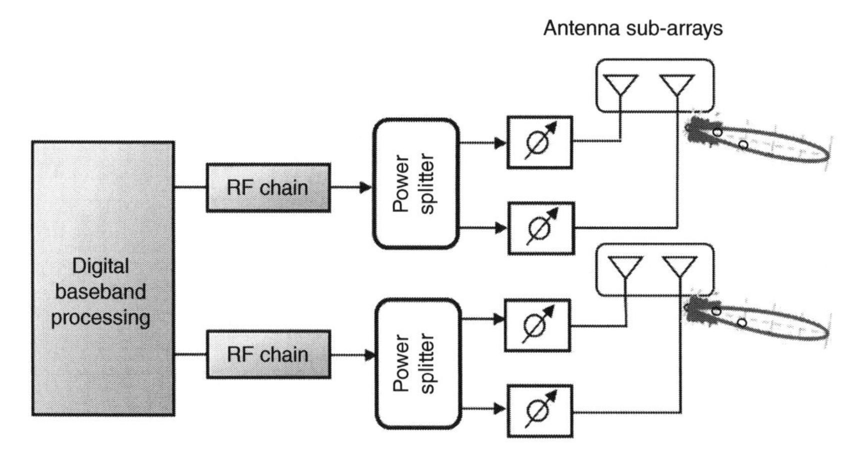

Figure 8.13 Illustration of a hybrid beamforming in a wireless transmitter.

the subarray has a phase shifter. Therefore, multiple beams can be sent with less RF chains, which reduces the system cost. Figure 8.13 shows a hybrid beamforming in a wireless transmitter. This type of beamforming does not have the full flexibility of digital beamforming.

Review Questions

8.1 List the advantages of 5G technology.

8.2 Explain the application areas of 5G technology.

8.3 Briefly describe the benefits of the future 6G technology.

8.4 Explain the advantages and challenges of using mmWave frequencies for 5G technology.

8.5 Explain and illustrate the 5G resource grid.

8.6 List the channel bandwidths of the FR1 and FR2 bands of 5G NR technology.

8.7 Draw and explain the frame structure of a 5G NR wireless system.

8.8 Describe and draw the resource grid of a 5G wireless transmission.

8.9 Briefly explain the advantages of using MIMO technology in 5G systems.

8.10 Write an equation for a radio channel matrix of a 2×2 MIMO configuration.

8.11 Illustrate and explain the basic concept of beamforming technology.

8.12 Compare and contrast analog, digital, and hybrid beamforming types.

References

1 5G System Overview, https://www.3gpp.org/technologies/5g-system-overview.
2 Erik Dahlman, Stefan Parkvall, Johan Skold, *5G NR: The Next Generation Wireless Access Technology*, 2nd ed., Academic Press, 2020.
3 Yu Heejung, Lee Howon, Jeon Hongbeom, "What is 5G? Emerging 5G Mobile Services and Network Requirements," *Sustainability*, 2017.
4 Theodore S. Rappaport, Robert W. Heath Jr., Robert C. Daniels, James N. Murdock, *Millimeter Wave Wireless Communications*, 1st ed., Pearson, 2014.
5 Wei Jiang, Fa-Long Luo, *6G Key Technologies: A Comprehensive Guide*, Wiley, 2022.
6 Borwen You, Ja-Yu Lu, *Terahertz Technology*, Intechopen, 2022.
7 Gerardus Blokdyk, *Massive MIMO A Complete Guide*, 5STARCooks, 2022.
8 Long Zhao, Hui Zhao, Kan Zheng, Wei Xiang, *Massive MIMO in 5G Networks: Selected Applications*, 1st ed. Springer Nature, 2018.
9 Robert W. Heath Jr., Angel Lozano, *Foundations of MIMO Communication*, Cambridge University Press, 2018.
10 Qurrat-Ul-Ain Nadeem, Abla Kammoun, et al., "Design of 5G Full Dimension Massive MIMO Systems," *IEEE Transactions on Communications*, Volume: 66, Issue: 2, 2018.
11 ShareTechnote, https://www.sharetechnote.com.html
12 Zhenyu Xiao, Lipeng Zhu, Lin Bai, Xiang-Gen Xia, *Array Beamforming Enabled Wireless Communications*, 1st ed., CRC Press, 2023.

Suggested Readings

A. Vasuki, E. Shyaam Sundhar, Karri Sowmya, Muthu Kannan M., "Performance Analysis of Massive MIMO Systems Under Wireless Fading Channels," *2nd International Conference on Vision Towards Emerging Trends in Communication and Networking Technologies (ViTECoN)*, 2023.

Aamir Rashid, Syed Shahid Shah, et al., "Mutual Coupling Reduction in MIMO Antenna for 5G Application by Self-Decoupled Method," *2nd International Conference on Vision Towards Emerging Trends in Communication and Networking Technologies (ViTECoN)*, IEEE, 2023.

Aidi Ren, Haoran Yu, et al., "A Broadband MIMO Antenna Based on Multimodes for 5G Smartphone Applications," *IEEE Antennas and Wireless Propagation Letters*, Volume: 22, Issue: 7, 2023.

Ao Liu, Taneli Riihonen, Weixing Sheng, "Full-Duplex Analog Beamforming Design for mm-Wave Integrated Sensing and Communication," *IEEE Radar Conference (RadarConf23)*, 2023.

Chouaib Bencheikh Lehocine, Fredrik Brännström, Erik G. Ström "Robust Analog Beamforming for Periodic Broadcast V2V Communication," *IEEE Transactions on Intelligent Transportation Systems*, Volume: 23, Issue: 10, 2022.

Deniz Dosluoglu, Kun-Da Chu, et al., "A Reconfigurable Digital Beamforming V-Band Phased-Array Receiver," *ESSCIRC 2022- IEEE 48th European Solid State Circuits Conference (ESSCIRC)*, 2022.

Dingyan Cong, Shuaishuai Guo, et al., "Vehicular Behavior-Aware Beamforming Design for Integrated Sensing and Communication Systems," *IEEE Transactions on Intelligent Transportation Systems*, Volume: 24, Issue: 6, 2023.

Felipe Augusto Pereira de Figueiredo, "An Overview of Massive MIMO for 5G and 6G," *IEEE Latin America Transactions*, 2022.

Ghanshyam Singh, Ajay Abrol, et al., "Isolation Enhancement in a Two-Element MIMO Antenna Using Electromagnetic Metamaterial," *International Conference on Device Intelligence, Computing and Communication Technologies, (DICCT)*, 2023.

Kefayet Ullah, Satheesh Bojja Venkatakrishnan, John L. Volakis, "Millimeter-Wave Digital Beamforming Receiver Using RFSoC FPGA for MIMO Communications," *IEEE 22nd Annual Wireless and Microwave Technology Conference (WAMICON)*, 2022.

Krishna Kanth Varma P, K. N. Nagesh, "Compact Two-Port MIMO Antenna Supporting the Sub-6 GHz 5G Spectrum Band of n77-n78," *International Conference on Computer Communication and Informatics (ICCCI)*, 2022.

Kwanghoon Lee, Jonghyun Kim, Eui Whan Jin, Kwang Soon Kim, "Extreme Massive MIMO for Upper-Mid Band 6G Communications," *13th International Conference on Information and Communication Technology Convergence (ICTC)*, 2023.

Markus Hofer, David Löschenbrand, et al., "Massive MIMO Channel Measurements for a Railway Station Scenario," *IEEE Wireless Communications and Networking Conference (WCNC)*, 2023.

Mohamed Rihan, Tarek Abed Soliman, Chen Xu, et al., "Taxonomy and Performance Evaluation of Hybrid Beamforming for 5G and Beyond Systems," *IEEE Access*, 8, 2020.

Mohammed A. AlQaisei, Abdel-Fattah A. Sheta, Ibrahim Elshafiey, "Hybrid Beamforming for Multi-User Massive MIMO Systems at Millimeter-Wave Networks," *39th National Radio Science Conference (NRSC)*, 2022.

Nishan Das, Md Sharif Mia, et al., "A Dual-Band MIMO Antenna for 5G Sub-6 GHz/WiFi/WiMAX/WLAN/Bluetooth/C-Band Applications," *Internationeal Conference on Electrical, Computer and Communication Engineering (ECCE)*, 2023.

Satoshi Suyama, Tatsuki Okuyama, et al., "Recent Studies on Massive MIMO Technologies for 5G Evolution and 6G," *IEEE Radio and Wireless Symposium (RWS)*, 2022.

Swarnima Jain, Anhad Markan, C. M. Markan, "Performance Evaluation of a Millimeter Wave MIMO Hybrid Beamforming System," *IEEE Latin-American Conference on Communications (LATINCOM)*, 2020.

Zhiwen Qin, Zhiming Yi, et al., "Analog Beamforming Transceiver for Millimeter-Wave Communication," *IEEE International Students' Conference on Electrical, Electronics and Computer Science (SCEECS)*, 2022.

9

RF Performance Verification of 5G NR Transceivers

This chapter provides the necessary knowledge for characterizing the RF performance of 5G NR base station transmitters and receivers. The key performance parameters of conducted and radiated (i.e., over-the-air (OTA)) conformance tests of 5G NR base station transmitters and receivers are covered. A brief description of the vector signal generator and vector spectrum analyzer which are essential for the RF performance verification of the 5G NR base station is presented. Test setups for characterizing RF performance of the 5G NR base-station transmitters are discussed. The transmitter parameters include the output power, total power dynamic range, transmit ON/OFF power, frequency error, error vector magnitude (EVM), time alignment error, occupied bandwidth, adjacent channel leakage power ratio, (ACLR), operating band unwanted emissions, transmitter spurious emissions, and transmitter intermodulation. The test setups for characterizing the RF performance of the receiver are also presented and discussed. These parameters include the reference sensitivity, dynamic range, adjacent channel selectivity, in-band and out-of-band blocking, receiver spurious emissions, and receiver intermodulation. Review questions are provided at the end of the chapter to help the readers understand the chapter's contents. Also, a list of references related to the covered topics and suggested reading are provided.

9.1 Chapter Objectives

After completing this chapter, the reader will be able to:

- Describe the test setups for the characterizing the RF performance verifications of 5G NR base station transmitters and receivers and describe the difference between conducted and over-the-air test methods.
- Explain and illustrate the error vector magnitude (EVM) of a digitally modulated signal and the cause of EVM in radio transmitters.
- Explain and illustrate the following performance parameters:
  - occupied bandwidth (OBW),
  - adjacent channel leakage ratio (ACLR),
  - operating band unwanted emission (OBUE),
  - transmitter's conducted spurious emission, and the transmitter's intermodulation.

Essentials of RF Front-end Design and Testing: A Practical Guide for Wireless Systems, First Edition. Ibrahim A. Haroun.

- Define the receiver reference sensitivity, receiver dynamic range, receiver's adjacent channel selectivity (ACS), and receiver's in-band blocking.
- Explain the test setup for measuring the over-the-air (OTA) receiver's sensitivity.

9.2 Test Instruments for Radio Performance Verification

The most common test instruments for performing conducted and radiated OTA testing of wireless systems such as 5G NR radios transceivers [1–4] are RF signal generators [5] and vector spectrum analyzers [6]. These instruments are described in the following subsections.

9.2.1 RF Signal Generators

RF signal generators are key test instruments for testing and characterizing 5G NR base station receivers and wireless communication systems. They generate RF signals with a set of properties to characterize the radio transceivers of wireless systems.

There are two types of RF signal generators, analog signal generators and vector signal generators. Analog signal generators generate CW signals and can provide AM, FM, and PM signals. Their applications include generating high-quality signal to be used as an LO signal for texting RF mixers, testing the gain, linearity, etc. of RF components, and receiver testing (e.g., two tones, generating an interferer signal for in-band blocking test). The advantages of analog signal generators are the cost and the signal quality (i.e., signal purity). The main specifications of the analog signal generators are the output power, frequency range, harmonics, and phase noise.

Vector signal generators generate all digital modulation schemes (i.e., QAM, QPSK, PSK, OFDM, etc.) that are needed for testing communication systems. Their application includes generating signals for testing 5G NR systems and MIMO and beam-forming subsystems. The signals that are used to test the 5G NR systems are called test models [2], which are specified by the 3GPP. Test models are defined signals with fixed configuration to carry out signal quality and spectrum measurements like EVM, spectrum emission mask, and ACLR. The models are defined for for both frequency ranges FR1 and FR2, for different numerology, and different bandwidths. Each model is defined for a specific test. The models for FR1 are called TM1.1, TM1.2, TM2, TM2a, TM2b, TM3.1a, TM3.1b, TM3.2, TM3.3. The test models for FR2 are FR2-TM1.1, FR2-TM2, FR2-TM2a, FR2-TM3.1, FR2-TM3.1a, FR2-TM3.1b, FR2-TM3.2, FR2-TM3.3. Figure 9.1 shows a photograph of a vector signal generator.

9.2.2 Vector Spectrum Analyzer

Vector spectrum analyzers (VSA) are essential test instruments for the conformance testing of 5G NR transceivers. They measure the amplitude, frequency, and phase of an input signal and

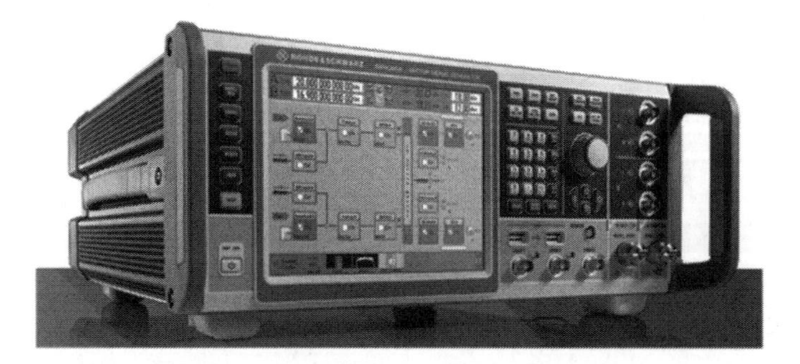

Figure 9.1 Vector signal generator (R&S SMW200A) [5].

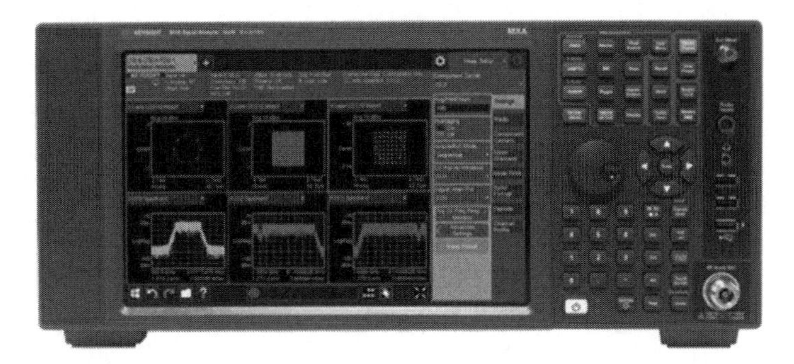

Figure 9.2 Vector spectrum analyzer (Keysight Technologies) [6].

demodulate digitally modulated signals. They are capable of displaying the signal spectrum, constellation diagram, and error-vector-magnitude. Figure 9.2 shows a photograph of a vector spectrum analyzer.

9.3 RF Performance Verification of 5G NR Transmitters

The RF performance verification of 5G NR base stations transmitters includes conducted [2, 3] and radiated (i.e., over-the-air, OTA) testing [4]. The conducted testing of the transmitter is performed at its antenna port connector (i.e., the antenna is not part of the test). However, in the OTA testing, the antenna is part of the test. Figure 9.3 shows the radiated and conducted reference points [3] for a base station type 1-H. The conducted characteristics are defined at the transceiver array boundary (TAB) connectors, which is the conducted interface between the transceiver unit array (TRXUA) and the composite antenna as shown in Figure 9.3. The transceiver unit array contains a number of transmitters and number of receivers, which have the ability to transmit/receive parallel independent modulated symbol streams. The composite antenna contains a radio distribution network (RDN) and an antenna array. The RDN is a linear passive network, which distributes the RF power generated by the transceiver unit array to the antenna array, and distributes the radio signals collected by the antenna array to the transceiver unit array. The key conformance tests of the 5G NR base station transmitters are listed in Table 9.1.

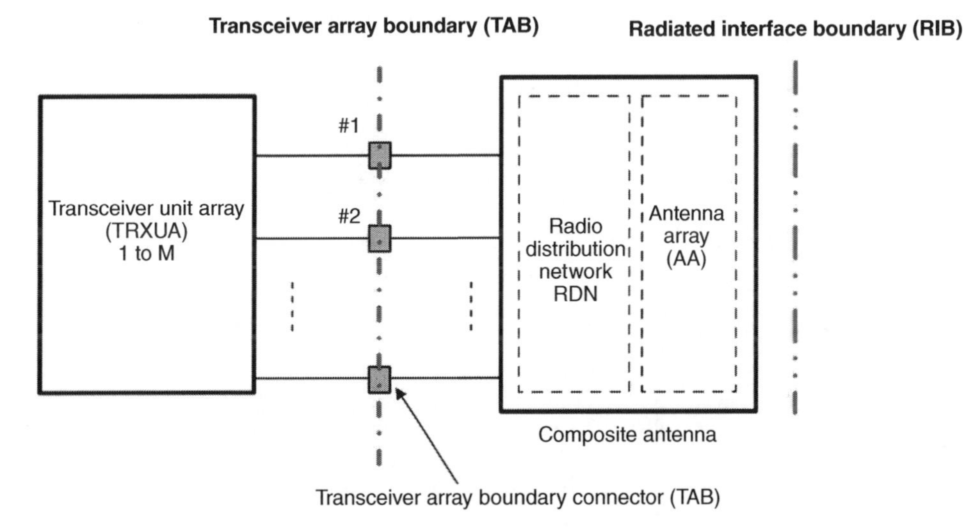

Figure 9.3 Radiated and conducted reference points of base station type 1-H.

Table 9.1 Conducted conformance tests of 5G NR base station transmitters.

Test case	Test name
1	Output power
2	Total power dynamic range
3	Transmit ON/OFF power
4	Frequency error
5	Modulation quality
6	Time alignment error
7	Occupied bandwidth
8	Adjacent channel leakage power ratio (ACLR)
9	Operating band unwanted emissions
10	Transmitter spurious emissions
11	Transmitter intermodulation

9.3.1 Transmitter Output Power

The purpose of the output power measurement test is to verify that the transmitter's maximum output power does not exceed the requirement specification across the frequency range of operation under the operating test conditions. The conducted transmitter output power of a wireless 5G base station [2, 3] is specified at the antenna connector of the transceiver's front-end (i.e., the *rated carrier output power per antenna connector*, $P_{\text{rated,c,AC}}$). Figure 9.4 shows the test setup of measuring the transmitter's output power at its antenna connector port.

In Figure 9.4, the transmitter's output signal at the Tx antenna connector port is connected to a spectrum analyzer through a variable attenuator to adjust the power level and to protect the spectrum analyzer. Transmit antenna ports that are not under test must be terminated. 3GPP defined signals called test models to carry conformance testing, the models are defined for both FR1 and FR2. In this test setup, the spectrum analyzer should be configured to enable testing TM1.1 signals. Table 9.2 [2] lists the output power limits for different classes of BS type 1-C.

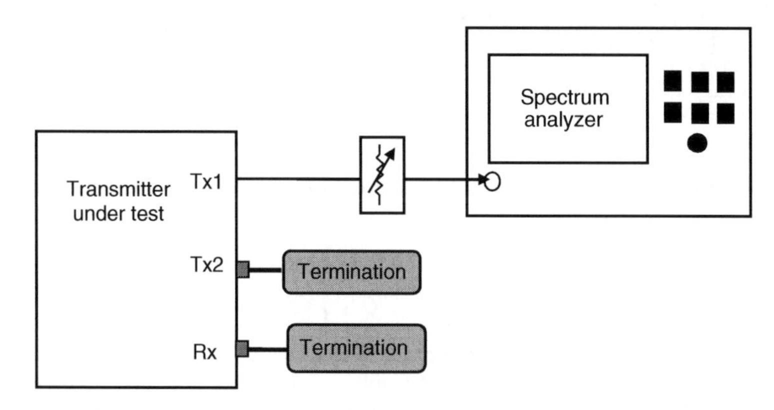

Figure 9.4 Test setup for measuring transmitter output power.

Table 9.2 Output power limits for BS type 1-C.

BS class	Prated (dBm)
Medium range BS	≤38
Local area BS	≤24

Table 9.3 5G NR BS total power dynamic range.

NR channel bandwidth (MHz)	Total power dynamic range (dB)		
	15 kHz SCS	30 kHz SCS	60 kHz SCS
5	13.5	10	N/A
10	16.7	13.4	10
15	18.5	15.3	12.1
20	19.8	16.6	13.4
25	20.8	17.7	14.5
30	21.6	18.5	15.3
35	22.7	19.6	16.4
40	22.9	19.8	16.6
45	23.8	20.7	17.6
50	23.9	20.8	17.7
60	N/A	21.6	18.5
70	N/A	22.3	19.2
80	N/A	22.9	19.8
90	N/A	23.4	20.4
100	N/A	23.9	20.9

9.3.2 Transmitter Total Power Dynamic Range

The purpose of the total power dynamic range measurement test [2] is to verify that the difference between the maximum and minimum transmit power is within the system performance requirement. Table 9.3 is an example of the specifications of the total power dynamic range for a 5G NR base station [2]. The test setup for measuring the transmitter's total power dynamic range is the same as that shown in Figure 9.4.

9.3.3 Transmit ON/OFF Power

The purpose of the transmit ON/OFF power measurement test is to verify that the transmitter's OFF power level, and the transient time length meets the standard requirement [1]. Transmit OFF power is the mean power measured over 70 μs when the transmitter is OFF and centered on the required channel. Figure 9.5 illustrates the transmitter's ON and OFF power levels and the time periods during which the transmitter is changing from OFF period to ON period and vice versa.

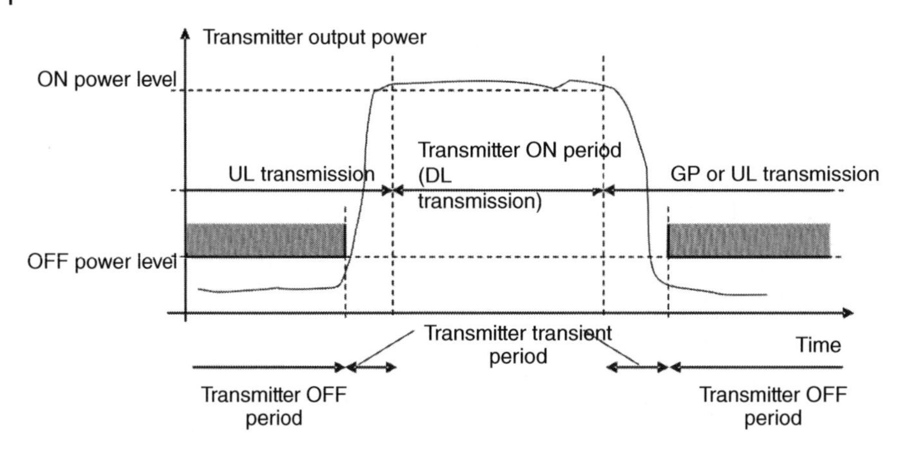

Figure 9.5 Illustration of transmitter ON and OFF periods [2].

Table 9.4 Transmit OFF power.

BS type	BS output power
1-C	≤−85 dBm/MHz per antenna connector

Table 9.5 Transient period.

Transition	Transient period length (μs)
OFF to ON	10
ON to OFF	10

Table 9.4 lists the specifications of transmit OFF power for radio base station type 1-C, and Table 9.5 lists the specification for the transient period.

The test setup for measuring the transmitter ON and OFF power is the same as that shown in Figure 9.4. In this test setup, the spectrum analyzer should be configured to enable testing TM1.1 signals. Figure 9.6 [7] shows an image of a measured transmitter OFF power.

From Figure 9.6, the filter bandwidth is 100 MHz, and the average power is −63.95 dBm, which is equivalent to −83.95 dBm in 1 MHz and meets the requirements of Table 9.4.

9.3.4 Transmitter Frequency Error

The purpose of the transmitter's frequency error measurement test is to verify that the difference between the actual transmitter's frequency and the assigned frequency meets the standard specification requirements under the test conditions. The frequency error is specified in parts per million (ppm); 1 ppm means $1/10^6$ part of a nominal frequency. Table 9.6 [2] shows the 3GPP frequency error specification for different base station classes.

The test setup shown in Figure 9.4 can be used for measuring the frequency error. In this test, the spectrum analyzer should be configured to enable testing models TM2/TM2a/TM3.1/TM3.2

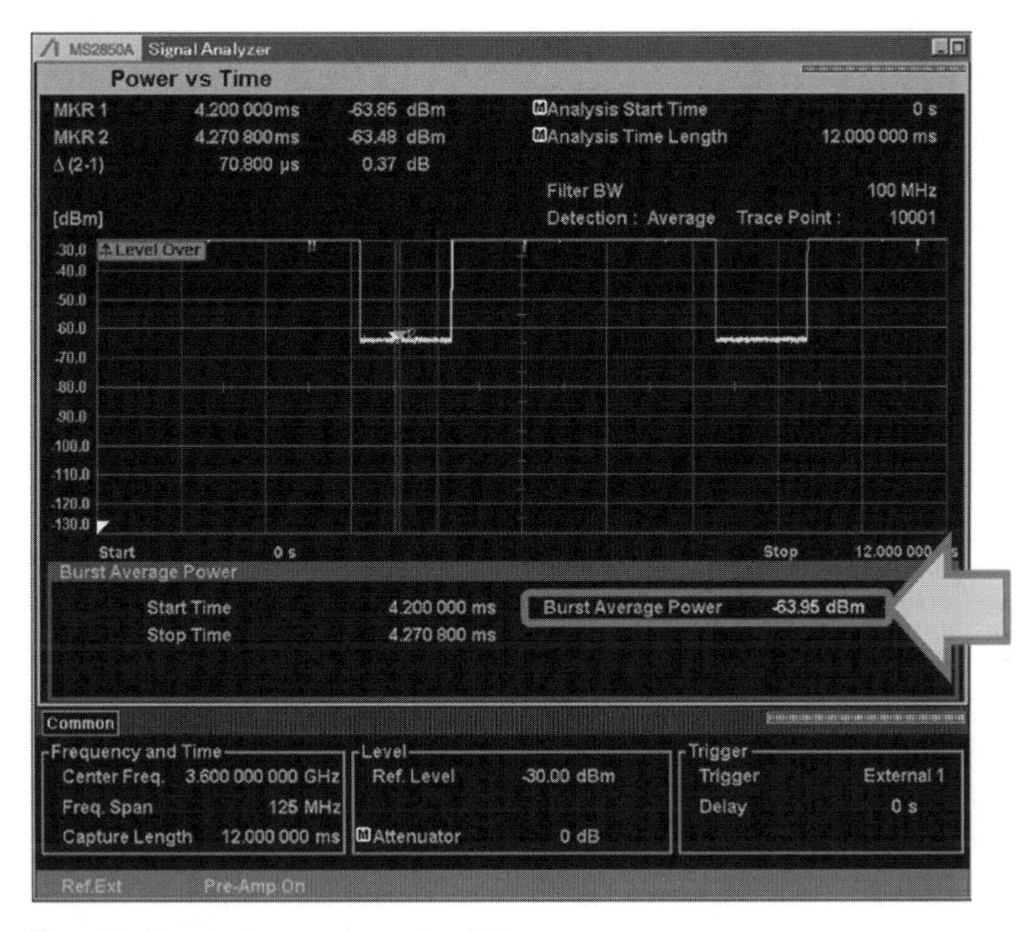

Figure 9.6 Example of measured transmitter OFF power.

Table 9.6 Frequency error specifications for different BS classes.

BS class	Frequency error
Wide area BS	±(0.05 ppm + 12 Hz)
Medium range BS	±(0.1 ppm + 12 Hz)
Local area BS	±(0.1 ppm + 12 Hz)

signals. Figure 9.7 [7] shows an image of measured frequency error for base station transmitter that transmits TM3.1a signal. From Figure 9.6, the frequency error is 0.16 Hz (0.000 ppm) and meets the requirements of Table 9.6.

9.3.5 Transmitter's Error Vector Magnitude (EVM)

The error vector magnitude (EVM) is a figure of merit for assessing the quality of digitally modulated signals. It is the vector difference between the ideal transmit signal and the actual measured

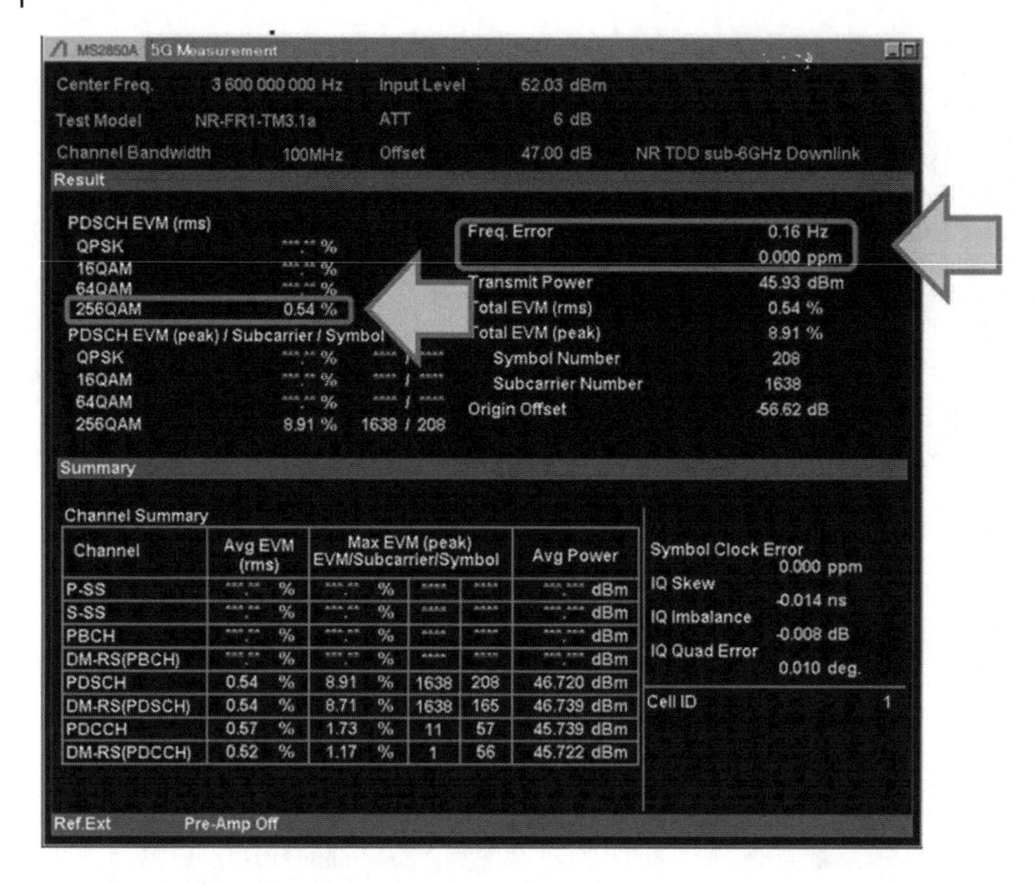

Figure 9.7 An image of measured frequency error [7].

Table 9.7 EVM requirements for 5G base station.

Modulation scheme	Required EVM (%)
QPSK	18.5
16 QAM	13.5
64 QAM	9
256 QAM	4.5

received signal at a given time and is expressed as a percentage (%) or dB. Figure 9.8 illustrates the concept of EVM.

The purpose of the EVM test is to verify the difference between the measured carrier signal and the ideal signal that meets standard specifications under the test conditions. The EVM measurements are helpful because they contain information about the digital signals' amplitude and phase errors. Table 9.7 shows the EVM requirements for BS type 1-C and BS type 1-H.

The test setup in Figure 9.4 can be used for performing the EVM testing; in this test, the vector spectrum analyzer should be configured to enable testing signals of different test models. Table 9.8 lists test models for testing the EVM 5G NR base station.

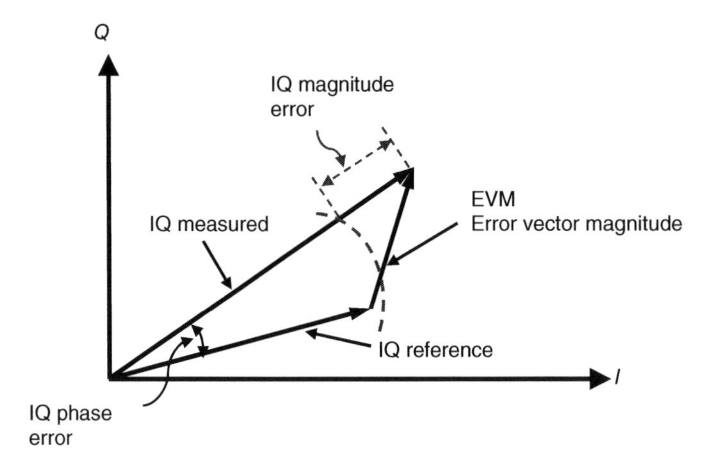

Figure 9.8 Illustration of signal's error vector magnitude.

Table 9.8 FR1Test models for testing EVM of 5G NR base station.

Test model	Modulation type
TM2	64 QAM (at min power)
TM2a	256 QAM (at min power)
TM2b	1024 QAM (at min power)
TM3.1	64 QAM (at max power)
TM3.1a	256 QAM (at max power)
TM3.1b	1024 QAM (at max power)
TM3.2	16QAM (at max power)
TM3.3	QPSK

Figure 9.9 shows an image [7] of measured EVM for FR1 5G NR base station transmitter that was tested for TM3.1a (i.e., 64QAM at max power). From Figure 9.9, the measured EVM for 256QAM is 0.54%, which meets the requirements of Table 9.7.

9.3.6 Time Alignment Error (TAE)

In MIMO systems, the transmitted frames of the signals present at the transmitter's antenna connectors are not perfectly aligned in time and may experience certain timing differences in relation to each other. Thus, it is essential to measure the time alignment error (TAE) to verify it is within the limit of the minimum requirement specified by the standard [2]. The 3GPP standard specifies TAE of less than 90 ns for MIMO systems. Figure 9.10 shows the test set up for measuring the TAE between two antenna connectors of a base station transmitter.

In this test, antenna ports that are not part of the test must be terminated to prevent any radiated coupling with the ports under test. The spectrum analyzer should be configured to enable testing model TM1.1 signals (i.e., test signal is QPSK). Figure 9.11 shows an image of measured TAE [7]. From Figure 9.11, the test mode that is applied for measuring the TAE is FR1-TM1.1 and the measured TAE is 22.9 ns, which meets the required specification of less than 90 ns.

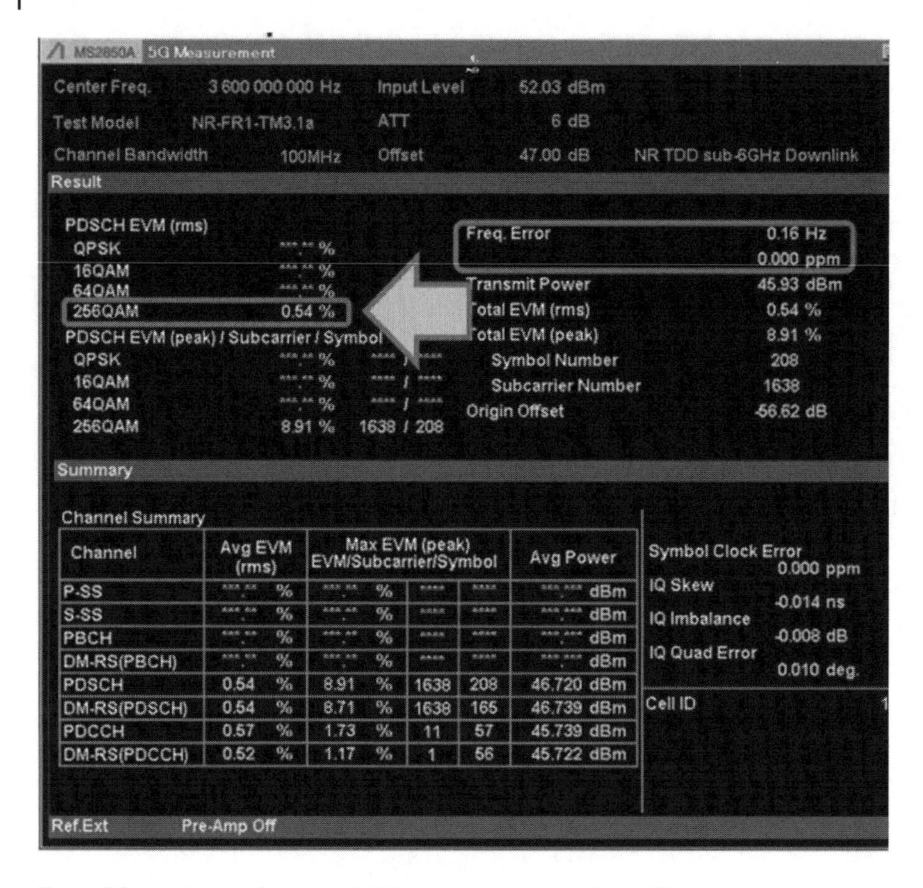

Figure 9.9 An image of measured EVM (error vector magnitude) [7].

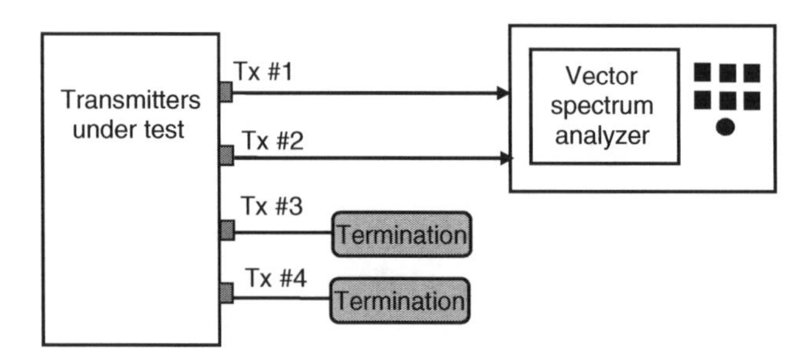

Figure 9.10 Test setup for measuring the TAE between two antenna connectors of a radio transmitter.

9.3.7 Occupied Bandwidth (OBW) Measurement

The occupied bandwidth (OBW) [1, 2] is defined by the 3GPP as the bandwidth containing 99% of the total integrated power of the transmitted spectrum, centered on the assigned channel frequency. The objective of an OBW measurement test is to verify the transmitted signal does not occupy excessive bandwidth. Figure 9.12 [7] shows an image of a measured OBW for a single carrier.

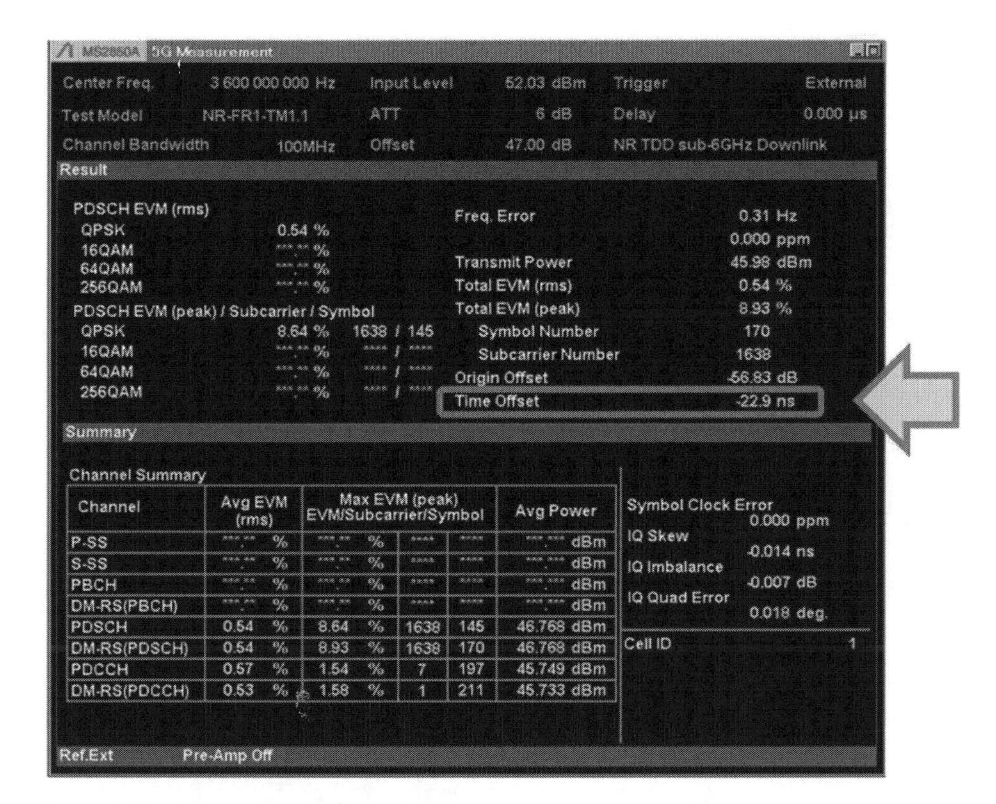

Figure 9.11 An image of measured TAE [7].

From Figure 9.12, the OBW is 96.81 MHz and contains 99% of the total transmitted power, which meets the 3GPP requirements. The test setup in Figure 9.4 can be used for measuring the OBW; in this test, the vector spectrum analyzer should be configured to enable measuring test model TM1.1 signals.

9.3.8 Adjacent Channel Leakage Power Ratio (ACLR)

The adjacent channel leakage power ratio (ACLR) is defined as the ratio of the filtered mean power centered on the assigned channel frequency to the filtered mean power centered on an adjacent channel frequency. The purpose of the ACLR test is to verify that the adjacent channel leakage power ratio meets the minimum requirement defined by the standard specifications. Table 9.9 lists the specification requirements of ACLR [2].

The ACLR measurement can be performed using the test setup in Figure 9.4 with the spectrum analyzer configured to test model TM1.1. Figure 9.13 shows measured ACLR of a 5G base station transmitter.

From Figure 9.13, the L1 and U1 are the adjacent channel leakage power ratios in the lower and upper first adjacent channels, and the L2 and U2 are the adjacent channel leakage power ratios in the lower and upper second adjacent channels. Figure 9.13 shows that the measured ACLR in the lower and upper adjacent channels are −55.3 dB, −54.6 dB, −56.16 dB, and −57.07 dB; the measured values are lower than the required specifications of Table 9.9 (i.e., the results pass with good margin).

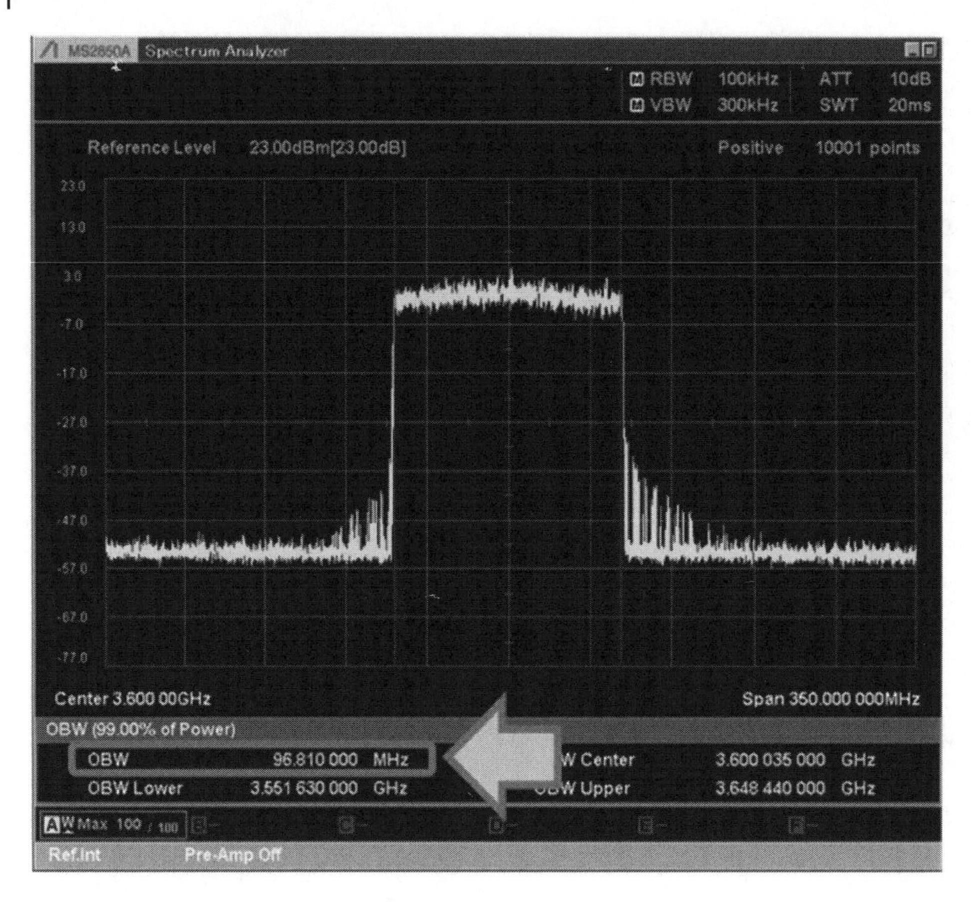

Figure 9.12 An image of measured OBW of a single carrier [7].

Table 9.9 ACLR requirements for different channel BW.

BS channel BW	ACLR limit (dB)
5, 10, 15, 20 MHz	44.2
25, 30, 40, 50, 60, 70, 80, 90, 100 MHz	43.8

9.3.9 Operating Band Unwanted Emissions (OBUE) Measurement

The operating band unwanted emission (OBUE) is defined by the 3GPP as all unwanted emissions in the operating band plus the frequency ranges of 10 MHz above and 10 MHz below the operating band [2, 3]. The purpose of this test is to verify that the unwanted emissions are under the mask's limit defined by the 3GPP standard. Figure 9.14 illustrates the OBUE range.

The test setup for measuring the OBUE is shown in Figure 9.15; in this test setup, the spectrum analyzer is configured to enable testing signals of models TM1.1 and TM1.2. All connectors that are not under test must be terminated.

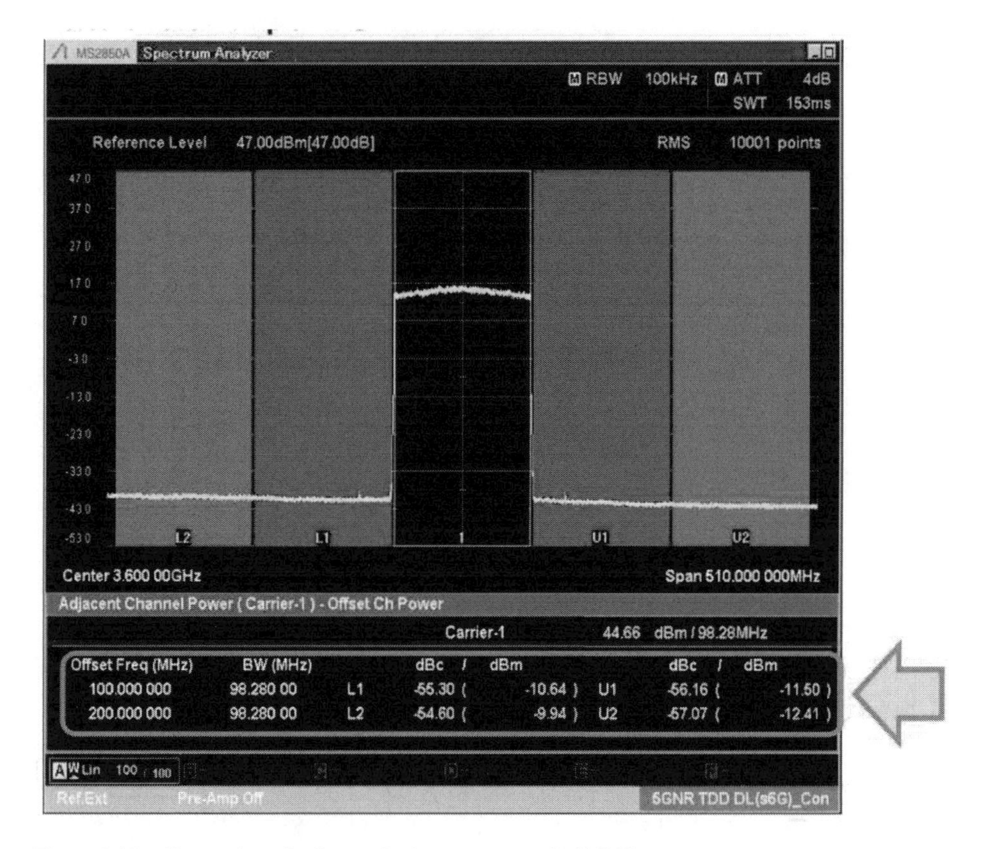

Figure 9.13 Illustration of adjacent leakage power ratio (ACLR).

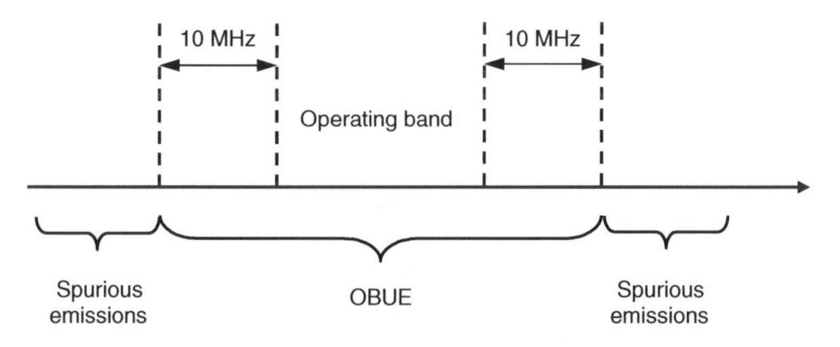

Figure 9.14 Illustration of operating band unwanted emission (OBUE).

9.3.10 Conducted Spurious Emission

The transmitter spurious emissions are emissions at frequencies that are outside the assigned channel. The purpose of the conducted spurious emission measurement test is to verify that the spurious emissions are under the limits from 9 kHz to 12.75 GHz as specified by 3GPP, excluding the operating band. The frequency range should be extended to the fifth harmonic of the operating band's edge if it is higher than 12.75 GHz. These emissions are caused by unwanted transmitter effects

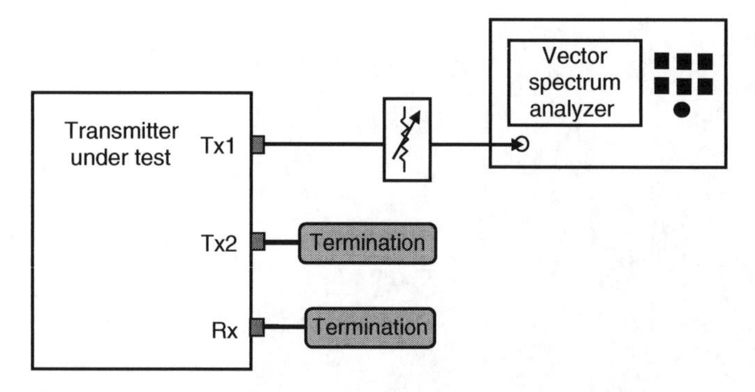

Figure 9.15 Test setup for measuring OBUE.

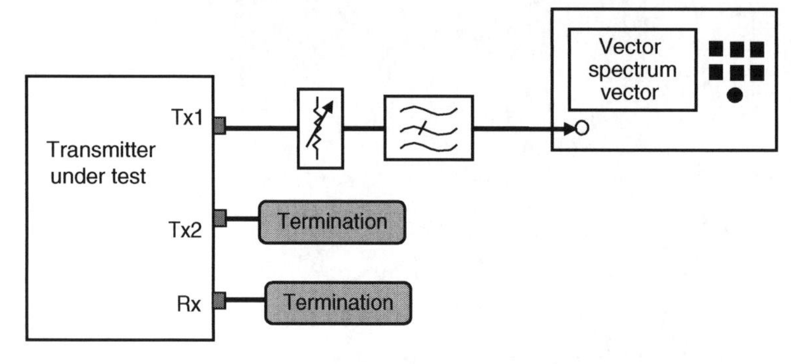

Figure 9.16 Test setup for measuring transmitter spurious emissions.

such as harmonics emission, parasitic emission, intermodulation products, and frequency conversion products, but exclude out-of-band (OOB) emissions. Regulatory agencies such as the Federal Communications Commission (FCC) in the United States and the European Telecommunications Standards Institute (ETSI) in Europe also specify spurious emissions limits. Figure 9.16 shows the test setup for measuring transmitter spurious emissions. In this test setup, a band rejection filter (also called notch filter) is used to reject the carrier signal but passes the spurious; the spectrum analyzer should be configured to enable testing model TM1.1 signals.

Tables 9.10 and 9.11 are examples of the general base station spurious emission limits specification in FR1, Category A and Category B. Figure 9.17 shows an image of measured transmitter spurious emissions.

Table 9.10 General BS transmitter spurious emission limits in FR1, Category A.

Spurious emission range	Basic limits	Measurement bandwidth
9 kHz–150 kHz	−13 dBm	1 kHz
150 kHz–30 MHz		10 kHz
30 MHz–1 GHz		100 kHz
1 GHz–12.75 GHz		1 MHz
12.75 GHz–fifth harmonic of the upper frequency edge of the DL operating band in GHz		1 MHz

Table 9.11 General BS transmitter spurious emission limits in FR1, Category B.

Spurious frequency range	Basic limits (dBm)	Measurement bandwidth
9 kHz–150 kHz	−36	1 kHz
150 kHz–30 MHz		10 kHz
30 MHz–1 GHz		100 kHz
1 GHz–12.75 GHz		1 MHz
12.75 GHz–fifth harmonic of the upper frequency edge of the DL operating band in GHz	−30	1 MHz

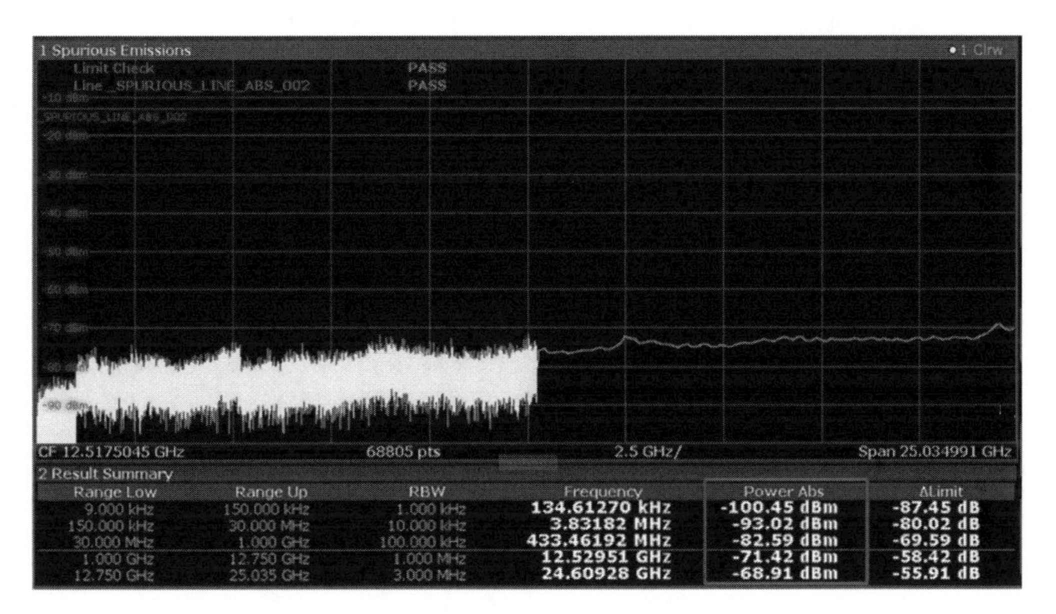

Figure 9.17 An example of measured transmitter spurious emissions.

9.3.11 Transmitter Intermodulation

The transmitter intermodulation requirement is a measure of the transmitter's ability to perform in the presence of a wanted signal and an interfering signal at its antenna port. The purpose of the transmitter's intermodulation measurement test is to verify the intermodulation products caused by the wanted signal and the interfering signal are below the specified requirements. The requirement includes the third and fifth-order intermodulation products excluding the interfering signal frequencies. Also, it should not exceed the emission limits for ACLR, OBUE, or spurious emission, and Figure 9.18 shows the test setup for measuring the transmitter's intermodulation.

In this test setup, a circulator is used to inject the interfering signal into the transmitter's antenna port. The test is performed to capture the critical intermodulation products, including the 3rd, 5th, and 7th orders. The signal generator is configured to test model TM1.1 signals. The intermodulation

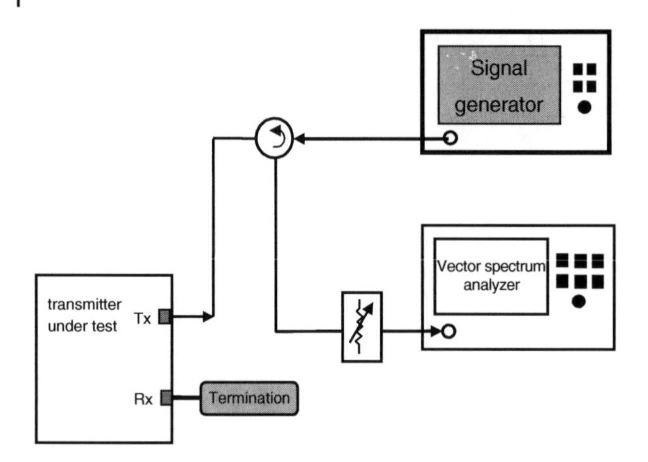

Figure 9.18 Test setup for measuring transmitter intermodulation.

Table 9.12 Interfering and wanted signals for transmitter intermodulation requirement.

Parameter	Value
Wanted signal type	NR single carrier, or multicarrier, or multiple intraband contiguously or noncontiguously aggregated carriers, with NB-IoT operation in NR in-band if supported
Interfering signal type	NR signal, the minimum BS channel bandwidth (BWChannel) with 15 kHz SCS of the band defined
Interfering signal level	Rated total output power (Prated, *t*, AC) in the operating band −30 dB
Interfering signal center frequency offset from lower/upper edge of wanted signal or edge of subblock inside subblock gap	$f_{\text{offset}} = \pm\, \text{BW}_{\text{channel}}\,(n - 1/2)$, for $n = 1, 2$ and 3

level should not exceed the emission limits of ACLR, OBUE, or spurious emissions. Table 9.12 lists the interfering and wanted signal specifications for transmitter intermodulation [2, 3].

9.4 RF Performance Verification of 5G NR Receivers

The purpose of the RF receiver tests is to verify the ability of the receiver to demodulate signals under various conditions and configurations correctly. Table 9.13 [2, 3] lists the key conformance test of 5G NR base station receivers.

Conducted conformance tests of 5G NR base station receivers are performed at the receiver's antenna port connectors (i.e., the antenna is not included in the test). The receiver performance requirements specify a minimum throughput or maximum bit-error rate (BER) that should be achieved at a specific signal-to-noise ratio (SNR) or Eb/No (energy per bit to noise power spectral density ratio). The receiver test cases of Table 9.13 are described in the following subsections.

Table 9.13 Key conformance tests of 5G NR base station receivers.

Test case	Test name
1	Reference sensitivity level
2	Dynamic range
3	Adjacent channel selectivity ACS
4	In-band and out-of-band blocking
5	Receiver spurious emissions
6	Receiver intermodulation

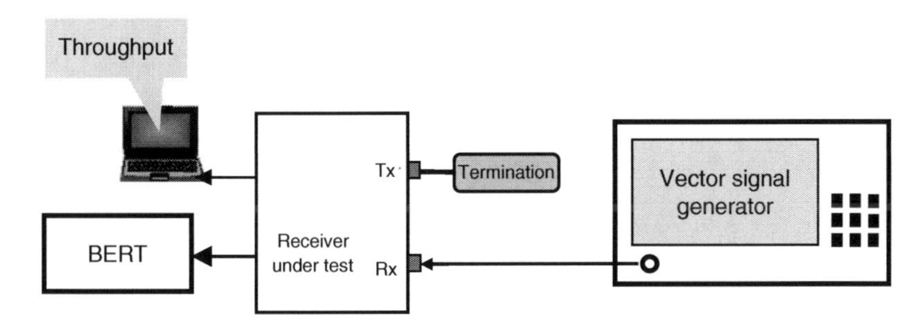

Figure 9.19 Test setup for measuring the receiver reference sensitivity.

9.4.1 Receiver Reference Sensitivity

The receiver reference sensitivity power level P_{REFSENS} is the minimum mean power received at the antenna port connector, where the throughput or the BER meets the standard requirement for a specified reference measurement channel. The reference sensitivity is given by

$$P_{\text{REFSENS}}\,(\text{dBm}) = -174 + 10\,\log\,(\text{RX BW}) + \text{NF} + \text{SNR} \tag{9.1}$$

where RX BW is the receiver bandwidth in Hz, NF is the receiver noise figure in dB, and SNR is the base band demodulation signal-to-noise ratio in dB. Figure 9.19 shows the test setup for the receiver reference sensitivity measurement.

In this test setup, the vector signal generator is configured to enable measuring the required test model signals; the signal generator's output is applied to the receiver's input via the antenna port connector. The receiver under test should be set to the desired channel. The BER is measured using a BERT tester, also some vector signal generators have BER measurement capabilities and can be used for performing this test. The test should be done at the bottom, middle, and top of the channel to be tested. The 3GPP specifications for 5G NR receivers require throughput $\geq$95% of the maximum throughput of the reference channel. Table 9.14 [2] is an example of the reference sensitivity-level specifications for a NR wide-area base station.

9.4.2 Receiver Dynamic Range

The dynamic range of a radio receiver is the range of input levels (including wanted and interference signals) at the antenna port connector and over which the receiver's performance meets the

Table 9.14 NR wide-area BS reference sensitivity levels.

BS channel bandwidth (MHz)	Subcarrier spacing (kHz)	Reference sensitivity power level, PREFSENS (dBm)		
		$f \leq 3.0$ GHz	3.0 GHz $< f \leq 4.2$ GHz	4.2 GHz $< f \leq 6.0$ GHz
5, 10, 15	15	−101	−100.7	−100.5
		−101 (Note)	−100.7 (Note)	−100.5 (Note)
10, 15	30	−101.1	−100.8	−100.6
10, 15	60	−98.2	−97.9	−97.7
20, 25, 30, 35, 40, 45, 50	15	−94.6	−94.3	−94.1
		−94.6 (Note)	−94.3 (Note)	−94.1 (Note)
20, 25, 30, 35, 40, 45, 50, 60, 70, 80, 90, 100	30	−94.9	−94.6	−94.4
20, 25, 30, 35, 40, 45, 50, 60, 70, 80, 90, 100	60	−95	−94.7	−94.5

Note: The requirements apply to BS that supports NB-IoT operation in NR in-band.

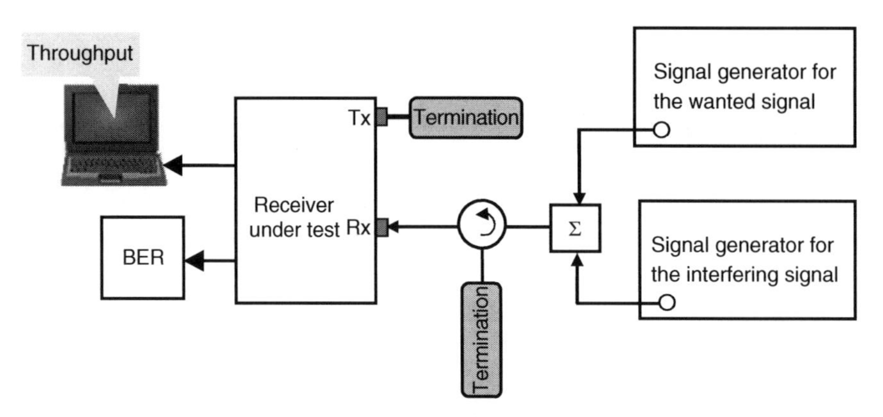

Figure 9.20 Test setup for measuring receiver dynamic range.

system requirements of the throughput or BER. The low end of the dynamic range is governed by the receiver sensitivity, whereas the high end is governed by its third-order intercept point.

The receiver dynamic range test measures the receiver's ability to receive a wanted signal when there is an interferer inside the channel bandwidth. Figure 9.20 shows the receiver dynamic range test setup; in this setup, the vector signal generators are configured to provide the required wanted and interference signals. The measured throughput must meet the requirement of ≥95% of the maximum throughput. Table 9.15 [2] lists the dynamic range specification for a wide-area base station.

Table 9.15 Receiver dynamic range requirements for a wide-area BS.

BS channel bandwidth (MHz)	Subcarrier spacing (kHz)	Reference measurement channel	Wanted signal mean power (dBm)	Interfering signal mean power (dBm)/BWConfig	Type of interfering signal
5	15	G-FR1-A2-1	−70.4	−82.5	AWGN
	30	G-FR1-A2-2	−71.1		
10	15	G-FR1-A2-1	−70.4	−79.3	AWGN
	30	G-FR1-A2-2	−71.1		
	60	G-FR1-A2-3	−68.1		
15	15	G-FR1-A2-1	−70.4	−77.5	AWGN
	30	G-FR1-A2-2	−71.1		
	60	G-FR1-A2-3	−68.1		
20	15	G-FR1-A2-4	−64.2	−76.2	AWGN
	30	G-FR1-A2-5	−64.2		
	60	G-FR1-A2-6	−64.5		
25	15	G-FR1-A2-4	−64.2	−75.2	AWGN
	30	G-FR1-A2-5	−64.2		
	60	G-FR1-A2-6	−64.5		
30	15	G-FR1-A2-4	−64.2	−74.4	AWGN
	30	G-FR1-A2-5	−64.2		
	60	G-FR1-A2-6	−64.5		
35	15	G-FR1-A2-4	−64.2	−73.7	AWGN
	30	G-FR1-A2-5	−64.2		
	60	G-FR1-A2-6	−64.5		
40	15	G-FR1-A2-4	−64.2	−73.1	AWGN
	30	G-FR1-A2-5	−64.2		
	60	G-FR1-A2-6	−64.5		
45	15	G-FR1-A2-4	−64.2	−72.6	AWGN
	30	G-FR1-A2-5	−64.2		
	60	G-FR1-A2-6	−64.5		
50	15	G-FR1-A2-4	−64.2	−72.1	AWGN
	30	G-FR1-A2-5	−64.2		
	60	G-FR1-A2-6	−64.5		
60	30	G-FR1-A2-5	−64.2	−71.3	AWGN
	60	G-FR1-A2-6	−64.5		
70	30	G-FR1-A2-5	−64.2	−70.7	AWGN
	60	G-FR1-A2-6	−64.5		
80	30	G-FR1-A2-5	−64.2	−70.1	AWGN
	60	G-FR1-A2-6	−64.5		
90	30	G-FR1-A2-5	−64.2	−69.5	AWGN
	60	G-FR1-A2-6	−64.5		
100	30	G-FR1-A2-5	−64.2	−69.1	AWGN
	60	G-FR1-A2-6	−64.5		

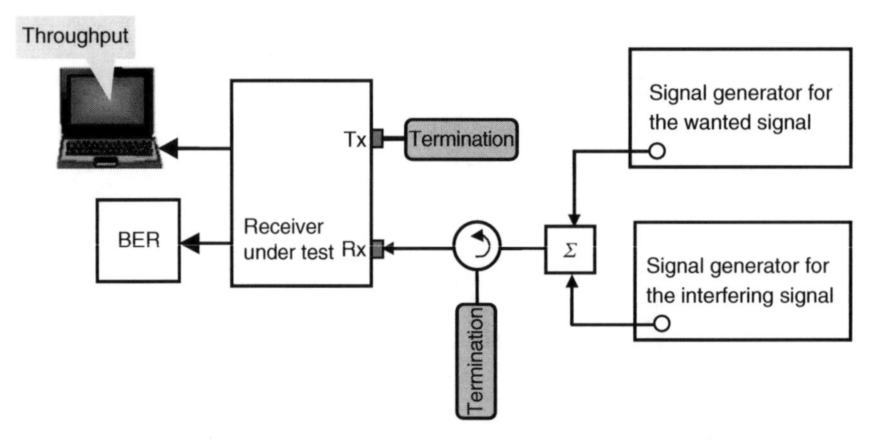

Figure 9.21 Test setup for measuring adjacent channel selectivity (ACS).

9.4.3 Adjacent Channel Selectivity (ACS) Measurement

The receiver's adjacent channel selectivity (ACS) [3] is a measure of the receiver's ability to receive a wanted signal at its assigned channel frequency in the presence of an adjacent channel signal with a specified center frequency offset from the band edge of the wanted channel. Figure 9.21 shows the test setup for measuring ACS.

In this test setup, the signal generator of the wanted signal is configured to provide test signals with specifications as per the standard requirements. The signal generator of the interfering signal is set to a frequency offset as specified by the standard specifications [3]. For 5G NR receivers, the measured throughput must meet the requirement of ≥95% of the maximum throughput of the reference channel. Figure 9.22 illustrates the relationship between the wanted and interference signals.

Table 9.16 [2, 3] shows an example of the ACS requirements for 5G NR base station, and Table 9.17 shows the base station ACS interferer frequency offset values.

9.4.4 In-band Blocking

In-band blockers [2, 3] are strong signals that fall inside the receiving band creating nonlinear distortion and interfering with weak RF carrier signals. Such blockers get mixed in the radio front end

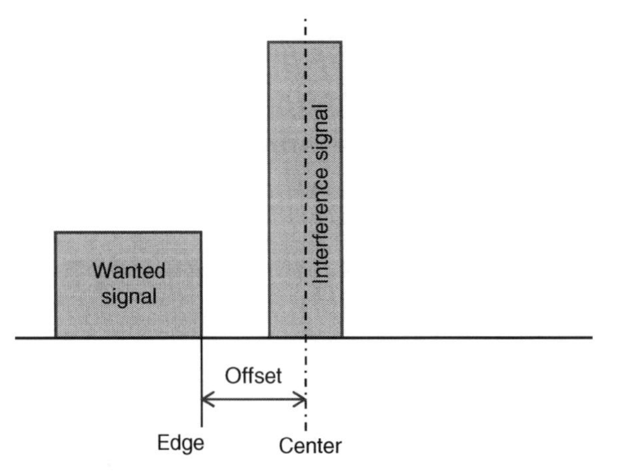

Figure 9.22 Relationship between wanted and interference signals.

Table 9.16 Base station ACS requirement.

BS channel bandwidth of the lowest/highest carrier received (MHz)	Wanted signal mean power (dBm)	Interfering signal mean power (dBm)
5, 10, 15, 20, 25, 30, 35, 40, 45, 50, 60, 70, 80, 90, 100	$P_{REFSENS} + 6\,dB$	Wide area BS: −52 Medium range BS: −47 Local area BS: −44

Note: The SCS for the lowest/highest carrier received is the lowest SCS supported by the BS for that bandwidth.

Table 9.17 Base station ACS interferer frequency offset values.

BS channel bandwidth of lowest/highest carrier received (MHz)	Interfering signal center frequency offset from the lower/upper base station RF bandwidth edge or subblock edge inside a subblock gap (MHz)	Type of interfering signal
5	±2.5025	5 MHz DFT-s-OFDM NR signal, SCS: 15 kHz, 25 RB
10	±2.5075	
15	±2.5125	
20	±2.5025	
25	±9.4675	20 MHz DFT-s-OFDM NR signal, SCS: 15 kHz, 100 RB
30	±9.4725	
40	±9.4625	
45	±9.4725	
50	±9.4625	
60	±9.4725	
70	±9.4675	
80	±9.4625	
90	±9.4625	
100	±9.4675	

and produce unwanted in-band frequency products. The purpose of the in-band blocking test is to verify the ability of the receiver to withstand high levels of in-band interferences from unwanted signals at specified frequency offsets without degradation of its sensitivity. Figure 9.23 illustrates the in-band blocking, and Figure 9.24 shows the test setup for measuring receiver's in-band blocking.

In this test setup, the throughput should be ≥95% of the maximum throughput of the reference measurement channel, with a wanted and an interfering signal coupled to BS antenna connector. The in-band blocking requirement applies from $F_{UL\_low} - \Delta f_{OOB}$ to $F_{UL\_high} + \Delta f_{OOB}$, ($F_{UL}$ is the

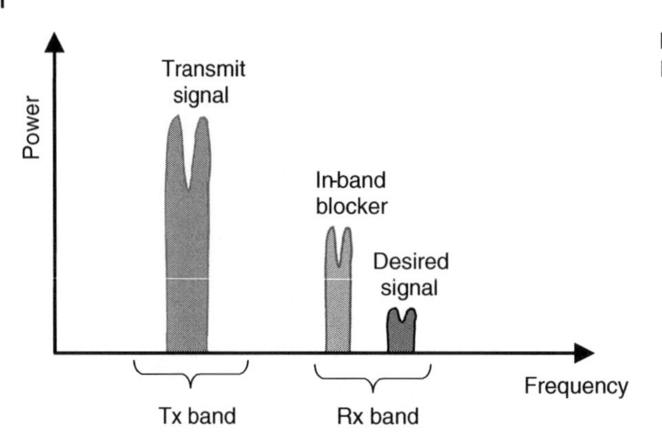

Figure 9.23 Illustration of in-band blocking.

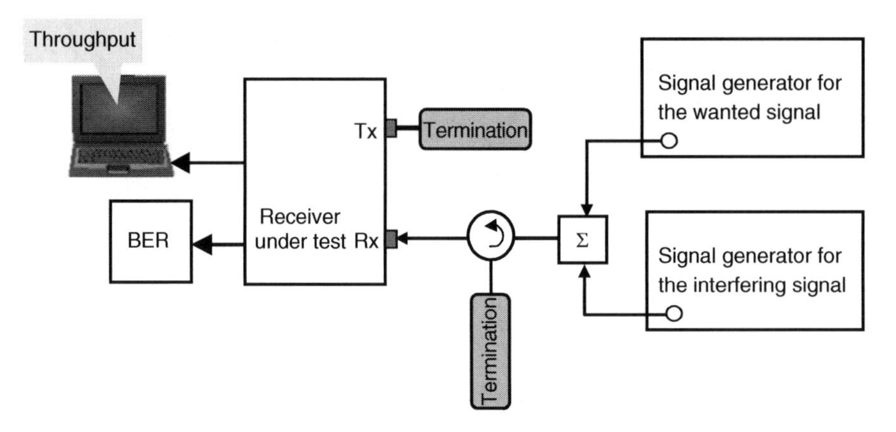

Figure 9.24 Test setup for measuring in-band blocking.

Table 9.18 Δf_{OOB} offset for NR operating bands.

BS type	Operating band characteristics	Δf_{OOB} (MHz)
BS type 1-C	$F_{UL\_high} - F_{UL\_low} \leq 200$ MHz	20
	200 MHz $\leq F_{UL\_high} - F_{UL\_low} \leq 900$ MHz	60
BS type 1-H	$F_{UL\_high} - F_{UL\_low} \leq 100$ MHz	20
	100 MHz $\leq F_{UL\_high} - F_{UL\_low} \leq 900$ MHz	60

uplink frequency) excluding the downlink frequency range of the operating band. Table 9.17 shows the Δf_{OOB} offset for NR operating bands, and Table 9.18 shows the general blocking requirement for base station's receiver. Table 9.19 [2, 3] lists the base station narrowband blocking requirements.

The interfering DFT-spread OFDM (discrete Fourier transform spread OFDM) signal is a form of single-carrier technique that possesses almost all advantages of the multicarrier OFDM technique (i.e., high spectral efficiency, flexible bandwidth allocation, low sampling rate, and low-complexity equalization).

Table 9.19 Base station general blocking requirement.

BS channel bandwidth of the lowest/highest carrier received (MHz)	Wanted signal mean power (dBm)	Interfering signal mean power (dBm)	Interfering signal center frequency minimum offset from the lower/upper base station RF bandwidth edge or subblock edge inside a subblock gap (MHz)	Type of interfering signal
5, 10, 15, 20	$P_{REFSENS}$ + 6 dB	Wide area BS: −43 Medium range BS: −38 Local area BS: −35	±7.5	5 MHz DFT-s-OFDM NR signal, 15 kHz SCS, 25 RBs
25, 30, 35, 40, 45, 50, 60, 70, 80, 90, 100	$P_{REFSENS}$ + 6 dB	Wide area BS: −43 Medium range BS: −38 Local area BS: −35	±30	20 MHz DFT-s-OFDM NR signal, 15 kHz SCS, 100 RBs

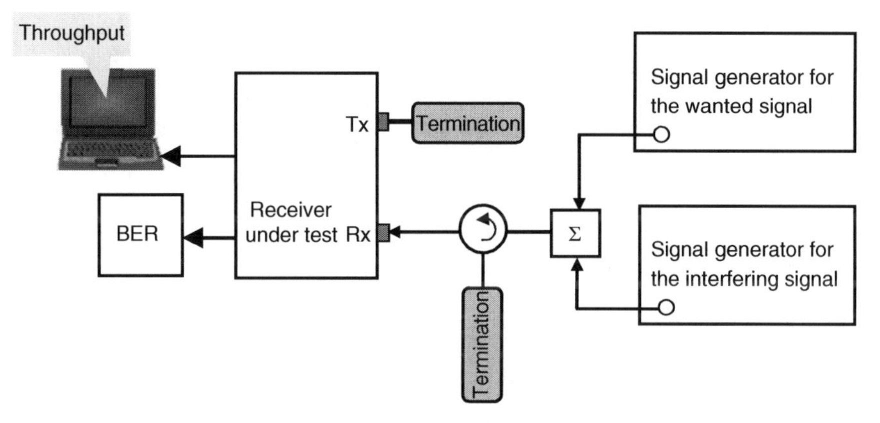

Figure 9.25 Test setup for measuring the out-of-band blocking.

9.4.5 Out-of-Band Blocking

The out-of-band blocking characteristics [3] is a measure of the receiver ability to receive a wanted signal at its assigned channel at the antenna connector in the presence of an unwanted interferer out of the operating band, which is a CW signal. Figure 9.25 shows the test setup for measuring the out-of-band blocking, and Figure 9.26 illustrates the out-of-band blocking.

In this test setup, the throughput is measured at the center of frequency of the allocated frequency and should be ≥95% of the maximum throughput of the reference measurement channel. Table 9.20 shows an example of the out-of-band blocking performance requirement.

9.4.6 Receiver Spurious Emissions

The receiver spurious emissions are the emissions generated or amplified in the receiver chain and appear at the receiver's antenna connector [3]. The purpose of a receiver spurious emissions test is

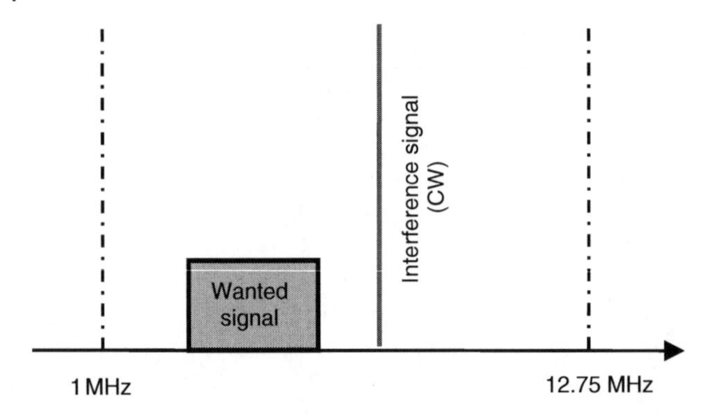

Figure 9.26 Illustration of out-of-band blocking.

Table 9.20 Base station narrowband blocking requirement.

BS channel bandwidth of the lowest/ highest carrier received (MHz)	Wanted signal mean power (dBm)	Interfering signal mean power (dBm)
5, 10, 15, 20, 25, 30, 35, 40, 45, 50, 60, 70, 80, 90, 100	$P_{REFSENS}$ + 6 dB	Wide area BS: −49 medium Range BS: −44 Local Area BS: −41

Note: The SCS for the lowest/highest carrier received is the lowest SCS supported by the BS for that BS channel bandwidth.

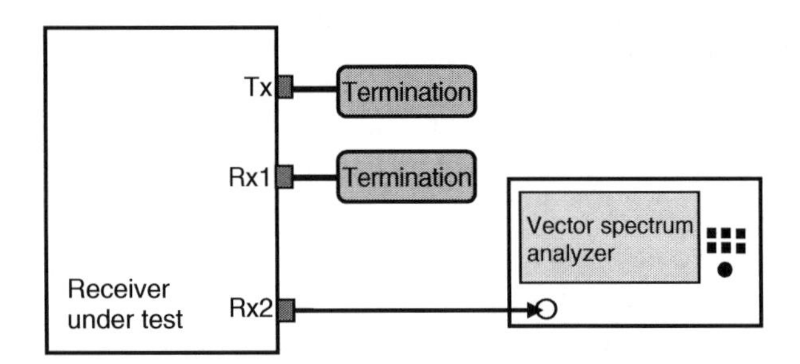

Figure 9.27 Test setup for measuring receiver spurious emissions.

to verify that the receiver is not generating any harmful interferences. Figure 9.27 shows the test setup for measuring the receiver's spurious emissions.

In this test setup, the spectrum analyzer is configured to measure NR-FR1-TM1.1 signal. If the receiver is a subsystem of a transceiver system, the receiver should be tested with both the Tx (i.e., transmitter) and the Rx (i.e., receiver) ON, and the Tx antenna port is terminated. Table 9.21 [3] shows the general BS receiver spurious emissions limits.

9.4.7 Receiver Intermodulation

Third and higher-order mixing of the two interfering RF signals can produce an interfering signal in the band of the desired channel. Intermodulation response rejection is a measure of the capability of

Table 9.21 Out-of-band blocking performance requirement.

Wanted signal mean power (dBm)	Interfering signal mean power (dBm)	Type of interference	Interference signal frequency range
PREFSENS + 6 dB	−15	CW carrier	1 MHz to $F_{UL,low} - \Delta f_{OOB}$ and from $F_{UL,high} + \Delta f_{OOB}$ up to 12750 MHz

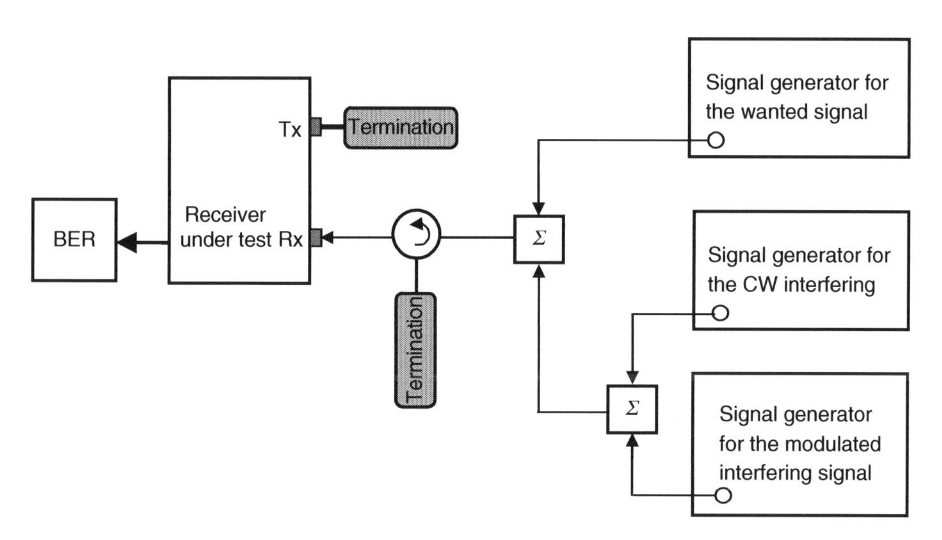

Figure 9.28 Test setup for receiver intermodulation characterization.

the receiver to receive a wanted signal on its assigned channel frequency at the antenna connector port and meets the performance requirements of the BER or throughput in the presence of two interfering signals, which have a specific frequency relationship to the wanted signal.

The purpose of the receiver intermodulation rejection test [3] is to verify the receiver's ability to receive a wanted signal on its assigned channel frequency in the presence of two interfering signals. Figure 9.28 shows the test setup of receiver intermodulation characterization.

In this test setup, the receiver under test should be tested with both the Tx (transmitter) and the Rx (receiver) are ON, and the Tx antenna port is terminated. The setting of the power levels of the wanted and interfering signals generated by the signal generators are set according to the standard requirements. Figure 9.29 shows an illustration of interference signals that cause receiver intermodulation. Tables 9.22 and 9.23 show the general BS receiver spurious emissions limits and the general intermodulation requirement, respectively. The specifications of the interfering signal for intermodulation test are listed in Table 9.24 [3].

9.4.8 In-channel Selectivity

In-channel selectivity (ICS) [3] is a base station receiver measurement, which characterizes the ability to receive a wanted signal in the presence of an adjacent channel signal with a specified center frequency offset. Figure 9.30 shows the test setup for measuring the in-channel selectivity of base station receiver. For BS type 1-C and BS type 1-H, the throughput should be ≥95% of the maximum throughput of the reference measurement channel. Table 9.25 shows the requirements for a wide-area BS in-channel selectivity testing.

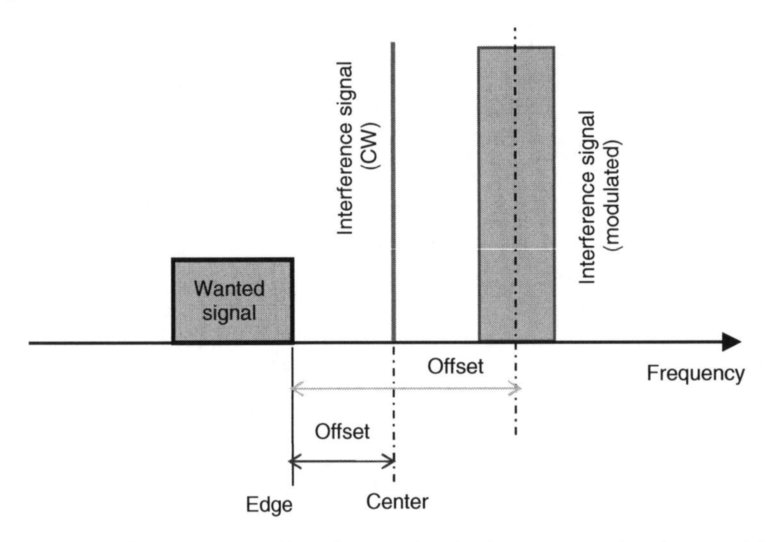

Figure 9.29 illustration of interference signals that cause receiver intermodulation.

Table 9.22 General BS receiver spurious emissions limits.

General BS receiver spurious emissions limits	Basic limit (dBm)	Measurement bandwidth
30 MHz–1 GHz	−57	100 kHz
1 GHz–12.75 GHz	−47	1 MHz
12.75 GHz—5th harmonic of the upper frequency edge of the UL operating band in GH	−47	1 MHz
12.75 GHz–26 GHz	−47	1 MHz

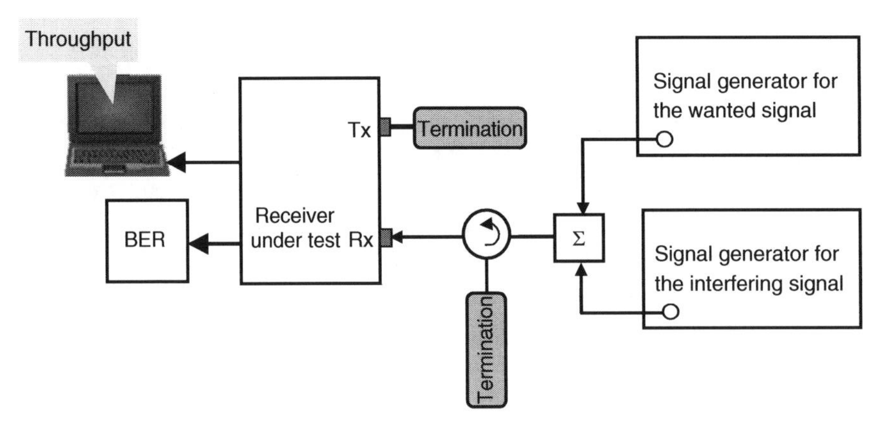

Figure 9.30 Test setup for in-channel selectivity measurement.

Table 9.23 General intermodulation requirement.

Bas station type	Wanted signal mean power (dBm)	Mean power of interfering signals (dBm)
Wide-area BS	$P_{REFSENS}$ + 6 dB	−52
Medium range BS	$P_{REFSENS}$ + 6 dB	−47
Local area BS	$P_{REFSENS}$ + 6 dB	−44

Table 9.24 Interfering signal for intermodulation requirement.

Channel bandwidth of the lowest/highest carrier received (MHz)	Interfering signal center frequency offset from the lower/upper RF bandwidth edge (MHz)	Type of interfering signal
5	±7.5	CW
	±17.5	5 MHz DFT-s-OFDM NR signal, (Note 1)
10	±7.465	CW
	±17.5	5 MHz DFT-s-OFDM NR signal, (Note 1)
15	±7.43	CW
	±17.5	5 MHz DFT-s-OFDM NR signal, (Note 1)
20	±7.395	CW
	±17.5	5 MHz DFT-s-OFDM NR signal, (Note 1)
25	±7.465	CW
	±25	20 MHz DFT-s-OFDM NR signal, (Note 2)
30	±7.43	CW
	±25	20 MHz DFT-s-OFDM NR signal, (Note 2)
40	±7.45	CW
	±25	20 MHz DFT-s-OFDM NR signal, (Note 2)
50	±7.35	CW
	±25	20 MHz DFT-s-OFDM NR signal, (Note 2)
60	±7.49	CW
	±25	20 MHz DFT-s-OFDM NR signal, (Note 2)
70	±7.42	CW
	±25	20 MHz DFT-s-OFDM NR signal, (Note 2)
80	±7.44	CW
	±25	20 MHz DFT-s-OFDM NR signal, (Note 2)
90	±7.46	CW
	±25	20 MHz DFT-s-OFDM NR signal, (Note 2)
100	±7.48	CW
	±25	20 MHz DFT-s-OFDM NR signal, (Note 2)

Note 1: For the 15 kHz subcarrier spacing, the number of RB is 25. For the 30 kHz subcarrier spacing, the number of RB is 10.
Note 2: For the 15 kHz subcarrier spacing, the number of RB is 100. For the 30 kHz subcarrier spacing, the number of RB is 50. For the 60 kHz subcarrier spacing, the number of RB is 24.

Table 9.25 Wide-area BS in-channel selectivity.

NR channel bandwidth (MHz)	Subcarrier spacing (kHz)	Reference measurement channel	Wanted signal mean power (dBm)			Interfering signal mean power (dBm)	Type of interfering signal
			$f \leq 3.0$ GHz	3.0 GHz $< f \leq 4.2$ GHz	4.2 GHz $< f \leq 6.0$ GHz		
5	15	G-FR1-A1-7	−99.2	−98.8	−98.5	−81.4	DFT-s-OFDM NR signal, 15 kHz SCS, 10 RBs
10, 15, 20, 25, 30, 35	15	G-FR1-A1-1	−97.3	−96.9	−96.6	−77.4	DFT-s-OFDM NR signal, 15 kHz SCS, 25 RBs
40, 45, 50	15	G-FR1-A1-4	−90.9	−90.5	−90.2	−71.4	DFT-s-OFDM NR signal, 15 kHz SCS, 100 RBs
5	30	G-FR1-A1-8	−99.9	−99.5	−99.2	−81.4	DFT-s-OFDM NR signal, 30 kHz SCS, 5 RBs
10, 15, 20, 25, 30, 35	30	G-FR1-A1-2	−97.4	−97	−96.7	−78.4	DFT-s-OFDM NR signal, 30 kHz SCS, 10 RBs
40, 45, 50, 60, 70, 80, 90, 100	30	G-FR1-A1-5	−91.2	−90.8	−90.5	−71.4	DFT-s-OFDM NR signal, 30 kHz SCS, 50 RBs
10, 15, 20, 25, 30, 35	60	G-FR1-A1-9	−96.8	−96.4	−96.1	−78.4	DFT-s-OFDM NR signal, 60 kHz SCS, 5 RBs
40, 45, 50, 60, 70, 80, 90, 100	60	G-FR1-A1-6	−91.3	−90.9	−90.6	−71.6	DFT-s-OFDM NR signal, 60 kHz SCS, 24 RBs

Note: Wanted and interfering signal are placed adjacently around F_c, where the F_c is defined for *BS channel bandwidth* of the wanted signal according to Table 5.4.2.2-1 in TS 38.104 [2]. The aggregated wanted and interferer signal shall be centered in the BS channel bandwidth of the wanted signal.

9.5 Over-The-Air (OTA) Testing of Radio Systems

OTA testing [4] is a method of connecting a device to the test equipment, such as spectrum analyzer and vector signal generator, via the free space using two antennas (i.e., transmit and receive antennas). Figure 9.31 illustrates the conducted and OTA test methods.

In a conducted testing, the device under test (DUT) is connected directly to the test equipment, whereas in an OTA testing, the device is connected to the test equipment indirectly. The OTA testing is performed in an anechoic chamber that represents a free-space environment; the chamber is the replacement of the cable connectors in the conducted test method. The DUT must be placed in the far-field zone with respect to the source antenna, and Figure 9.32 illustrates the near/far field boundary. In the far-field, one needs to measure only the signal's amplitude because the radiated signals are plane waves.

The far-field distance, R, is given by,

$$R \geq \frac{2D^2}{\lambda} \tag{9.2}$$

where D is the largest dimension (in meter) of device's antenna, and λ is the wavelength of the radiating signal. Figure 9.33 illustrates an OTA test setup, in this setup, the transmit and receive antennas at the ends of radio link are placed inside the chamber, whereas the test equipment and control devices are placed outside the chamber.

In Figure 9.33, the angle $\boldsymbol{\phi}$ represents the direction in the azimuth plane of the unit under test which is placed on the turn table. The angel $\boldsymbol{\theta}$ represents the elevation angle of the source antenna mounted on the positioner. This test setup is used to verify the performance of a receiver for specified angles of arrival. In such a setup, the vector signal generator is set to the required channel and is connected to a horn antenna that is placed inside an anechoic chamber. The position of the

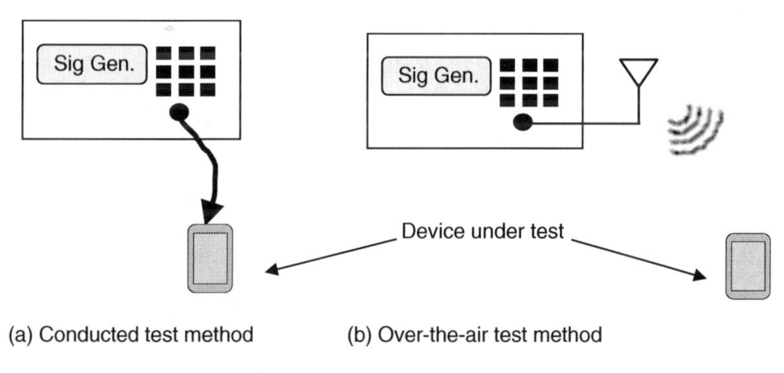

Figure 9.31 Illustration of conducted and over-the-air test methods.

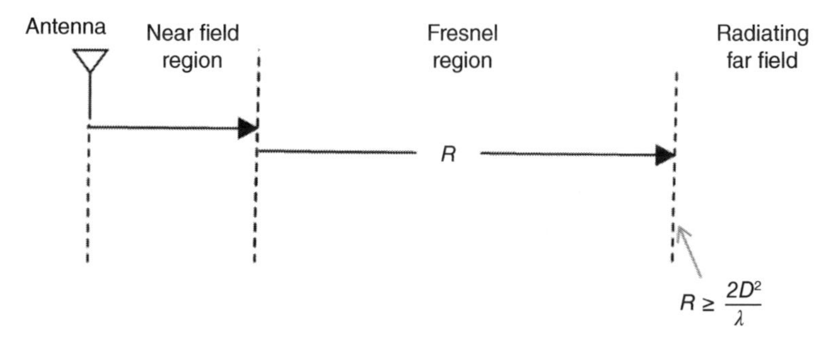

Figure 9.32 Illustration of near/far filed boundary.

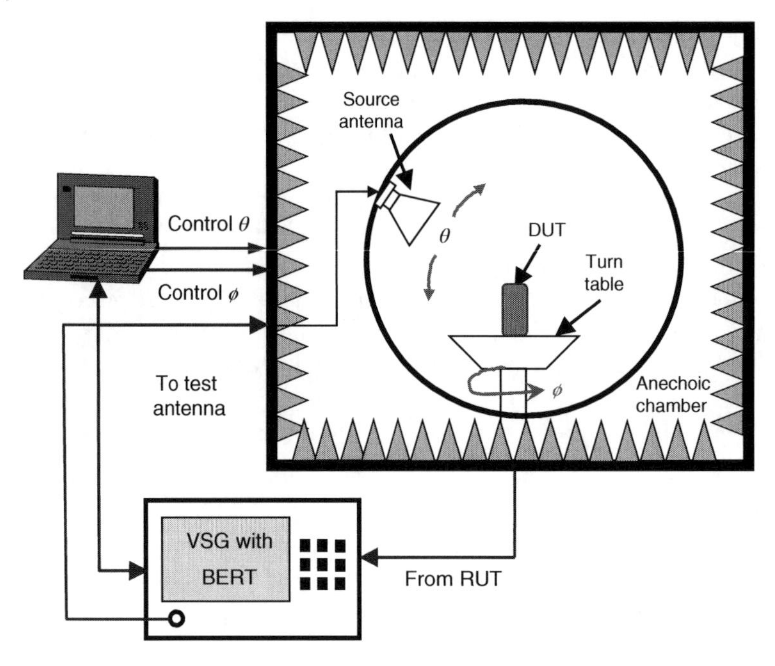

Figure 9.33 Illustration of an OTA test setup.

horn antenna for certain elevation angel $\boldsymbol{\theta}$ is controlled by a computer placed outside the chamber. The DUT is placed on a turn table, which is located in the far-field region from the horn antenna and can be rotated from 0 to 360 degrees. The output of the unit under test is connected to the test equipment that is placed outside the chamber (e.g., BERT, spectrum analyzer, etc.)

The far-field condition requirement for large massive MIMO system requires a huge chamber, which is a big challenge for both the space and the system cost, this problem can be resolved by using *Compact Antenna Test Range* (CATR) system, which enables meeting the far-field condition in a small range. The CATR system [8–10] is explained in Chapter 7 of this book.

The purpose of OTA receiver testing is to validate the receiver performance in a certain angle of arrivals (AoA) in the azimuth and elevation planes (i.e., $\boldsymbol{\theta}, \boldsymbol{\phi}$). In OTA testing [4], the receiver under test must be placed in the far-field region from the source test antenna in an anechoic chamber. Table 9.26 [2] lists the OTA conformance tests for 5G NR base station receivers.

Table 9.26 OTA conformance tests for 5G Nr base station receivers.

Test case	Test name
1	OTA sensitivity
2	OTA reference sensitivity level
3	OTA dynamic range
4	OTA adjacent channel selectivity
5	OTA in-band blocking
6	OTA out-of-band blocking
7	OTA receiver spurious emissions
8	OTA receiver intermodulation
9	OTA in-channel selectivity

Review Questions

9.1 Describe the test setup for measuring the power at the output port of an RF transmitter and explain the purpose of this test.

9.2 Draw a diagram of the test setup for measuring the transmitter's output power dynamic range and define the purpose of the test.

9.3 Explain and illustrate the EVM of a digitally modulated signal.

9.4 Describe the test setup and the test condition for measuring a transmitter's EVM and explain the purpose of the test and the cause of EVM.

9.5 Define the meaning of OBW and how to measure it.

9.6 Explain how to measure the ACLR and its impact on the system performance.

9.7 Illustrate and explain the meaning of the OBUE and describe the test setup for measuring the OBUE of an RF transmitter.

9.8 Describe the test setup for measuring the transmitter's conducted spurious emission and the causes of such emissions.

9.9 Draw a diagram of the test setup for measuring the transmitter's intermodulation, and what is the purpose of the test and the cause of intermodulation in RF transmitters.

9.10 Define the meaning of the receiver reference sensitivity and write an expression for the receiver sensitivity.

9.11 Explain the impact of the receiver sensitivity on the receiver's performance.

9.12 Describe the test setup for measuring the receiver dynamic range and define the meaning of receiver dynamic range.

9.13 Draw a diagram for measuring the receiver's ACS and define the term ACS.

9.14 Illustrate and explain the in-band blocking in a receiver band and describe the test setup for in-band blocking characterization.

9.15 Explain how to measure the receiver spurious emission and the purpose of performing such a measurement.

9.16 Describe the test setup for measuring the receiver intermodulation and explain the impact of intermodulation on the receiver's performance.

9.17 Explain why OTA testing is used for validating the performance of 5G NR wireless systems.

9.18 Draw a diagram of the test setup for measuring OTA receiver sensitivity and describe the test setup.

References

1 5G System Overview, https://www.3gpp.org/technologies/5g-system-overview.
2 ETSI TS 138 141-1 V17.5.0 (2022-04), Base Station (BS) conformance testing Part 1: Conducted conformance testing (3GPP TS 38.141-1 version 17.5.0 Release 17).
3 3GPP Technical Specification Group Radio Access Network, NR Base Station Conformance Testing, Part 1: Conducted Conformance Testing, Release 15; TS 38.141-1, V15.6.0, 2020.
4 3GPP Technical Specification Group Radio Access Network, NR Base Station (BS) Conformance Testing Part 2: Radiated Conformance Testing, Release 15; TS 38.141-2 V.15.6.0, 2020.
5 Rohde & Schwarz, *Vector Signal Generators*, https://www.rohde-schwarz.com.
6 Keysight Technologies, *Vector Spectrum Analyzers*, www.keysight.com.
7 5G NR Sub-6 GHz Measurement Methods, Anritsu Application Note, https://dl.cdn-anritsu.com/en-en/test-measurement/files/Application-Notes/Application-Note/ms2850a-5gnr-ef1200.pdf.
8 Vitawat Sittakul, Sarinya Pasakawee, "Design of millimeter-wave Antenna for Compact Antenna Test Range (CATR)," *International Symposium on Antennas and Propagation*, IEEE, 2022.
9 Michael D. Foegelle, "Validation of CATRs for 5G mmWave OTA Testing Applications," *15th European Conference on Antennas and Propagation (EuCAP)*, 2021.
10 Corbett Rowell, Benoit Derat, Adrian Cardalda-Garcia, "Multiple CATR Reflector System for 5G Radio Resource Management Measurements," *15th European Conference on Antennas and Propagation (EuCAP),* IEEE, 2021.

Suggested Readings

Chuanting Liu, Jonas Fridén, et al., "OTA Test Method in Extreme Temperature for 5G Massive MIMO Devices," *17th European Conference on Antennas and Propagation (EuCAP)*, 2023.

Corbett Rowell, Benoit Derat, Adrian Cardalda-Garcia, "Multiple CATR Reflector System for 5G Radio Resource Management Measurements," *15th European Conference on Antennas and Propagation (EuCAP)*, IEEE, 2021.

David Reyes Paredes, Mark Beach, Moray Rumney, "Over-The-Air Test Method for Evaluation of 5G Millimeter Wave Devices Under 3D Spatially Dynamic Environment from Single Feeder," *IEEE/MTT-S International Microwave Symposium*, 2022.

Fengchun Zhang, Lassi Hentilä, Pekka Kyösti, et al., "Millimeter-Wave New Radio Test Zone Validation for MIMO Over-the-Air Testing," *IEEE Transactions on Antennas and Propagation*, 2022.

Hans Andersson, Stefan Nilsson, et al., "Investigation on Simplified Test Environment of OTA In-Band Blocking for 5G Millimeter-Wave Radio Base Stations," *2022 16th European Conference on Antennas and Propagation (EuCAP)*, 2022.

Jonas Fridén, Sam Agneessens, Aidin Razavi, et al., "5G Over-the-Air Conformance Testing," *14th European Conference on Antennas and Propagation (EuCAP)*, 2020.

Michael D. Foegelle, "Validation of CATRs for 5G mmWave OTA Testing Applications," *15th European Conference on Antennas and Propagation (EuCAP)*, 2021.

Nan Ma, Jinyu Wang, et al., "Dynamic MmWave channel reconstruction of MIMO OTA testing for handover scenarios," *China Communications*, Volume: 20, Issue: 4, 2023.

Penghui Shen, Yihong Qi, et al., "A Directly Connected OTA Measurement for Performance Evaluation of 5G Adaptive Beamforming Terminals," *IEEE Internet of Things Journal*, Volume: 9, Issue: 16, 2022.

Vitawat Sittakul, Sarinya Pasakawee, "Design of Millimeter-Wave Antenna for Compact Antenna Test Range (CATR)," *International Symposium on Antennas and Propagation*, IEEE, 2022.

Wei Fan, Fengchun Zhang, Gert F. Pedersen, "Channel Spatial Profile Validation for FR2 New Radio Over-the-air Testing," *IEEE 4th International Conference on Electronic Information and Communication Technology (ICEICT),* 2021.

Xiaohang Yang, Hao Sun, et al., "Standardization Progress and Challenges for 5G MIMO OTA Performance Testing," *IEEE 5th International Conference on Electronic Information and Communication Technology (ICEICT)*, 2022.

Yanpu Hu, Shouyuan Wang, Shaogeng An, "Over the Air Testing and Error Analysis of 5G Active Antenna System Base Station in Compact Antenna Test Range," *Photonics & Electromagnetics Research Symposium*, 2019.

Index

a

ABCD parameters 98–102
absolute gain method 200
absorptive switch 137
additive white Gaussian noise (AWGN) 216
adjacent channel leakage power ratio (ACLR) 143, 164, 223, 233–234
adjacent channel selectivity (ACS) 242
admittance chart 88
Advanced Design System (ADS) 63
advanced mobile phone system (AMPS) 6
aliasing 43
amplitude modulation (AM) 1, 11, 19, 20–23
analog beamforming 219
analog communication systems 20
  AM 20–23
  AM demodulation 26
  FM demodulation 32–34
  noise suppression, in FM systems 32
  PM 34–35
  single-sideband modulation 23–25
analog-to-digital converters (ADC) 5, 39, 145
angle of arrivals (AoA) 252
angular frequency deviation 27
antenna anechoic chamber 169
antenna array 191, 225
antenna efficiency 176, 181
antennas 6, 170
  aperture efficiency 176–178
  bandwidth 178
  beamwidth 176
  characterization
    CATR 198–199
    RF antenna anechoic chamber 198
  E-plane radiation pattern 187
  ERP 183–184
  gain 175–176
  gain measurements
    absolute gain method 200
    antenna radiation pattern 201
    directivity measurement 201
    gain-transfer method 200–201
    input impedance measurement 201–202
  gain-to-noise-temperature (G/T) 190–191
  impedance mismatch 189
  input impedance 178
  multielement antenna (array) 191–193
  noise temperature 190
  polarization 178–179
  polarization mismatch 189
  radiation pattern and parameters 173–174
  rectangular patch 187
  short antenna fields 180–181
  types
    dipole antenna 184–185
    microstrip antenna 186–187
antenna temperature 190
Australia Communication and Media Authority (ACMA) 13

b

balanced modulator 24
bandpass filter (BPF) 123, 145
bandstop filters 123
baseband signals 2, 11
beamforming technology 191
  analog beamforming 219
  antenna beam 216

beamforming technology (*contd.*)
  digital beamforming 219
  hybrid beamforming 219–220
  3D beamforming 217
BER *see* bit-error rate (BER)
Bessel function 28
bi-directional communication 4
binary frequency-shift keying (BFSK) 45
binary phase shift keying (BPSK) 39
bit-error rate (BER) 46, 126, 144, 194
Boltzmann's constant 147, 190

c

capture range 126
Carson's rule 28
cascaded noise figure 113
cavity resonator filters 123–124
cellular phone systems 1
characteristic impedance 64, 67
chip rate 54
circulators/isolators 4, 132–133
coax connectors 106
coaxial transmission lines 61, 62
code division multiple access (CDMA) 1, 8, 52
commercial attenuator products 134
commercial mixer module 119
commercial off-the-shelf (COTS)
  connectorized modules 112
compact antenna test range (CATR) 169, 198–199, 252
continuous-phase frequency-shift keying (CPFSK) 46
conversion loss 121
coplanar waveguide (CPW) 61
CPFSK *see* continuous-phase frequency-shift keying (CPFSK)
CPW *see* coplanar waveguide (CPW)
cut-off frequency 32
cyclic prefix (CP) 211

d

damped oscillation 127
data conversion
  analog-to-digital conversion 42–44
  digital-to-analog conversion 44–45
delay spread 194
differential phase-shift keying (DPSK) modulator 39, 49
digital beamforming 219
digital communication systems
  data conversion
    analog-to-digital conversion 42–44
    digital-to-analog conversion 44–45
  digital modulation
    FSK 45–46
    PSK 47–49
    QAM 49–51
  spectral efficiency and noise 51–52
  wideband modulation
    DSSS 54–56
    FHSS 52–54
    OFDM 56–57
digital down-converter (DDC) 158
digital modulation 39
  FSK 45–46
  PSK 47–49
  QAM 49–51
digital phase shifters 137
digital-signal-processing (DSP) 5, 158
digital-to-analog converters (DAC) 5, 39
digital up-converter (DUC) 158
direct conversion (zero-IF) receiver 155, 157–158, 162
direct-conversion transmitters 162
direct digital synthesizer (DSS) 125
directional couplers 133–134
direct sequence spread spectrum (DSSS) 39, 54–56
dispersion phenomenon 76
double-sideband suppressed carrier (DSB SC) 19, 24
double-stub matching 91–98
down converter mixer 6
dual-conversion superheterodyne receivers 156–157
duplexers 4, 124
dynamic range (DR) 121, 152

e

effective radiated power (ERP) 183–184
8-PSK modulation 39
electric field intensity 181
electromagnetic (EM) spectrum 9

electromagnetic (EM) waves 2, 6, 40, 145, 170, 171
enhanced mobile broadband (eMBB) 9, 208
ENR *see* excess noise ratio (ENR)
equivalent noise temperature 147
error vector magnitude (EVM) 143, 223, 229
European Telecommunications Standards Institute (ETSI) 164, 236
EVM *see* error vector magnitude (EVM)
excess noise ratio (ENR) 114

f

far-field distance 172
fast frequency hopping (FFH) 53
Federal Communication Commission (FCC) 13, 164, 236
feedforward linearization 131–132
FFH *see* fast frequency hopping (FFH)
fifth-generation new radio-access (5G NR) 8, 207
  cellular phone systems 13
  frequency ranges 209–210
  numerology 211
  radio frame structure 210–211
  resource grid 212–213
  RF performance verification of 5G NR receivers 246–247
    ACS measurement 242
    ICS 247–250
    in-band blocking 242–245
    out-of-band blocking 245
    receiver dynamic range 239–242
    receiver reference sensitivity 239
    receiver spurious emissions 245–246
  RF performance verification of 5G NR transmitters
    ACLR 233–234
    conducted spurious emission 235–237
    OBUE measurement 234–235
    OBW 232–233
    TAE 231–232
    transmit ON/OFF power 227–228
    transmitter frequency error 228–229
    transmitter intermodulation 237–238
    transmitter output power 226–227
    transmitter's EVM 229–231
    transmitter total power dynamic range 227
5G mobile communication 9
fixed wireless access (FWA) 209
flat fading 194
FM *see* frequency modulation (FM)
4G long term evolution (LTE) 52
Fraunhofer region 170
free-space wave impedance 181
frequency discrimination 32
frequency division multiple access (FDMA) 1, 6
frequency division multiplexing (FDM) 6, 209
frequency drift 129
frequency error 143, 223
frequency-hopping spread spectrum (FHSS) 39, 52–54, 195
frequency modulation (FM) 1, 12
  demodulation 32–34
frequency pushing 129
frequency range 1 (FR1) 1, 209
frequency range 2 (FR2) 1, 14, 209
frequency selective fading 194
frequency-shift keying (FSK) 39, 45–46
frequency spectrum allocation 1
Fresnel region 171
Friis equation 190
full dimension multiple-input multiple-output (FD-MIMO) 214
full-duplex radio system 3
full-duplex superheterodyne radios 155

g

gain-to-noise-temperature (G/T) 190–191
gain-transfer method 200–201
Gaussian minimum shift keying (GMSK) 46
group delay 125

h

half-duplex radio system 1, 3
harmonic distortion (HD) 116, 151
HFSS *see* high-frequency structure simulator (HFSS)
high-frequency transmission lines 61, 62
  ABCD parameters 98–102
  impedance matching, using Smith chart
    double-stub matching 91–98
    single-stub matching 89–91
  input impedance, of transmission line 70–73
  planar transmission lines, in radio systems

high-frequency transmission lines (*contd.*)
microstrip line structure 76
quarter-wave transformer 73–74
RF transmission lines 62–63
Smith chart 76–88
S-parameters 102–106
transmission line
analysis 63–67
connectors 106
VSWR 68–69
high-pass filter (HPF) 118, 122–123
homodyne 157
hybrid beamforming 219–220

i

ICS *see* in-channel selectivity (ICS)
image frequency 118–119, 155
immittance chart 79, 80
impedance matching
double-stub matching 91–98
single-stub matching 89–91
in-band blocking 242–245
in-channel selectivity (ICS) 247–250
incident normalized voltage 103
incident power 73
Innovation Science and Economic Development Canada (ISED) 13
in-phase/quadrature (I/Q) demodulator 145
input reflection coefficient 113
insertion loss 125
intermediate frequency (IF) 2, 118, 144
intermodulation distortion 115–117
intermodulation products 115, 150
International Telecommunication Union (ITU) 13, 208
internet-of-things (IoT) 9
intersymbol interference (ISI) 56, 163
inverse fast Fourier transform (IFFT) 56
isolator 6, 111, 132, 133
isotropic antenna 179–181

k

Kirchhoff's voltage and current laws 63

l

left-hand circular (LHC) polarization 179
line-of-sight (LOS) 194, 197
load pulling 129
local oscillators (LO) 5, 118, 145
long-term evolution (LTE) 8
lower sideband (LSB) 13, 25, 118
low-noise amplifiers (LNA) 5
intermodulation distortion 115–117
NF 112–113
NF measurement methods 114–115
performance parameters 117
low-pass filters (LPF) 50, 121–122, 145

m

magnetic fields 65, 75, 127, 172, 181
massive machine-type communication (mMTC) 9, 208
microstrip antenna 186–187
millimeter-wave systems 124
minimum shift keying (MSK) 46
mixer modules 120
modulation index 21, 27
modulator 5, 10, 11, 20, 24, 32, 40
monopole antennas 184
MSK *see* minimum shift keying (MSK)
multielement antenna (array) 191–193
multilevel frequency-shift keying (FSK) 46
multipath propagation 169
flat fading 194
radio Fresnel zones 197
radio wave propagation path loss 195–197
multiple-input multiple-output (MIMO) 1, 8, 10, 135, 207
massive MIMO 5G systems 213–216
MU-MIMO 213
multi-user multiple-input multiple-output (MU-MIMO) 213

n

narrowband frequency modulation (NBFM) 28
N-element microstrip antenna array 191, 192
noise factor 120
noise figure (NF) 112–113, 121
measurement methods
cold noise method 115
Y-Factor method 114–115
noise suppression, in FM systems 32
nonreturn to zero (NRZ) signals 39, 41
normalized impedance 83

normalized load admittance 93, 95
Nyquist frequency 43

o
OBUE *see* operating band unwanted emissions (OBUE)
occupied bandwidth (OBW) 143, 162, 164, 232–233
OFDM *see* orthogonal frequency division multiplexing (OFDM)
one dB compression point 115, 150
operating band unwanted emissions (OBUE) 143, 164, 234–235
orthogonal frequency division multiple access (OFDMA) 1, 8
orthogonal frequency division multiplexing (OFDM) 39, 56–57, 195
out-of-band (OOB)
  blocking 245
  emissions 164, 236
over-the-air (OTA) testing 223, 251–252

p
pattern multiplication 192
permeability 65
permittivity 65
phase constant 65, 67
phase departure 34
phase discrimination 32
phase-locked loop (PLL)
  demodulator 19
  FM demodulator 33–34
  frequency synthesizer 125–127
phase modulation (PM) 12, 34–35
phase sensitivity 34
phase shifters 135
phase-shift keying (PSK) 39, 47–49
phase-shift SSB modulator 24
phase velocity 65
planar transmission lines 62
PLL *see* phase-locked loop (PLL)
polarization loss factor (PLF) 189
power amplifiers (PA) 5, 129, 130
  linearization techniques
    feedforward linearization 131–132
    performance parameters 132
power splitter/combiner 134
processing gain 54
pseudorandom noise (PN) 52
PSK *see* phase-shift keying (PSK)

q
quadrature amplitude modulation (QAM) 39, 49–51
quantization error 43
quarter-wave transformer 73–74

r
radiation absorbent material (RAM) 198
radio distribution network (RDN) 225
radiofrequency (RF)
  spectrum allocation 13–14
  switches 2
    performance parameters 138
    pin diode RF switch 137–138
  transceivers 5
    direct conversion (zero-IF) receiver 157–158
    direct-conversion transmitters 162
    dual-conversion superheterodyne receivers 156–157
    receiver system parameters 146–157
    RF block-level budget analysis 159–162
    RF transmitters system parameters 162–165
    SDR 158–159
    superheterodyne receiver architecture 144–145
radiofrequency (RF) subsystem blocks
  attenuators 134–135
  circulators/isolators 132–133
  directional couplers 133–134
  duplexers 124
  filters
    bandstop filters 123
    cavity resonator filters 123–124
    HPF 122–123
    LPF 121–122
    RF filter's specifications 125
  frequency synthesizers 125–127
  LNA
    intermodulation distortion 115–117
    measurement methods, NF 114–115
    NF 112–113

radiofrequency (RF) subsystem blocks (*contd.*)
performance parameters 117
power amplifier linearization techniques
feedforward linearization 131–132
performance parameters 132
power splitter/combiner 134
RF building blocks 111–112
RF mixers
double-balanced mixers 119–120
frequency conversion 119
image frequency 118–119
image rejection mixers 119
mixer's conversion LOS 120
mixer's NF 120
mixer's performance parameters 120–121
RF oscillators 127
frequency stability 128
performance parameters, VCO 129
VCO 128–129
RF phase shifters 135–137
RF power amplifiers 129–130
RF switches
performance parameters 138
pin diode RF switch 137–138
radio Fresnel zones 169
radio performance verification
RF signal generators 224
VSA 224–225
radio system bandwidth 190
radio transceivers 3–6, 132
front-end, using duplexer 3
with RF switch 4
using circulator 4
receiver intermodulation 246–247
receiver reference sensitivity 239
receiver selectivity 155
receiver spurious emissions 245–246
receiver spurious response 154–155
receiver system parameters
intermediate frequency and images 155–156
intermodulation 150–154
NF 147–150
receiver selectivity 155
receiver sensitivity 146–147
receiver spurious response 154–155
SFDR 150–154
reflection coefficient 67, 69
resource element (RE) 212
return loss 104
return to zero (RTZ) signals 39, 41
right-hand circular (RHC) polarization 179

s

scattering matrix 105
scattering parameters *see* S-parameters
settling time 45, 126
SFDR *see* spurious-free dynamic range (SFDR)
signal-to-noise ratios (SNR) 44, 112, 144, 190, 193
simplex radio system 3
simplified mathematical model, MIMO 214–216
simulink simulation 159
single-pole/double-throw (SPDT or 1:2) switch 137
single-pole/multiple-throw (SPnT) switch 137
single-sideband (SSB) 19, 127
modulation
filtering method 24
phase method 24–25
single-stub matching 89–91
sinusoidal modulating signal 28
16-quadrature amplitude modulation (QAM) modulator 39, 49, 50
slow frequency hopping (SFH) 53
SMA connectors 106
Smith chart 76–88
software-defined radios (SDR) 143, 158–159
S-parameters 102–106
spectrum allocations 14
spread-spectrum (SS) 52
spurious emission 143
spurious-free dynamic range (SFDR) 44, 117, 150–154
subcarrier spacing (SCS) 210
submillimeter-wave 9
surface-mount mixer package 119

t

telegrapher's equations 63
terahertz (THz), for 6G wireless technology 1, 9–10
thermal noise 112
third generation partnership project (3GPP) organization 9, 207

third-order intercept point (IP3) 116, 121, 150, 151
third-order intermod (IM) 150
third-order intermodulation distortion 116, 151
time alignment error (TAE) 223, 231–232
time-division duplexing (TDD) systems 13, 137
time division multiple access (TDMA) multiplexing 1, 7
time-division multiplexing transmission 209
transceiver array boundary (TAB) connectors 225
transmission, cascade, and chain parameters *see* ABCD parameters
transmit/receive (T/R) switch 4, 138
transmitters
  EVM 163–164
  frequency error 162
  intermodulation 143, 165, 237–238
  output power DR 162
  spurious emission 164
  spurious emissions 235
transverse electromagnetic (TEM) waves 62, 172
two-port network 100–102

u
ultra-massive multiple-input multiple-output (UM-MIMO) systems 10
ultra reliable low-latency communications (URLLC) 9, 208
upper sideband (USB) 13, 118
user equipment (UE) 209

v
vector network analyzer (VNA) 103, 113
vector signal generators 224
vector spectrum analyzers (VSA) 224–225
VNA *see* vector network analyzer (VNA)
Voice over Internet Protocol (VoIP) 8
voltage-controlled oscillator (VCO) 33, 125, 128–129
voltage-standing wave ratio (VSWR) 61, 68–69

w
wideband modulation
  DSSS 54–56
  FHSS 52–54
  OFDM 56–57
wireless communication model 1, 2
wireless systems 1
  cellular phone systems 6–9
  MIMO 10
  radiofrequency spectrum allocation 13–14
  radio systems classification 3–6
  THz, for 6G wireless technology 9–10
  wireless communications 2

y
Y-Factor 111, 114–115

Printed and bound by CPI Group (UK) Ltd, Croydon, CR0 4YY
17/04/2025
14658877-0001